RNA Polymerases as Molecular Motors

On the Road

2nd Edition

Chemical Biology

Editor-in-chief:
Kira J. Weissman, *Lorraine University, France*

Titles in the series:

1: High Throughput Screening Methods: Evolution and Refinement
2: Chemical Biology of Glycoproteins
3: Computational Tools for Chemical Biology
4: Mass Spectrometry in Chemical Biology: Evolving Applications
5: Mechanisms of Primary Energy Transduction in Biology
6: Cyclic Peptides: From Bioorganic Synthesis to Applications
7: DNA-targeting Molecules as Therapeutic Agents
8: Protein Crystallography: Challenges and Practical Solutions
9: Oxidative Folding of Proteins: Principles, Biological Regulation and Design
10: Chemical and Biological Synthesis: Enabling Approaches for Understanding Biology
11: Synthetic Glycomes
12: The Chemical Biology of Sulfur
13: The Chemical Biology of Phosphorus
14: DNA Damage, DNA Repair and Disease: Volume 1
15: DNA Damage, DNA Repair and Disease: Volume 2
16: The Discovery and Utility of Chemical Probes in Target Discovery
17: Inhibitors of Protein–Protein Interactions
18: The Chemical Biology of Nitrogen
19: RNA Polymerases as Molecular Motors: On the Road

How to obtain future titles on publication:
A standing order plan is available for this series. A standing order will bring delivery of each new volume immediately on publication.

For further information please contact:
Book Sales Department, Royal Society of Chemistry, Thomas Graham House, Science Park, Milton Road, Cambridge, CB4 0WF, UK
Telephone: +44 (0)1223 420066, Fax: +44 (0)1223 420247
Email: booksales@rsc.org
Visit our website at www.rsc.org/books

RNA Polymerases as Molecular Motors

On the Road

2nd Edition

Edited by

Robert Landick
University of Wisconsin–Madison, USA
Email: landick@bact.wisc.edu

Terence Strick
École normale supérieure, France
Email: strick@biologie.ens.fr

and

Jue Wang
University of Wisconsin–Madison, USA
Email: wang@bact.wisc.edu

ROYAL SOCIETY OF CHEMISTRY

Chemical Biology No. 19

Print ISBN: 978-1-78801-365-9
PDF ISBN: 978-1-83916-056-1
EPUB ISBN: 978-1-83916-066-0
Print ISSN: 2055-1975
Electronic ISSN: 2055-1983

A catalogue record for this book is available from the British Library

The Royal Society of Chemistry is a charity, registered in England and Wales, Number 207890, and a company incorporated in England by Royal Charter (Registered No. RC000524), registered office: Burlington House, Piccadilly, London W1J 0BA, UK, Telephone: +44 (0) 20 7437 8656.

For further information see our website at www.rsc.org

Printed in the United Kingdom by CPI Group (UK) Ltd, Croydon, CR0 4YY, UK

Preface to the Second Edition

Ten years ago, Henri Buc enjoined us to view RNA polymerases as molecular motors, and to focus on understanding how the motor starts and runs, but also how it stops. The Royal Society of Chemistry volume which emerged from this perspective, 'RNA Polymerases as Molecular Motors', provided readers with a complete and detailed view of the key aspects of RNA polymerase structure and function, bringing together the perspectives of genetics, biochemistry, structural biology, biophysics, and modelling.

The absolutely central place of transcription in our contemporary understanding of biology cannot be underestimated, and this is true for both historical as well as conceptual reasons. From a historical standpoint, RNA polymerase is most probably the best-understood molecular system we have. The study of transcription has proceeded unabated since the earliest days of the molecular biological revolution, and a tremendous amount of fundamental knowledge has been obtained in this period which brings together numerous subfields of biology, but also has attracted the interest of numerous fields of science, including physics, chemistry, and mathematics. Whereas the fundamental study of other molecular systems has at times been waylaid by promises and dreams of translational applications and drug discovery, the study of RNA polymerase has remained predominantly fundamental, and the ongoing accumulation of this class of knowledge has permitted appropriate translational developments to emerge from the fundamental studies, rather than to push aside the fundamental studies. RNA polymerase stands out among molecular systems as one for which our knowledge is at its most advanced state. As a result, the ongoing discoveries made at the frontiers of this knowledge continue to irrigate all of the fields of biology.

Chemical Biology No. 19
RNA Polymerases as Molecular Motors: On the Road, 2nd Edition
Edited by Robert Landick, Terence Strick and Jue Wang

Published by the Royal Society of Chemistry, www.rsc.org

Today we are pleased to provide you with the second installment of 'RNA Polymerases as Molecular Motors' which we hope will capture some of the exciting developments in our understanding of transcription over the last ten years. In essence we will take the motor for a ride around town to see how it deals with fuel shortages, traffic, and assorted obstacles it may typically encounter while on the road. The central premise of the book will be to explore how RNA polymerase carries out transcription in the crowded context of the cell. One of the main developments we hope to communicate is the increasingly broad yet detailed view of RNA polymerase as a central hub for the entire organism, integrating *via* unanticipated chemical and physical interactions a range of signals of the cellular state including metabolic capacity, cell cycle state, and DNA damage, to name but a few.

Because one of the likely reasons for the success of the first edition was its determination to synthesize and centralize a vast amount of mechanistically-relevant information that had been accumulated over the past decade or two, we felt it was vital that this new edition continue to fulfill this task as well. So as to keep the second edition fresh and lively, we proposed to rely on the first edition for the foundations of transcription – indeed the first edition, although out of print, remains a popular download on the Royal Society of Chemistry webstore. In this second edition we asked our contributors to update key established mechanistic developments, taking as granted the information already provided in the first edition. The book is thus organized in two parts; a first part which goes into the new workings of the motor that have recently been discovered, and a second part which focuses on how the RNAP motor integrates into the cellular landscape.

We hope you will enjoy your ride with this latest of models.

Robert Landick
Terence Strick
Jue Wang

Foreword to the First Edition

The process of making RNA obliges the DNA-dependent RNA polymerases to perform a remarkable number of separate tasks:

- They scan duplex DNA to find the sites for initiating transcription.
- At these sites, the promoters, they select the DNA strand that will serve as the transcription template and expose it for base-paired binding of their ribonucleoside triphosphate substrates.
- They polymerize RNA chains through distinctive processes of initiation, promoter escape and elongation.
- They discriminate against the incorporation of deoxyribonucleotides into RNA.
- They recognize specific DNA sites for disassembling the elongating transcription complex, releasing the nascent transcript and disengaging from the DNA template.

They also do many things that are less obviously intrinsic to a nucleic acid polymerase:

- They abortively and repetitively initiate transcription at the promoter, generating short ribo-oligonucleotides.
- They elongate RNA chains at non-uniform rates, pausing at characteristic sites.
- They detect incorrect nucleotide addition to the elongating transcript and increase the accuracy of transcription by proofreading.
- They set intrinsically unequal rates of transcription at different promoters by DNA sequence-determined variations of binding affinity, rates of promoter opening and ease of escape from the promoter.

Chemical Biology No. 19
RNA Polymerases as Molecular Motors: On the Road, 2nd Edition
Edited by Robert Landick, Terence Strick and Jue Wang

Published by the Royal Society of Chemistry, www.rsc.org

- They exert force on transcription-obstructing DNA-bound proteins, clearing them out of their path as they elongate RNA chains.
- They exert force on their DNA templates, with diverse consequences for the architecture of the eukaryotic nucleus and the prokaryotic nucleoid/chromoid.
- They monitor their DNA templates for damage.
- They recruit accessory proteins for several of these activities, particularly transcript elongation, termination and DNA repair, and they also interact with other molecular machines in order to couple transcription with subsequent RNA processing.

This is an exciting time for understanding and admiring transcription in its machine-like and mechanistic terms. Within the past approximately 10 years, the determination of the structure of the multisubunit RNA polymerases has profoundly transformed the way in which ideas about transcription mechanisms are formulated and tests of these ideas are designed. The determination of the structure of the eukaryotic RNA polymerase II from budding yeast and, most recently, of an archaeal RNA polymerase has made the common evolutionary roots of the multisubunit enzymes vividly apparent. At the same time, comparison of structures of single-subunit nucleic acid polymerases with multisubunit RNA polymerases establishes the existence of catalytic mechanisms that are common to all nucleic acid polymerizations.

During the same period, the spectacular development of methods for examining single molecules of RNA polymerase in action has opened up entirely new possibilities for probing the mechanical and motor-like aspects of RNA polymerases. The process of directly observing RNA synthesis one molecule at a time reveals insights that are difficult to retrieve from, or are entirely obscured in, observations of ensembles in bulk solution. At the same time, technical advances have significantly increased the power of longer established analytical approaches (*e.g.*, fast reaction kinetics).

This book presents a synthesis of these streams of endeavor. Overview chapters that focus on the mechanism–structure interface and the structure–machine interface introduce the two sections of the book, while individual chapters within each section concentrate more specifically on particular processes – kinetic analysis, single-molecule spectroscopy, and termination of transcription, for example. Seen from the perspective of (nearly) 50 years ago, the detail in which every step of transcription is currently understood is remarkable, and the ways in which that detail illuminates every aspect of gene regulation is enormously satisfying. From the current perspective, the dominant sense is of unanswered questions, of experiments addressing key aspects of mechanism that remain open to competing interpretation, of further technical development that would yield insights currently just out of reach, of molecular computations that have not yet been done – *i.e.*, of a work in progress and a field of activity urgently engaged in finding fascinating new questions to answer.

The current understanding of the mechanistic and machine-like aspects of transcription has been formed primarily through work with the bacterial RNA polymerases. While the common evolutionary and mechanistic basis of all transcription can now be appreciated at the structural level, research on the eukaryotes has been dominated by the challenge of enumerating and understanding elaborations of the core transcription machinery with extrinsic initiation, elongation and termination factors and complexes (the core transcription initiation factors of budding yeast alone comprise nearly 50 polypeptide chains with an aggregate mass of more than 2.5×10^6 MDa), the dominant role of chromatin structure and modification in regulation of transcription, direct coupling of transcription with post-transcriptional RNA processing and, recently, the role of small RNAs in these processes. Many questions about the RNA polymerase machine that are specific to eukaryotic transcription, especially transcription of nucleosomal chromatin and the important role of elongation factors, are open to lines of experimentation and analysis that are described here for the bacterial enzymes. For these lines of inquiry, the work on bacterial transcription that is presented here points to, and lights up, the path.

A Foreword is a good place to relate how and where this very large endeavor started. The activity of DNA-dependent RNA polymerase was discovered at the University of Chicago's Argonne Cancer Research Hospital in 1959 and announced in a brief note in the *Journal of the American Chemical Society* in August of that year by S. B. Weiss and his assistant L. Gladstone. They showed that incorporation of ^{32}P-labeled CTP into acid-insoluble (*i.e.*, polymeric) form in a preparation of rat liver nuclei required all four ribo NTPs, ATP, GTP, CTP and UTP. The product of their synthesis was degraded by pancreatic ribonuclease but not by deoxyribonuclease I. Moreover, degradation with alkali of the radioactive product made with CTP ^{32}P-labeled in the alpha position distributed radioactivity to all four ribonucleoside 2′ and 3′ monophosphates. This implied the synthesis of RNA polymers of complex sequence, as opposed to the mere addition of CMP to the ends of nucleic acid chains, either adventitiously to DNA or as the matured CCA adduct of tRNAs. RNA synthesis was strongly inhibited by pyrophosphate but indifferent to orthophosphate, distinguishing the enzyme from polynucleotide phosphorylase. The role of DNA in RNA synthesis was, however, not resolved.

A year later, A. Stevens, then a postdoctoral fellow at NIH, and J. Hurwitz, A. Bresler and R. Diringer at the NYU School of Medicine separately announced the existence of a comparable activity in extracts of *Escherichia coli*. RNA synthesis by the abundantly active bacterial extracts was readily shown to be profoundly dependent on DNA. In contrast, Weiss could not separate the mammalian RNA polymerase activity from DNA and also turned to work with a bacterial enzyme (from *Micrococcus luteus*). With the bacterial preparations, J. J. Furth, Hurwitz and M. Goldman and also A. Stevens, followed by Weiss and T. Nakamoto soon showed the correspondence of the relative incorporation of (AMP+UMP) to (GMP+CMP) into synthesized RNA with the guanine- cytosine content of added DNA. Weiss and Nakamoto extended the analysis of the DNA template–RNA product relationship to the level of nearest neighbor nucleotide pairs by essentially copying an elegant analytical strategy devised for DNA polymerase (by J. Josse, A. D. Kaiser and A. Kornberg) that had been published just months before.

Using CsCl density centrifugation, B. D. Hall and S. Spiegelman had just provided the definitive proof that RNA made in phage T2-infected *E. coli* was specific to the infecting virus by showing that it was able to form DNA–RNA hybrid duplexes with phage DNA. Weiss, Nakamoto and I adopted this approach to show that the RNA synthesized *in vitro* by the bacterial RNA polymerase generated a polymeric product that was fully complementary to the eliciting double-stranded T2 phage DNA and corresponded, in that sense, to RNA made in the phage-infected cell. This (simple) experiment showed that DNA, which was still widely referred to as the "primer" of RNA synthesis, in fact was its template. The same series of experiments yielded the additional information that the newly synthesized RNA was released from its template and that transcription did not separate template DNA strands. However, both strands of T2 DNA were transcribed, yielding RNA that was self-complementary. In similar experiments with the RNA polymerase activity from *E. coli* and phage ϕX174 DNA, M. N. Hayashi and S. Spiegelman at the University of Illinois, as well as M. Chamberlin and P. Berg at Stanford, also found both strands of their phage DNA templates transcribed. In contrast, the RNA isolated from cells infected with diverse phages was soon found to be DNA-strand-selected, or "asymmetric," which is consistent with the requirements of messenger RNA function in instructing protein synthesis. Was something missing from the *in vitro* RNA synthesis system, or was it conceivable that mRNA might have to be selected after transcription by an additional process, with unusable transcripts rapidly disposed of? The first alternative implied an ability to select specific sites on DNA for starting transcription; the alternative hypothesis clearly failed on two counts: it was too elaborate and inelegant to be plausible, and it implied that transcription yields its functional products at the cost of an energy-consuming futile cycle. Thus, the hunt for DNA strand-selective "asymmetric" transcription was on. Of course, it can now be appreciated that the dichotomy was, to some extent, false. The human genome is pervasively transcribed, with both complementary DNA strands frequently serving as transcription templates. Moreover, cellular processes for very fast disposal of nonfunctional/unusable transcripts do exist.

In any case, the question of template strand selection in transcription was soon answered by J. Marmur's group at Brandeis, Hayashi and Spiegelman, as well as G. P. Tocchini-Valentini and co-workers at Chicago and the International Laboratory of Genetics and Biophysics in Naples, with bacterial RNA polymerase preparations that yielded strand-selective transcription, implying the ability of the enzyme to select specific DNA sites for production of RNA *in vitro*. In hindsight, the something missing from, or inactivated in, the polymerase preparations used for the initial experiments must have been σ, the initiation-specific subunit of bacterial RNA polymerases, discovered several years later by R. R. Burgess and A. A. Travers at Harvard, and E. K. F. Bautz and J. J. Dunn at Rutgers. Within the next year, partially purified bacterial RNA polymerase had been prepared in several laboratories, including those already referred to, W. Zillig's group and others. Those first experiments also provided what we now understand to have been a straightforward and simple demonstration of the existence of genes under positive transcriptional control: "asymmetric" transcription of phage T4 DNA, and the DNA of large-tailed phages infecting *Bacillus subtilis*, selectively yielded transcripts that correspond to RNA produced at the outset of phage infection, the so-called early RNA. This is, in a sense, the historical baseline for the "modern synthesis" represented by this book.

E. P. Geiduschek
Division of Biological Sciences, UCSD, La Jolla

Preface to the First Edition

It is now fashionable to view the cell as an ensemble of molecular machines or motors working in concert. What exactly do we mean by employing these terms normally reserved for industry and engineering? Up to what point is the analogy drawn between man-made motors and Nature's molecular motors justified? When does this analogy become counterproductive? If one considers RNA polymerase, the analogy is evidently useful for helping ask the right questions about its function. How does a given RNA polymerase molecule cope with the various biochemical and biomechanical tasks it has to accomplish – physically translocating along DNA, acting as a faithful chemical replicator, and properly targeting biological "start" and "stop" signals along a given DNA sequence? These are some of the naïve questions posed and addressed within the ten chapters of this book.

Tackling these challenges and concisely reporting the present achievements and bottlenecks encountered in this venture requires both clarity and humility. We are very grateful to all the authors for their illuminating contributions, and for their compliance with our numerous requests.

We are also extremely grateful to David Lilley for having offered us the opportunity to publish this book in the prestigious collection of the RSC Biomolecular Science series. It has been a pleasure working with Janet Freshwater and Annie Jacob of the Royal Society of Chemistry and we thank them for their support, patience and humor.

In Dante's Inferno there must exist a tiny, poorly-lit room where pale fellows frantically scratch at thick piles of papers while still receiving "final" versions, totally ruining their previous efforts. We have left Tantalus and Sisyphus there, thanks to our saviors, Marie-Héléne and Cary.

Henri Buc and Terence Strick

Chemical Biology No. 19
RNA Polymerases as Molecular Motors: On the Road, 2nd Edition
Edited by Robert Landick, Terence Strick and Jue Wang

Published by the Royal Society of Chemistry, www.rsc.org

Contents

Chemical Biology No. 19
RNA Polymerases as Molecular Motors: On the Road, 2nd Edition
Edited by Robert Landick, Terence Strick and Jue Wang

Published by the Royal Society of Chemistry, www.rsc.org

Structural Atlas

SETH DARST*

Rockefeller University, 1230 York Avenue, New York, NY 10065, USA
*E-mail: darst@rockefeller.edu

Chemical Biology No. 19
RNA Polymerases as Molecular Motors: On the Road, 2nd Edition
Edited by Robert Landick, Terence Strick and Jue Wang

Published by the Royal Society of Chemistry, www.rsc.org

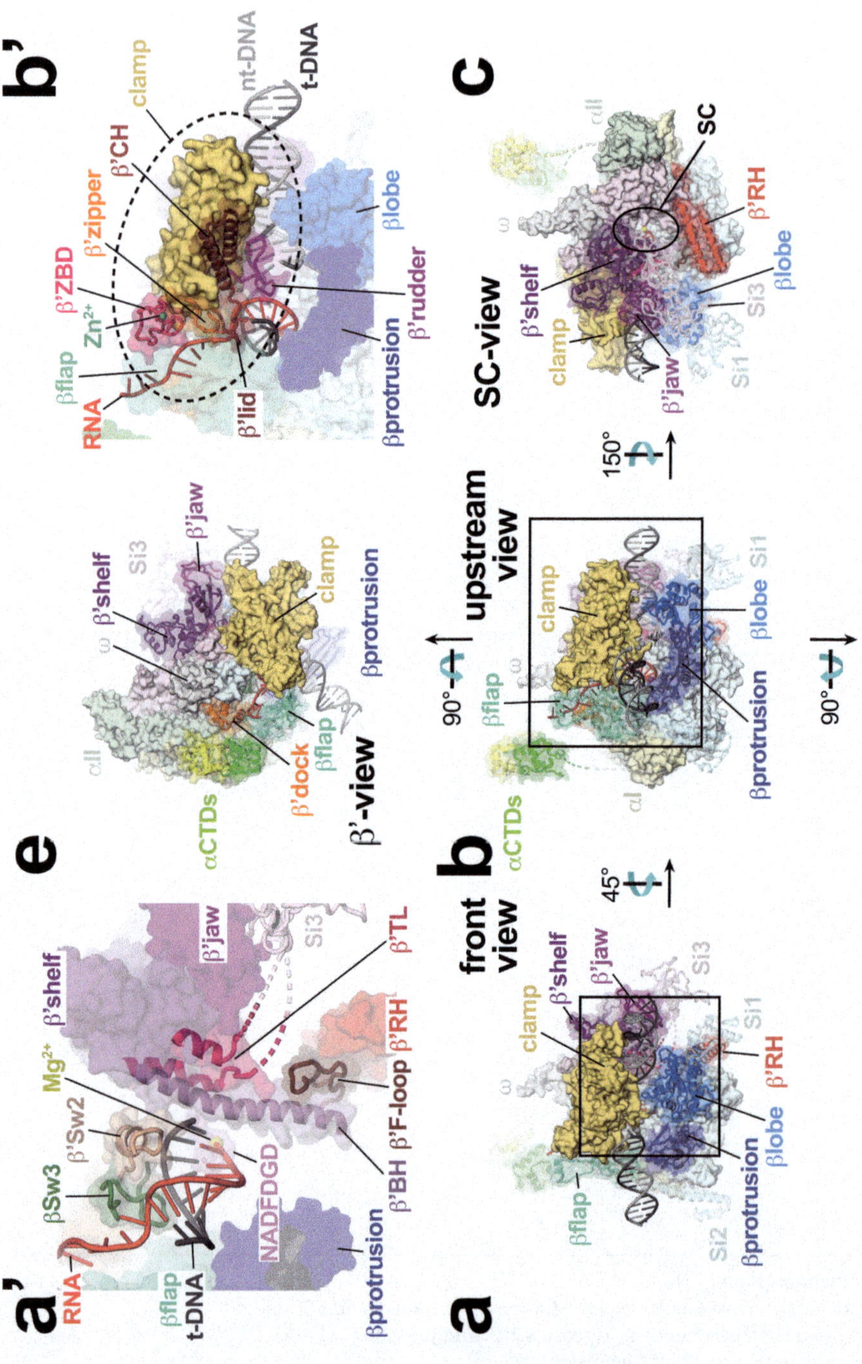
a'
RNA
βflap
t-DNA
βSw3
β'Sw2
Mg2+
NADFDGD
βprotrusion
β'BH
β'F-loop
β'RH
β'TL
β'shelf
β'jaw
Si3
e
αCTDs
αII
ω
β'shelf
Si3
β'jaw
clamp
β'dock
βflap
βprotrusion
β'-view
b'
RNA
βflap
Zn2+
β'ZBD
β'zipper
β'CH
clamp
nt-DNA
t-DNA
β'lid
βprotrusion
β'rudder
βlobe
a
front view
βflap
clamp
ω
β'shelf
β'jaw
Si3
Si1
β'RH
βlobe
βprotrusion
Si2
45°
b
upstream view
αCTDs
αI
βflap
ω
clamp
βprotrusion
βlobe
Si1
90°
90°
150°
c
SC-view
clamp
β'shelf
β'jaw
Si1
Si3
βlobe
β'RH
SC
ω
αII

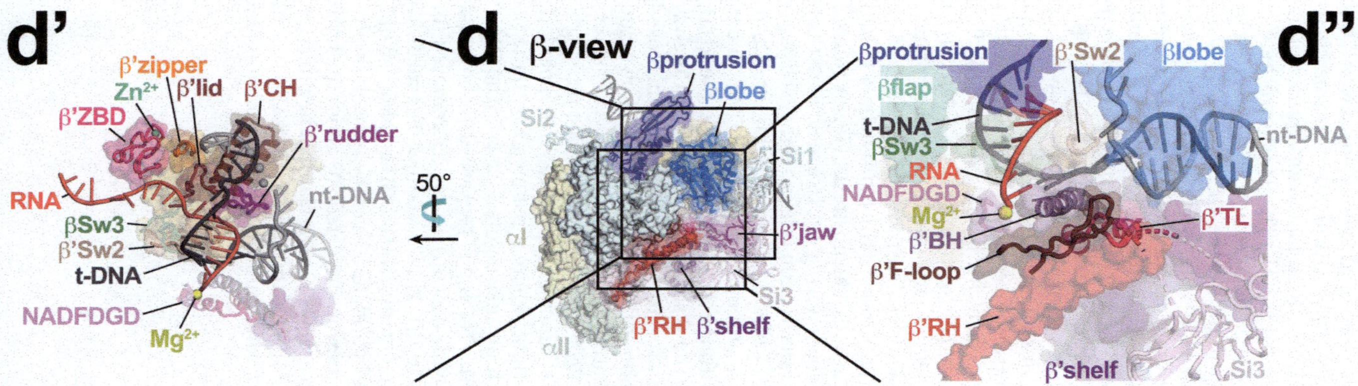

Figure 1 ***Escherichia coli*** **RNA polymerase structure**. Most of the structural features illustrated here were described by Zhang *et al.*[1] and Cramer *et al.*[2] Views of an *E. coli* (*Eco*) RNA polymerase (RNAP) transcription elongation complex (EC).[3] Proteins are shown as molecular surfaces. Some highlighted structural features are shown as backbone ribbons with transparent molecular surfaces. The nucleic acids (template-strand DNA, t-DNA; nontemplate-strand DNA, nt-DNA; RNA) are shown as cartoon worms. Lineage-specific inserts Si1, Si2, and Si3 (unique to *Eco* RNAP, also called βi4, βi9, and β′i6)[4] are shown as backbone ribbons without the molecular surface. The structural features illustrated here are also detailed in Table 1. (a) Front view of the EC, looking straight into the RNAP active site cleft between the clamp and βprotrusion–βlobe. The direction of transcription is from left to right. The downstream duplex DNA (on the right) is enclosed in a channel between the clamp and β′jaw, and the βlobe. (a′) Cutaway view (with obscuring features removed for clarity) showing the boxed region in (a), affording a view of structural features surrounding the RNAP active site. The active site is marked by the Mg^{2+}-ion chelated by the NADFDGD motif. The growing end of the RNA forms an RNA–DNA hybrid with the t-DNA, but upstream single-stranded RNA is directed up to the left in an RNA exit channel. The switch regions (βSw3 and β′Sw2) interact with the RNA–DNA hybrid, and β′Sw2 acts as a hinge connecting the mobile clamp domain to the rest of the RNAP. The bridge-helix (β′BH) bridges across the active site cleft. The BH and trigger-loop (TL)-helices interact to form a structural unit that is central to RNAP function, coordinating interactions with t-DNA and the incoming

Figure 1 (*continued*) substrate nucleoside triphosphate (NTP). The TL is a loop connecting the TL-helices, which tends to be unstructured in unliganded RNAP, but undergoes a folding transition to interact with the correct (matched) incoming nucleotide substrate.[5,6] The F-loop modulates transcription elongation by interacting with the folded TL. The TL shown here is unfolded. In *Eco* RNAP, the lineage-specific insertion Si3 is inserted by flexible linkers (dashed lines) in the middle of the TL. (b) The front view (a) is rotated 45° about the vertical axis as shown, giving rise to a view roughly parallel with the axis of the upstream duplex DNA (upstream view). The direction of transcription is from left to right. The α C-terminal domains (αCTDs, modeled) are linked by long, unstructured linkers (dashed lines) to the respective α N-terminal domains (αNTDs). (b′) Cutaway view (with obscuring features removed for clarity) showing the boxed region in (b) and focusing on structural elements of the clamp. The single-stranded RNA product can be seen extruding out the RNA exit channel underneath the βflap. The β′zinc-binding domain (β′ZBD) chelates a Zn^{2+}-ion (green sphere). (c) The upstream view is rotated 150° about the vertical axis as shown, giving a view into the secondary channel (SC-view) through which nucleotides can travel to the active site. The direction of transcription is right to left. The β′rim-helices (β′RH) form a rim at the entrance to the SC. (d) The upstream view is rotated 90° about the horizontal axis as shown, giving rise to a view onto the β subunit (β-view). The direction of transcription is left to right. The βlobe covers the downstream duplex DNA binding channel, while the βprotrusion covers the RNA–DNA hybrid and upstream duplex DNA. Due to the long linkers connecting Si3 with the TL, Si3 sits outside the RNAP active site cleft, interacting with the β′shelf and β′jaw. (d′). Cutaway view (with obscuring features removed for clarity) showing the boxed region in (d) but rotated by 50° about the vertical axis as shown and focusing primarily on structural features of the clamp. This view shows how the β′lid acts to disrupt the RNA–DNA hybrid, directing the single-stranded upstream RNA into the RNA exit channel towards the upper left, and the t-DNA towards the upper right, where it reanneals with the nt-DNA (grey spheres model the path of unstructured bases). The RNAP active site (NADFDGD and Mg^{2+}, BH, TL, and β′ jaw) are also shown (at the bottom). (d″). Cutaway view (with obscuring features removed for clarity) showing the boxed region in (d). This view shows the relationship between the RNAP active site elements (NADFDGD and Mg^{2+}, β′BH, β′TL, and β′F-loop), the RNA–DNA hybrid, and the downstream duplex DNA (on the right). (e) The upstream view is rotated 90° about the horizontal axis as shown, giving rise to a view onto the clamp. The direction of transcription is left to right. The single-stranded upstream RNA (red) can be seen extruding out the RNA exit channel between the clamp, the βflap, and the β′dock.

Table 1 RNAP structural features delineated in Figure 1.

RNAP subunit (color in Figure 1)	Structural module (color in Figure 1)	Approximates to residues (*Eco* RNAP numbering)	Visible in Figure 1	Comments	Reference
α (αI: pale yellow), (αII: pale green)	NTD	1–235	b, c, d, e	• Dimerize and form a platform for β and β′ assembly • αI: interacts primarily with β • αII: interacts primarily with β′	1,7–11
	CTD		a, b, c, e	• Interacts with a wide array of regulatory factors • Interacts with AT-rich DNA sequences upstream of the promoter −35 element (UP-elements)	8–10, 12–16
β (pale cyan)	protrusion (blue)	28–146; 456–519	a, a′, b, b′, d, d″, e	• Interacts with transcription activator CarD (*Thermus* and *Mycobacterium tuberculosis*) • Interacts with the mutation frequency decline (Mfd) [Transcription Repair Coupling Factor (TRCF)] protein	17–20
	lobe (marine blue)	153–225; 340–444	a, b, b′, c, d, d″	• Forms part of downstream duplex DNA binding channel	

(*continued*)

Table 1 (*continued*)

RNAP subunit (color in Figure 1)	Structural module (color in Figure 1)	Approximates to residues (*Eco* RNAP numbering)	Visible in Figure 1	Comments	Reference
	flap (green cyan)	829–937; 1040–1059	a, a′, b, b′, d″, e	• Interacts with σ domain 4 • Forms RNA exit channel • Plays a critical role in RNA hairpin-dependent transcriptional pausing • Interacts with many transcription regulators, including essential N-utilizing substance A (Nus)A	21–26
	Switch3 (Sw3) (forest green)	1248–1269	a′, d′, d″	• Conserved interactions with RNA–DNA hybrid	27
β′ (light pink)	Switch2 (Sw2) (wheat)	330–348	a′, d′, d″	• Hinge connecting clamp to rest of RNAP • Many bacterial RNAP inhibitors bind here and disrupt normal clamp dynamics, highlighting the importance of clamp motions to RNAP function	28–33
	dock (orange)	369–420	E	• Forms part of RNA exit channel	
	NADFDGD (violet)	458–464	a′, d′, d″	• Universally conserved motif of multi-subunit cellular RNAPs • Three aspartate residues (D's) chelate a Mg^{2+}-ion in the RNAP active site	1 and 34

Table 1 (*continued*)

RNAP subunit (color in Figure 1)	Structural module (color in Figure 1)	Approximates to residues (*Eco* RNAP numbering)	Visible in Figure 1	Comments	Reference
	rim-helices (RH) (red)	648–713	a, a′, c, d, d″	• Two anti-parallel α-helices form a rim at the entrance to the secondary channel (SC, see below) • Provide a binding platform for SC-regulatory factors, such as growth regulator (Gre)-factors, transfer region R (TraR), and DnaK suppressor A (DksA)	35–44
	F-loop (chocolate)	742–765	a′, d″	• Plays a critical role in modulating catalysis and elongation rate by interacting with folded TL • Interacts with many RNAP inhibitors	6,45–51
	Bridge helix (BH) (pale violet)	769–803	a′, d″	• Bridges β- and β′-pincers across RNAP active site cleft, connecting the F-loop with the shelf • Central to RNAP function, coordinating interactions with template DNA, the incoming substrate NTP, and other key structural elements such as the trigger-loop helices (TL-helices) • Observed in straight and kinked conformations	1,21, 52–55

(*continued*)

Table 1 (*continued*)

RNAP subunit (color in Figure 1)	Structural module (color in Figure 1)	Approximates to residues (*Eco* RNAP numbering)	Visible in Figure 1	Comments	Reference
	shelf (violet purple)	787–914; 1147–1150; 1216–1317	a, a′, c, d, d″, e	• Key structural module that forms a part of the SC and supports the BH and TL-helices	
	TL-helices/ TL (hot pink)	915–940; 1133–1146	a′, d′, d″	• TL-helices and BH interact to form a structural unit central to RNAP function, coordinating interactions with template DNA and the incoming substrate NTP • The TL loop connects the TL-helices. The TL tends to be unstructured in unliganded RNAP, but undergoes a folding transition to interact with the correct (matched) incoming nucleotide substrate. Correct folding of the TL is critical for substrate recognition, enzyme fidelity, and efficient catalysis • Many bacterial lineages contain a large lineage-specific insert in the middle of the TL, including *Eco* (Si3, see below)	5,21 and 52
	jaw (purple)	1151–1215	a, a′, c, d, d′, d″, e	• Conserved structural element unique to bacterial RNAPs • Interacts with downstream duplex DNA • Interacts with β′Si3 (when present)	

Table 1 (*continued*)

RNAP subunit (color in Figure 1)	Structural module (color in Figure 1)	Approximates to residues (*Eco* RNAP numbering)	Visible in Figure 1	Comments	Reference
Clamp					
β, β′	clamp (yellow orange)	β: 1319–1342 β′: 1–342; 1318–1344 Zn^{2+}–I	a, b, b′, c, d′, e	• Comprises one pincer of the RNAP crab-claw • Highly mobile structural element; opens and closes to clamp on nucleic acids (RNA–DNA hybrid and downstream duplex DNA) within RNAP active site cleft	27,38,56 and 57
	β′zipper (orange)	44–53	b′, d′	• Residues Y46 and R47 interact with the DNA backbone in the spacer between the −35 and −10 elements of RNAP holoenzyme open promoter complex	58 and 59
	β′ZBD (hot pink)	64–94	b′, d′	• Conserved structural element • Involved in diverse protein–protein and protein–nucleic acid interactions	
	β′lid (firebrick red)	249–265	b′, d′	• Plays a role in separating the RNA transcript from the RNA–DNA hybrid	3,21,60 and 61
	β′clamp helices (CH) (chocolate)	266–306	b′, d′	• Provides a platform for binding of regulatory factors such as σ or N-utilizing substance • G (NusG)	22,62 and 63
	β′rudder (magenta)	307–327	b′, d′	• Contributes to elongation complex stability	64

(*continued*)

Table 1 (*continued*)

RNAP subunit (color in Figure 1)	Structural module (color in Figure 1)	Approximates to residues (*Eco* RNAP numbering)	Visible in Figure 1	Comments	Reference
Lineage-Specific Inserts					
β	Si1(βi4)	226–339	a, b, b′, c, d	• Inserted in βlobe • Function unknown	4,65 and 66
	Si2(βi9)	938–1039	a, d,	• Inserted in βflap • Function unknown	4,66 and 67
β′	Si3(β′i6)	941–1132	a, a′, c, d″, e	• Inserted in the middle of the TL • Modulates transcriptional pausing by controlling TL folding	68 and 69

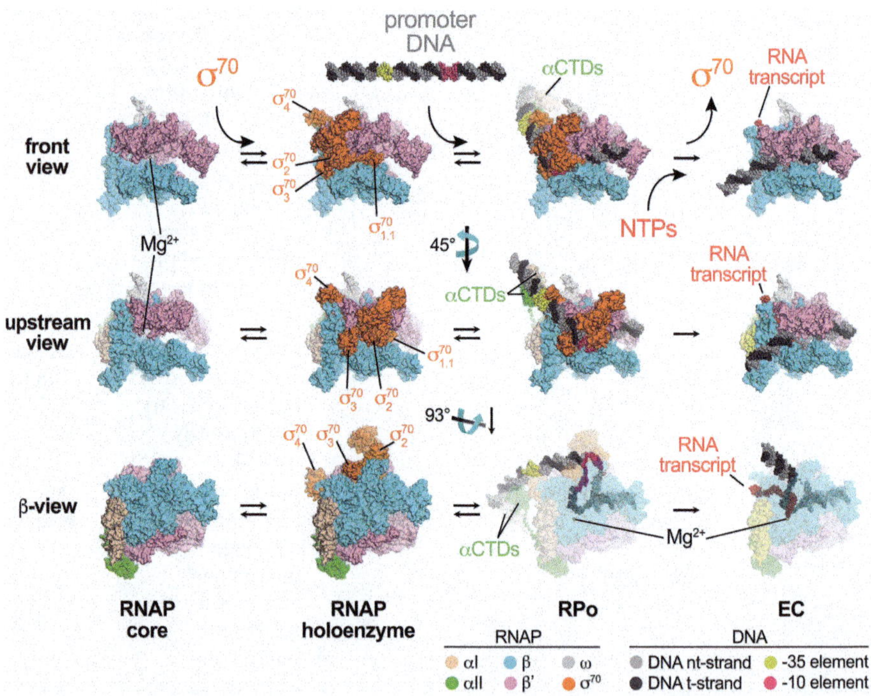

Figure 2 ***Eco* RNAP structure progressing through the transcription cycle.** Structures of *Eco* RNAP are shown in different stages of the transcription cycle. Proteins are shown as molecular surfaces; nucleic acids are shown as atomic spheres

(color-coding is shown in the key at lower right). Each row shows a different view of the complex (top row, front view; middle row, upstream view; bottom row, β-view). Each column shows a different complex in the transcription cycle. From left to right; RNAP core, RNAP holoenzyme (core + σ^{70}), RNAP holoenzyme open promoter complex (RPo), and finally the transcription elongation complex (RNAP core + DNA template + RNA transcript; EC). The RPo structure shown here was adapted from PDB 6OUL.[38] The RNAP holoenzyme and RNAP core models shown here were derived from 6OUL by removing the DNA and σ^{70}, respectively. The disposition of $\sigma^{70}_{1.1}$ in the RNAP holoenzyme structure was modeled from PDB 4LK1.[70] The EC structure was adapted from PDB 6ALF.[3] RNAP core: The front and upstream views illustrate the overall crab-claw-like shape of the RNAP. The top pincer in this view comprises primarily the β′ subunit, with the main part of the pincer comprising the clamp domain (see Figure 1a, b, b′, c, d′ and e). The bottom pincer comprises primarily the β subunit. The large open cleft between the pincers contains the RNAP active site (marked by a Mg^{2+}-ion) and will accommodate the nucleic acids in the RPo and EC structures. RNAP holoenzyme: The σ^{70} structure comprises a series of structured domains ($\sigma^{70}_{1.1}$, σ^{70}_{2}, σ^{70}_{3}, and σ^{70}_{4}) linked by flexible linkers.[71–74] Upon binding to the RNAP core, $\sigma^{70}_{1.1}$ situates in the RNAP active site cleft, possibly serving to protect the RNAP active site from engaging with non-specific nucleic acids.[70] The $\sigma^{70}_{1.1}$ is ejected from the RNAP active site cleft upon the entry of promoter DNA.[75] The main functional role of σ^{70}_{2}, which binds to the β′clamp-helices (β′CH; see Figure 1b′, d′), is to initiate transcription bubble formation and recognize the promoter −10 element.[76] The σ^{70}_{3} recognizes the promoter extended −10 element, when present.[59,77] The main functional role of σ^{70}_{4}, which comprises a canonical helix–turn–helix DNA binding domain, is to recognize the promoter −35 element.[72] The σ^{70}_{4} is perched on the RNAP βflap-tip helix (see Figure 1a, b, b′, e). Binding of σ^{70}_{2} to the β′CHs and σ^{70}_{4} to the βflap positions the two σ^{70} domains appropriately to engage the promoter −10 and −35 elements, that are typically separated by a 17 base pair spacer.[22,78,79] The σ^{70}_{3} and σ^{70}_{4} are linked by a long linker (the σ^{70}_{finger}) that snakes into the RNAP active site cleft near the RNAP active site, then back out through the RNA exit channel.[22] The σ^{70}_{finger} facilitates *de novo* initiation (initiation of RNA chain synthesis that does not require an oligonucleotide primer) by helping to properly position the t-strand DNA,[72,80,81] but upon RNA chain initiation and subsequent elongation, the σ^{70}_{finger} becomes an obstacle to RNA chain elongation that must be displaced.[6,22,82,83] RPo: Location of a duplex promoter sequence by the RNAP holoenzyme triggers a series of conformational changes as the enzyme unwinds 12 to 14 base pairs of DNA to generate the transcription bubble and loads the t-strand DNA into the RNAP active site.[30,84–87] This multi-step process culminates in the formation of the transcription competent RPo. In RPo, σ^{70}_{4} recognizes the duplex −35 element and σ^{70}_{2} engages the −10 element and stabilizes the upstream edge of the transcription bubble.[59,72,76] The single-stranded transcription start site is loaded into the RNAP active site, and the duplex DNA downstream of the start site is bound in the downstream duplex channel, displacing $\sigma^{70}_{1.1}$.[75] At many promoters, the αCTDs interact with AT-rich DNA upstream of the −35 element (UP-elements).[12,14–16] Additional promoter–RNAP holoenzyme interactions are detailed in Figure 3. EC: The formation of RPo allows NTP substrates to enter the RNAP active site (through the SC-channel; Figure 1c) and RNA chain extension to begin. The initial elongation of the RNA is characterized by fits and starts, marked by scrunching and abortive initiation (release of short RNA chains without release of RNAP holoenzyme from the promoter),[88–90] but ultimately the RNA product is extruded through the RNA exit channel and σ^{70} is released, giving rise to the EC.

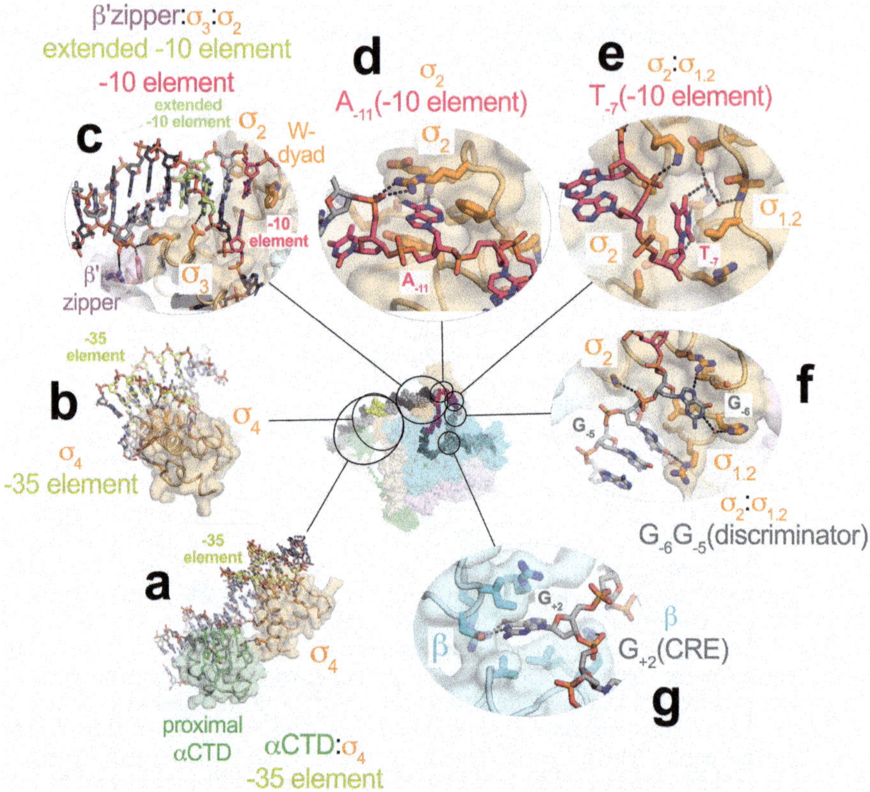

Figure 3 **Structural determinants of RNAP holoenzyme–promoter DNA interactions.** The center of the image shows the β-view of RPo (the same as Figure 2). The circled features are highlighted around the outside of the image. (a) UP-element: At some promoters, the αCTDs bind to an AT-rich sequence just upstream of the promoter −35 element.[12,14,15] The proximal αCTD (the one closest to RNAP) can make favorable interactions with σ_4^{70} [16] as shown here. (b) −35 element (consensus $^{-35}$TTGACA^{-30}; where the superscripts indicate the usual position of the element with respect to the transcription start site, which can vary):[78] In *Eco*, about 43% of σ^{70} promoters contain an identifiable −35 element,[91] making it the most prevalent promoter element after the −10 element. The −35 element sequence is recognized by σ_4^{70}, a canonical helix–turn–helix DNA binding domain.[72,79] (c) Extended −10 element ($^{-15}$TG^{-14}): In *Eco*, about 18% of σ^{70} promoters contain an extended −10 element; a TG sequence spaced one base pair upstream of the −10 element.[91,92] The presence of an extended −10 element can substitute for the absence of a −35 element.[93] The extended −10 element is recognized by residues of σ_3^{70}.[59,77] Upstream of the extended −10, residues from the β′zipper interact with the DNA phosphate backbone at the −17 position of the nt-strand.[58,59] (d) −10 element ($^{-12}$TATAAT^{-7}; A_{-11}): The −10 element is the most conserved promoter element, and the A at the second position (normally A_{-11}) is the most conserved base of the −10 element.[78] The A_{-11} is thought to be the first base that flips out of the duplex DNA stack to nucleate transcription bubble formation.[76,94–96] The flipped-out A_{-11} is captured in a cognate pocket of σ_2^{70}.[76] (e) −10 element

($^{-12}$TATAAT^{-7}; T_{-7}): The last T of the −10 element (normally T_{-7}) is nearly as conserved as the A_{-11}.[78] T_{-7} also flips out of the DNA base stack and is captured in a separate pocket of σ_2^{70}.[76] (f) Discriminator ($^{-6}$GG^{-5}): The 'discriminator' is defined as the sequence between the −10 element and the start site (to −6 to −1). Characteristics of the discriminator have a profound effect on RPo stability.[97–99] The two positions just downstream of the −10 element are particularly important, where a $^{-6}$GG^{-5} sequence can form specific interactions with σ_2^{70}.[80] (g) Core recognition element (CRE: G_{+2}): A G at the +2 position of the nucleotide strand can bind in a G-specific pocket of the β-subunit.[80]

References

1. G. Zhang, *et al.*, Crystal structure of *Thermus aquaticus* core RNA polymerase at 3.3 A resolution, *Cell*, 1999, **98**, 811–824.
2. P. Cramer, D. A. Bushnell and R. D. Kornberg, Structural basis of transcription: RNA polymerase II at 2.8 angstrom resolution, *Science*, 2001, **292**, 1863–1876.
3. J. Y. Kang, *et al.*, Structural basis of transcription arrest by coliphage HK022 nun in an *Escherichia coli* RNA polymerase elongation complex, *eLife*, 2017, **6**, e25478.
4. W. J. Lane and S. A. Darst, Molecular evolution of multisubunit RNA polymerases: sequence analysis, *J. Mol. Biol.*, 2010, **395**, 671–685.
5. D. Wang, D. A. Bushnell, K. D. Westover, C. D. Kaplan and R. D. Kornberg, Structural basis of transcription: role of the trigger loop in substrate specificity and catalysis, *Cell*, 2006, **127**, 941–954.
6. Y. Zuo and T. A. Steitz, Crystal structures of the *E. coli* transcription initiation complexes with a complete bubble, *Mol. Cell*, 2015, **58**, 534–540.
7. W. Zillig, P. Palm and A. Heil, Function and reassembly of subunits of DNA-dependent RNA polymerase, in *RNA Polymerase*, ed. R. Losick and M. Chamberlin, Cold Spring Harbor Laboratory, 1976, vol. 101.
8. K. Igarashi and Ak. Ishihama, Bipartite functional map of the *E. coli* RNA polymerase alpha subunit: involvement of the C-terminal region in transcription activation by cAMP-CRP, *Cell*, 1991, **65**, 1015–1022.
9. R. S. Hayward, K. Igarashi and A. Ishihama, Functional specialization within the α-subunit of *Escherichia coli* RNA polymerase, *J. Mol. Biol.*, 1991, **221**, 23–29.
10. E. E. Blatter, W. Ross, H. Tang, R. L. Gourse and R. H. Ebright, Domain organization of RNA polymerase alpha subunit: C-terminal 85 amino acids constitute a domain capable of dimerization and DNA binding, *Cell*, 1994, **78**, 889–896.
11. G. Zhang and S. A. Darst, Structure of the *Escherichia coli* RNA polymerase α subunit amino-terminal domain, *Science*, 1998, **281**, 262–266.
12. W. Ross, *et al.*, A third recognition element in bacterial promoters: DNA binding by the alpha subunit of RNA polymerase, *Science*, 1993, **262**, 1407–1413.

13. Y. H. Jeon, *et al.*, Solution structure of the activator contact domain of the RNA polymerase alpha subunit, *Science*, 1995, **270**, 1495–1497.
14. W. Ross, A. Ernst and R. L. Gourse, Fine structure of *E. coli* RNA polymerase-promoter interactions: alpha subunit binding to the UP element minor groove, *Genes Dev.*, 2001, **15**, 491–506.
15. B. Benoff, *et al.*, Structural basis of transcription activation: the CAP-alpha CTD-DNA complex, *Science*, 2002, **297**, 1562–1566.
16. W. Ross, D. A. Schneider, B. J. Paul, A. Mertens and R. L. Gourse, An intersubunit contact stimulating transcription initiation by *E. coli* RNA polymerase: interaction of the α C-terminal domain and sigma region 4, *Genes Dev.*, 2003, **17**, 1293–1307.
17. A. M. Deaconescu, *et al.*, Structural basis for bacterial transcription-coupled DNA repair, *Cell*, 2006, **124**, 507–520.
18. C. L. Stallings, *et al.*, CarD is an essential regulator of rRNA transcription required for *Mycobacterium tuberculosis* persistence, *Cell*, 2009, **138**, 146–159.
19. L. F. Westblade, *et al.*, Structural basis for the bacterial transcription-repair coupling factor/RNA polymerase interaction, *Nucleic Acids Res.*, 2010, **38**, 8357–8369.
20. B. Bae, *et al.*, CarD uses a minor groove wedge mechanism to stabilize the RNA polymerase open promoter complex, *eLife*, 2015, **4**, e08505.
21. D. G. Vassylyev, M. N. Vassylyeva, A. Perederina, T. H. Tahirov and I. Artsimovitch, Structural basis for transcription elongation by bacterial RNA polymerase, *Nature*, 2007, **448**, 157–162.
22. K. S. Murakami, S. Masuda and S. A. Darst, Structural basis of transcription initiation: RNA polymerase holoenzyme at 4 A resolution, *Science*, 2002, **296**, 1280–1284.
23. K. Kuznedelov, *et al.*, A role for interaction of the RNA polymerase flap domain with the sigma subunit in promoter recognition, *Science*, 2002, **295**, 855–857.
24. I. Toulokhonov and R. Landick, The flap domain is required for pause RNA hairpin inhibition of catalysis by RNA polymerase and can modulate intrinsic termination, *Mol. Cell*, 2003, **12**, 1125–1136.
25. X. Guo, *et al.*, Structural basis for NusA stabilized transcriptional pausing, *Mol. Cell*, 2018, **69**, 816–827.e4.
26. K. S. Ha, I. Toulokhonov, D. G. Vassylyev and R. Landick, The NusA N-terminal domain is necessary and sufficient for enhancement of transcriptional pausing via interaction with the RNA exit channel of RNA polymerase, *J. Mol. Biol.*, 2010, **401**, 708–725.
27. A. L. Gnatt, P. Cramer, J. Fu, D. A. Bushnell and R. D. Kornberg, Structural basis of transcription: an RNA polymerase II elongation complex at 3.3 A resolution, *Science*, 2001, **292**, 1876–1882.
28. W. Lin, *et al.*, Structural basis of transcription inhibition by fidaxomicin (Lipiarmycin A3), *Mol. Cell*, 2018, **70**, 60–71.e15.
29. H. Boyaci, *et al.*, Fidaxomicin jams *Mycobacterium tuberculosis* RNA polymerase motions needed for initiation via RbpA contacts, *eLife*, 2018, **7**, e34823.

30. H. Boyaci, J. Chen, R. Jansen, S. A. Darst and E. A. Campbell, Structures of an RNA polymerase promoter melting intermediate elucidate DNA unwinding, *Nature*, 2019, **565**, 382–385.
31. J. Mukhopadhyay, *et al.*, The RNA polymerase "switch region" is a target for inhibitors, *Cell*, 2008, **135**, 295–307.
32. G. A. Belogurov, *et al.*, Transcription inactivation through local refolding of the RNA polymerase structure, *Nature*, 2008, **457**, 332–335.
33. A. Srivastava, *et al.*, New target for inhibition of bacterial RNA polymerase: "switch region", *Curr. Opin. Microbiol.*, 2011, **14**, 532–543.
34. L. M. Iyer, E. V. Koonin and L. Aravind, Evolutionary connection between the catalytic subunits of DNA-dependent RNA polymerases and eukaryotic RNA-dependent RNA polymerases and the origin of RNA polymerases, *BMC Struct. Biol.*, 2003, **3**, 1–23.
35. O. Laptenko, J. Lee, I. Lomakin and S. Borukhov, Transcript cleavage factors GreA and GreB act as transient catalytic components of RNA polymerase, *EMBO J.*, 2003, **22**, 6322–6334.
36. N. Opalka, *et al.*, Structure and function of the transcription elongation factor GreB bound to bacterial RNA polymerase, *Cell*, 2003, **114**, 335–345.
37. E. Sosunova, *et al.*, Donation of catalytic residues to RNA polymerase active center by transcription factor Gre, *Proc. Natl. Acad. Sci. U. S. A.*, 2003, **100**, 15469–15474.
38. J. Chen, *et al.*, *E. coli* TraR allosterically regulates transcription initiation by altering RNA polymerase conformation, *eLife*, 2019, **8**, e49375.
39. V. Molodtsov, *et al.*, Allosteric effector ppGpp potentiates the inhibition of transcript initiation by DksA, *Mol. Cell*, 2018, **69**, 828–839.e5.
40. M. Abdelkareem, *et al.*, Structural basis of transcription: RNA polymerase backtracking and its reactivation, *Mol. Cell*, 2019, **75**, 298–309.e4.
41. S. Gopalkrishnan, W. Ross, A. Y. Chen and R. L. Gourse, TraR directly regulates transcription initiation by mimicking the combined effects of the global regulators DksA and ppGpp, *Proc. Natl. Acad. Sci. U. S. A.*, 2017, **114**, E5539–E5548.
42. A. Perederina, *et al.*, Regulation through the secondary channel--structural framework for ppGpp-DksA synergism during transcription, *Cell*, 2004, **118**, 297–309.
43. B. E. Nickels and A. Hochschild, Regulation of RNA polymerase through the secondary channel, *Cell*, 2004, **118**, 1–4.
44. A. Parshin, *et al.*, DksA regulates RNA polymerase in *Escherichia coli* through a network of interactions in the secondary channel that includes Sequence Insertion 1, *Proc. Natl. Acad. Sci. U. S. A.*, 2015, **112**, E6862–E6871.
45. N. Miropolskaya, I. Artsimovitch, S. Klimašauskas, V. Nikiforov and A. Kulbachinskiy, Allosteric control of catalysis by the F loop of RNA polymerase, *Proc. Natl. Acad. Sci. U. S. A.*, 2009, **106**, 18942–18947.
46. N. Miropolskaya, V. Nikiforov, S. Klimašauskas, I. Artsimovitch and A. Kulbachinskiy, Modulation of RNA polymerase activity through the trigger loop folding, *Transcription*, 2010, **1**, 89–94.

47. N. Miropolskaya, *et al.*, Interplay between the trigger loop and the F loop during RNA polymerase catalysis, *Nucleic Acids Res.*, 2014, **42**, 544–552.
48. B. Bae, *et al.*, CBR antimicrobials inhibit RNA polymerase via at least two bridge-helix cap-mediated effects on nucleotide addition, *Proc. Natl. Acad. Sci. U. S. A.*, 2015, **112**, E4178–E4187.
49. Y. Feng, *et al.*, Structural basis of transcription inhibition by CBR hydroxamidines and CBR pyrazoles, *Structure*, 2015, **23**, 1470–1481.
50. N. R. Braffman, *et al.*, Structural mechanism of transcription inhibition by lasso peptides microcin J25 and capistruin, *Proc. Natl. Acad. Sci. U. S. A.*, 2019, **116**, 1273–1278.
51. D. Degen, *et al.*, Transcription inhibition by the depsipeptide antibiotic salinamide A, *eLife*, 2014, **3**, e02451.
52. D. G. Vassylyev, *et al.*, Structural basis for substrate loading in bacterial RNA polymerase, *Nature*, 2007, **448**, 163–168.
53. A. Weixlbaumer, K. Leon, R. Landick and S. A. Darst, Structural basis of transcriptional pausing in bacteria, *Cell*, 2013, **152**, 431–441.
54. S. Tagami, *et al.*, Crystal structure of bacterial RNA polymerase bound with a transcription inhibitor protein, *Nature*, 2010, **468**, 978–982.
55. S. Tagami, S. Sekine and S. Yokoyama, A novel conformation of RNA polymerase sheds light on the mechanism of transcription, *Transcription*, 2011, **2**, 162–167.
56. A. Chakraborty, *et al.*, Opening and closing of the bacterial RNA polymerase clamp, *Science*, 2012, **337**, 591–595.
57. A. Feklistov, *et al.*, RNA polymerase motions during promoter melting, *Science*, 2017, **356**, 863–866.
58. Y. Yuzenkova, V. R. Tadigotla, K. Severinov and N. Zenkin, A new basal promoter element recognized by RNA polymerase core enzyme, *EMBO J.*, 2011, **30**, 3766–3775.
59. B. Bae, A. Feklistov, A. Lass-Napiorkowska, R. Landick and S. A. Darst, Structure of a bacterial RNA polymerase holoenzyme open promoter complex, *eLife*, 2015, **4**, e08504.
60. T. Naryshkina, K. Kuznedelov and K. Severinov, The role of the largest RNA polymerase subunit lid element in preventing the formation of extended RNA-DNA hybrid, *J. Mol. Biol.*, 2006, **361**, 634–643.
61. I. Toulokhonov and R. Landick, The role of the lid element in transcription by *E. coli* RNA polymerase, *J. Mol. Biol.*, 2006, **361**, 644–658.
62. D. G. Vassylyev, *et al.*, Crystal structure of a bacterial RNA polymerase holoenzyme at 2.6 Å resolution, *Nature*, 2002, **417**, 712–719.
63. J. Y. Kang, *et al.*, Structural basis for transcript elongation control by NusG family universal regulators, *Cell*, 2018, 1650–1662.e14.
64. K. Kuznedelov, N. Korzheva, A. Mustaev and K. Severinov, Structure-based analysis of RNA polymerase function: the largest subunit's rudder contributes critically to elongation complex stability and is not involved in the maintenance of RNA–DNA hybrid length, *EMBO J.*, 2002, **21**, 1369–1378.

65. K. Severinov, *et al.*, A non-essential domain of *Escherichia coli* RNA polymerase required for the action of the termination factor Alc, *J. Biol. Chem.*, 1994, **269**, 14254–14259.
66. I. Artsimovitch, Co-overexpression of *Escherichia coli* RNA polymerase subunits allows isolation and analysis of mutant enzymes lacking lineage-specific sequence insertions, *J. Biol. Chem.*, 2003, **278**, 12344–12355.
67. S. Borukhov, *et al.*, Mapping of trypsin cleavage and antibody-binding sites and delineation of a dispensable domain in the beta subunit of *Escherichia coli* RNA polymerase, *J. Biol. Chem.*, 1991, **266**, 23921–23926.
68. M. Chlenov, *et al.*, Structure and function of lineage-specific sequence insertions in the bacterial RNA polymerase β′ subunit, *J. Mol. Biol.*, 2005, **353**, 138–154.
69. J. Y. Kang, *et al.*, RNA polymerase accommodates a pause RNA hairpin by global conformational rearrangements that prolong pausing, *Mol. Cell*, 2018, **69**, 802–815.e1.
70. B. Bae, *et al.*, Phage T7 Gp2 inhibition of *Escherichia coli* RNA polymerase involves misappropriation of σ70 domain 1.1, *Proc. Natl. Acad. Sci. U. S. A.*, 2013, **110**, 19772–19777.
71. M. Lonetto, M. Gribskov and C. A. Gross, The sigma 70 family: sequence conservation and evolutionary relationships, *J. Bacteriol.*, 1992, **174**, 3843–3849.
72. E. A. Campbell, *et al.*, Structure of the bacterial RNA polymerase promoter specificity sigma subunit, *Mol. Cell*, 2002, **9**, 527–539.
73. T. M. Gruber and C. A. Gross, Multiple sigma subunits and the partitioning of bacterial transcription space, *Annu. Rev. Microbiol.*, 2003, **57**, 441–466.
74. A. Feklistov, B. D. Sharon, S. A. Darst and C. A. Gross, Bacterial sigma factors: a historical, structural, and genomic perspective, *Annu. Rev. Microbiol.*, 2014, **68**, 357–376.
75. V. Mekler, *et al.*, Structural organization of bacterial RNA polymerase holoenzyme and the RNA polymerase-promoter open complex, *Cell*, 2002, **108**, 599–614.
76. A. Feklistov and S. A. Darst, Structural basis for promoter -10 element recognition by the bacterial RNA polymerase σ Subunit, *Cell*, 2011, **147**, 1257–1269.
77. K. A. Barne, J. A. Bown, S. J. W. Busby and S. D. Minchin, Region 2.5 of the *Escherichia coli* RNA polymerase sigma70 subunit is responsible for the recognition of the "extended-10" motif at promoters, *EMBO J.*, 1997, **16**, 4034–4040.
78. R. K. Shultzaberger, Z. Chen, K. A. Lewis and T. D. Schneider, Anatomy of *Escherichia coli* sigma 70 promoters, *Nucleic Acids Res.*, 2007, **35**, 771–788.
79. K. S. Murakami, S. Masuda, E. A. Campbell, O. Muzzin and S. A. Darst, Structural basis of transcription initiation: an RNA polymerase holoenzyme-DNA complex, *Science*, 2002, **296**, 1285–1290.

80. Y. Zhang, *et al.*, Structural basis of transcription initiation, *Science*, 2012, **338**, 1076–1080.
81. A. Kulbachinskiy and A. Mustaev, Region 3.2 of the sigma subunit contributes to the binding of the 3′-initiating nucleotide in the RNA polymerase active center and facilitates promoter clearance during initiation, *J. Biol. Chem.*, 2006, **281**, 18273–18276.
82. L. Li, V. Molodtsov, W. Lin, R. H. Ebright and Y. Zhang, RNA extension drives a stepwise displacement of an initiation-factor structural module in initial transcription, *Proc. Natl. Acad. Sci. U. S. A.*, 2020, **117**, 5801–5809.
83. R. S. Basu, *et al.*, Structural basis of transcription initiation by bacterial RNA polymerase holoenzyme, *J. Biol. Chem.*, 2014, **289**, 24549–24559.
84. R. M. Saecker, M. T. Record Jr and P. L. deHaseth, Mechanism of bacterial transcription initiation: RNA polymerase - promoter binding, isomerization to initiation-competent open complexes, and initiation of RNA synthesis, *J. Mol. Biol.*, 2011, **412**, 754–771.
85. E. F. Ruff, M. T. Record Jr and I. Artsimovitch, Initial events in bacterial transcription initiation, *Biomolecules*, 2015, **5**, 1035–1062.
86. E. A. Hubin, *et al.*, Structure and function of the mycobacterial transcription initiation complex with the essential regulator RbpA, *eLife*, 2017, **6**, e22520.
87. J. Chen, *et al.*, Stepwise promoter melting by bacterial RNA polymerase, *Mol. Cell*, 2020, **78**, 275–288.e6.
88. A. Revyakin, C. Liu, R. H. Ebright and T. R. Strick, Abortive initiation and productive initiation by RNA polymerase involve DNA scrunching, *Science*, 2006, **314**, 1139–1143.
89. A. N. Kapanidis, *et al.*, Initial transcription by RNA polymerase proceeds through a DNA-scrunching mechanism, *Science*, 2006, **314**, 1144–1147.
90. L. M. Hsu, Promoter clearance and escape in prokaryotes, *Biochim. Biophys. Acta*, 2002, **1577**, 191–207.
91. E. A. Hubin, M. Lilic, S. A. Darst and E. A. Campbell, Structural insights into the mycobacteria transcription initiation complex from analysis of X-ray crystal structures, *Nat. Commun.*, 2017, **8**, 16072.
92. S. Keilty and M. Rosenberg, Constitutive function of a positively regulated promoter reveals new sequences essential for activity, *J. Biol. Chem.*, 1987, **262**, 6389–6395.
93. A. Kumar, *et al.*, The minus 35-recognition region of *Escherichia coli* sigma 70 is inessential for initiation of transcription at an "extended minus 10" promoter, *J. Mol. Biol.*, 1993, **232**, 406–418.
94. Y. F. Chen and J. D. Helmann, The *Bacillus subtilis* flagellar regulatory protein sigma D: overproduction, domain analysis and DNA-binding properties, *J. Mol. Biol.*, 1995, **249**, 743–753.
95. H. M. Lim, H. J. Lee, S. Roy and S. Adhya, A "master" in base unpairing during isomerization of a promoter upon RNA polymerase binding, *Proc. Natl. Acad. Sci. U. S. A.*, 2001, **98**, 14849–14852.

96. E. Heyduk, K. Kuznedelov, K. Severinov and T. Heyduk, A consensus adenine at position -11 of the nontemplate strand of bacterial promoter is important for nucleation of promoter melting, *J. Biol. Chem.*, 2006, **281**, 12362–12369.
97. S. P. Haugen, *et al.*, rRNA promoter regulation by nonoptimal binding of σ region 1.2: an additional recognition element for RNA polymerase, *Cell*, 2006, **125**, 1069–1082.
98. S. P. Haugen, W. Ross, M. Manrique and R. L. Gourse, Fine structure of the promoter–region 1.2 interaction, *Proc. Natl. Acad. Sci. U. S. A.*, 2008, **105**, 1–6.
99. K. L. Henderson, *et al.*, Mechanism of transcription initiation and promoter escape by *E. coli* RNA polymerase, *Proc. Natl. Acad. Sci. U. S. A.*, 2017, **114**, E3032–E3040.

CHAPTER 1

The Transition from Transcription Initiation to Transcription Elongation: Start-site Selection, Initial Transcription, and Promoter Escape

JARED T. WINKELMAN[a,b], BRYCE E. NICKELS*[b] AND RICHARD H. EBRIGHT*[a]

[a]Department of Chemistry and Waksman Institute, Rutgers University, Piscataway, NJ 08854, USA; [b]Department of Genetics and Waksman Institute, Rutgers University, Piscataway, NJ 08854, USA
*E-mail: bnickels@waksman.rutgers.edu, ebright@waksman.rutgers.edu

1.1 Introduction

In transcription initiation, RNA polymerase (RNAP) binds to promoter DNA, unwinds a turn of promoter DNA to yield an RNAP–promoter open complex containing an unwound "transcription bubble," and selects a transcription start site (TSS). In the next step of initiation, termed "initial transcription," RNAP remains bound to the promoter and synthesizes an RNA product of a threshold length of approximately 11–15 nucleotides (nt). In the final step of initiation,

Chemical Biology No. 19
RNA Polymerases as Molecular Motors: On the Road, 2nd Edition
Edited by Robert Landick, Terence Strick and Jue Wang

Published by the Royal Society of Chemistry, www.rsc.org

termed "promoter escape," RNAP breaks free of the promoter to yield a transcription elongation complex that synthesizes the rest of the RNA product. As a result of research over the last two decades, we now have a detailed mechanistic understanding of TSS selection, and we now understand the broad outlines of initial transcription and promoter escape. Here we review the current understanding of TSS selection, initial transcription, and promoter escape, focusing on these processes as they occur in the best characterized example, transcription initiation by *Escherichia coli* RNAP-σ^{70} holoenzyme, but also summarizing these processes as they occur in eukaryotic RNAP I, II, and III.

1.2 Transcription Start-Site (TSS) Selection

1.2.1 Mechanism

In contrast to DNA polymerases, RNAP can initiate nucleic acid synthesis using both primer-independent and primer-dependent mechanisms.[1–7] In primer-independent transcription initiation, RNAP uses an initiating nucleoside triphosphate (NTP) and an extending NTP or, alternatively, uses a non-canonical initiating nucleotide[8,9] [NCIN; an NTP-related compound, such as nicotinamide adenine dinucleotide (NAD^+) or its reduced form (NADH)] and an extending NTP. In primer-dependent transcription initiation, RNAP uses a short, 2 to ~5 nucleotide, RNA primer ("nanoRNA") and an extending NTP.[10–13]

In both primer-independent and primer-dependent transcription initiation, TSS selection entails two steps: (1) placing the start-site nucleotide (position +1) and the next nucleotide (position +2) of the template DNA strand into the RNAP active-center product site ("P site") and addition site ("A site"), respectively; and (2) placing the initiating entity—the initiating NTP or NCIN in primer-independent initiation, or the 3′ nucleotide of the RNA primer in primer-dependent initiation—in the P site and the extending NTP in the A site, respectively.

The position of the TSS relative to promoter core elements (promoter −35 element and −10 element for *E. coli* RNAP-σ^{70} holoenzyme) is variable.[14–23] The results of a comprehensive analysis of TSS selection by RNAP-σ^{70} holoenzyme indicate that TSS selection most frequently entails placing the template-strand position located 7 nucleotides downstream of the promoter −10 element into the RNAP active-center P site to serve as position +1[14,22,24] (see Figure 1.1). However, the TSS can vary over a 5 bp window, allowing placement of template-strand positions located 6, 7, 8, 9, or 10 nucleotides downstream of the promoter −10 element into the RNAP active-center P site to serve as position +1.[14,22,24]

The variability in TSS selection raises the structural question of how placement of the template strand relative to the RNAP active center can vary by up to 5 nucleotides, which corresponds to variation by up to at least ~17 Å (5 nucleotides × ~3.4 Å per nucleotide). In principle, this variability could be accommodated by differences in RNAP conformation, differences in DNA conformation, or both.

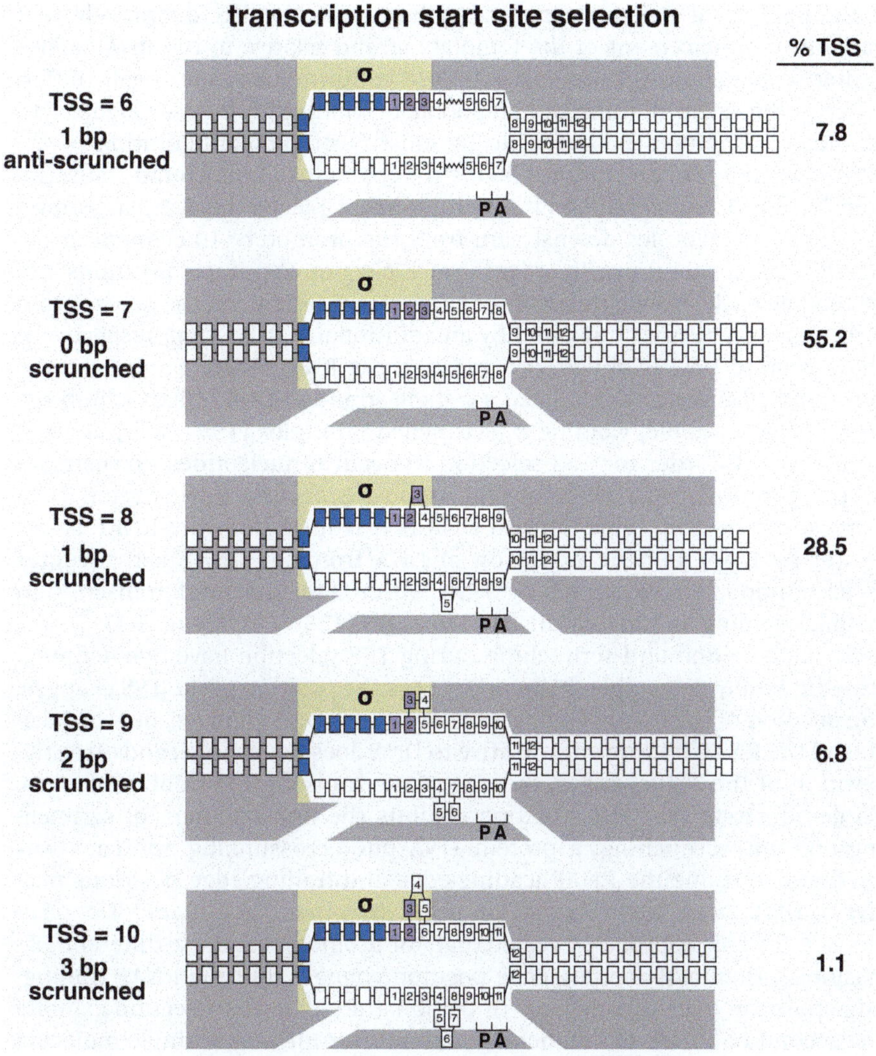

Figure 1.1 **Mechanism of transcription start site (TSS) selection.** Changes in TSS selection result from changes in DNA scrunching. The left column shows the TSS positions expressed in terms of distance downstream of the promoter −10 element (6, 7, 8, 9, or 10 nt downstream). The centre column shows the structures of anti-scrunched (TSS = 6), unscrunched (TSS = 7), and scrunched (TSS = 8, 9, 10) RNAP–promoter open complexes. Anti-scrunching is indicated by stretched lines between nucleotides; scrunching is indicated by bulged-out nucleotides. Positions of scrunched and anti-scrunched nucleotides are speculative. Gray, RNAP; yellow, σ; blue, −10-element nucleotides; purple, discriminator nucleotides; P and A, RNAP active center NTP-binding sites; boxes, DNA nucleotides (nontemplate-strand nucleotides above template-strand nucleotides; nucleotides downstream of the −10 element are numbered). The right column shows the observed percentage of transcription start sites at each position as measured by Vvedenskaya *et al.*[14]

Results from a series of experiments have now revealed, definitively, that variability in placement of the template strand relative to the RNAP active center is mediated by differences in DNA conformation, specifically differences in the extent of transcription-bubble unwinding[22,23,25,26] (see Figure 1.1). At most promoters, the energetically most favorable configuration of the RNAP–promoter open complex is one that contains an unwound transcription bubble 13 nucleotides in length and that places the template-strand position 7 nucleotides downstream from the promoter −10 element in the RNAP active-center P site. In order for TSS selection to occur at positions 8, 9, or 10 nucleotides downstream of the promoter −10 element, the downstream DNA duplex is further unwound by an additional 1, 2, or 3 bp, respectively; the unwound DNA is pulled into and past the RNAP active center, and the unwound DNA is accommodated as single-stranded DNA bulges within the transcription bubble, yielding a "scrunched" complex (TSS = 8, 9, or 10 in Figure 1.1). In order for TSS selection to occur 6 nucleotides downstream of the promoter −10 element, the opposite occurs: downstream DNA is rewound by 1 bp, downstream DNA is extruded from the RNAP active center by 1 bp, and the extrusion of DNA from the RNAP active center is accommodated by stretching DNA within the unwound transcription bubble, yielding an "anti-scrunched" complex (TSS = 6 in Figure 1.1).

Scrunching and anti-scrunching during TSS selection have two defining, experimentally detectable, hallmarks: (1) as the position of the TSS changes, the position of the RNAP leading edge relative to DNA changes, but the position of the RNAP trailing edge relative to DNA does not change; and (2) as the position of the TSS changes, the size of the transcription bubble changes. To demonstrate that TSS selection exhibits the first hallmark of scrunching and anti-scrunching, a protein–DNA photocrosslinking approach was used to determine the RNAP leading-edge and trailing-edge positions relative to DNA, *in vitro* and *in vivo*, on promoters that use different TSS positions.[22,23,25] This approach revealed that, for promoters that use different TSS positions, the RNAP leading-edge position changes, but the RNAP trailing-edge position does not change. To demonstrate that TSS selection exhibits the second hallmark of scrunching and anti-scrunching, a single-molecule magnetic-tweezers approach was employed that allowed direct assessment of transcription-bubble sizes for promoters that use different TSS positions.[25] This approach revealed a one-for-one correlation between TSS position and transcription-bubble size.

Following the placement of the template-strand +1 and +2 in the RNAP active-center P and A sites, the initiating entity and the extending NTP bind, through complementary base pairing interactions, to template-strand positions +1 and +2.[27,28] In primer-independent transcription initiation, binding of the initiating entity to the RNAP active-center P site is facilitated by a module of the σR3–σR4 linker termed the "σ finger",[29–31] which reaches into the RNAP active-center cleft and interacts with the DNA template strand.[29,30,32–34] As a result of interactions with the σ finger, the DNA template strand in the immediate vicinity of the RNAP active centre adopts an A-form helical conformation that facilitates binding of NTPs or NCINs.[29]

In primer-dependent transcription, a 2 to ~5 nucleotide RNA binds to the template strand, with its 3′ nucleotide base-paired to template-strand position +1 in the RNAP active-center P site. Primer-dependent transcription initiation can bypass the requirement for the σ finger, presumably because base pairing of an RNA primer to the DNA template strand pre-organizes the template strand as one strand of an RNA–DNA double helix and constrains the template strand to interact with the RNAP active center, facilitating binding of the extending NTP.[31–33]

1.2.2 Sequence Determinants

It has been known for over four decades that promoter DNA sequence in the TSS region is a key determinant of TSS selection in primer-independent transcription: specifically, there is a strong preference for a purine:pyrimidine base pair (R:Y, nontemplate strand:template strand) at position +1 in primer-independent transcription,[35–41] and there is a moderate preference for a pyrimidine:purine base pair (Y:R) at position −1 in primer-independent transcription.[35,42] The strong preference for R:Y at position +1 is attributable to a strong preference of the enzyme active-center P site for a purine NTP or purine-containing NCIN as initiating entity.[35–38,43,44] The moderate preference for Y:R at position −1 is attributable to the energetically favorable inter-chain base-stacking interaction that occurs between R at template-strand position −1 and an initiating NTP base-paired to template-strand position +1.[34] An analogous preference for Y:R, and an analogous energetically favorably inter-chain base-stacking interaction by template-strand R, are observed in primer-dependent initiation;[13] in this case, inter-chain base-stacking occurs between R at the template-strand position immediately upstream of the primer binding site and a primer base-paired to the primer binding site.[13]

A comprehensive analysis of TSS selection on a promoter library containing all 4^7 (~16 000) sequences at positions 4–10 bp downstream of the −10 element of a consensus σ^{70}-dependent promoter confirmed and quantified the preferences for R:Y at position +1 and Y:R at position −1 in primer-independent transcription.[14]

An analogous comprehensive study of TSS selection, in this case on a promoter library containing all 4^{10} (~1 000 000) sequences at positions 1–10 bp downstream of the −10 element, revealed that a nontemplate-strand segment immediately downstream of the promoter −10 element that makes sequence-specific protein–DNA interactions with σ,[29,45–47] termed the "discriminator element," also contributes to TSS selection.[22] Purine-rich discriminator nontemplate-strand sequences favor TSS selection at upstream positions; pyrimidine-rich discriminator nontemplate-strand sequences favor TSS selection at downstream positions.[22] The effect of the discriminator element on TSS selection explains the atypical TSS selection at the *E. coli* ribosomal RNA promoter *rrn*B P1.[23] The results of protein–DNA photo-crosslinking studies mapping RNAP leading-edge and RNAP trailing-edge positions relative to DNA, magnetic-tweezers studies measuring transcription-bubble size, and crystal structures of complexes containing purine-rich and pyrimidine-rich discriminator elements indicate that discriminator elements manifest their effects on TSS selection by modulating the energetics of scrunching and anti-scrunching.[22,23,25]

At least one other aspect of promoter sequence contributes to TSS selection: namely, a nontemplate-strand segment at the downstream part of the transcription bubble, termed the "core-recognition element (CRE)".[29,48] The CRE makes sequence-specific protein–DNA interactions with RNAP core that affect formation and stability of the RNAP-promoter open complex.[29,49,50] The role of the CRE in TSS selection[48] can be understood in terms of effects of RNAP–CRE interactions in defining the downstream edge of the transcription bubble in the RNAP–promoter open complex, which, in turn, defines the extent of scrunching or anti-scrunching in the RNAP–promoter open complex.

1.2.3 Bacterial RNAP *vs.* Eukaryotic RNAP

Eukaryotes possess three nuclear RNAPs—RNAP I, RNAP II, and RNAP III—each of which is a member of the same multi-subunit RNAP family as bacterial RNAP.[51] RNAP I and RNAP III show the same pattern of TSS variation, over a 5 bp window, as does bacterial RNAP, and most likely employ the same scrunching–anti-scrunching mechanism as bacterial RNAP.[52–57] Eukaryotic RNAP II, in the absence of accessory factors, also shows the same pattern of TSS variation, over a 5 bp window, as does bacterial RNAP. However, in the presence of the ATP-dependent DNA translocase transcription factor IIH (TFIIH), RNAP II can exhibit TSS variation over a >100 bp window.[58,59] The results of optical-tweezers and cryo-electron microscopy (cryo-EM) studies indicate that long-range TSS variation in the presence of TFIIH likely involves a scrunching–anti-scrunching mechanism analogous to that of bacterial RNAP, but having greater range due to the ability to access energy from TFIIH-dependent ATP hydrolysis in addition to energy available from the thermal bath.[60,61] Magnetic-tweezers studies, in contrast to the optical-tweezers and cryo-EM studies, have not detected long-range scrunching by RNAP II in the presence of TFIIH,[62] but differences in reagents and reaction conditions preclude direct comparison of the different studies.

Eukaryotic RNAP I, RNAP II, and RNAP III, like bacterial RNAP, are able to perform both primer-independent and primer-dependent transcription initiation.[51–53] In primer-independent transcription initiation by RNAP I, a structural module of a transcription initiation factor—the Rrn7 zinc ribbon and B-reader—carries out a function analogous to that of the σ finger in bacterial primer-independent transcription initiation, reaching into the RNAP active-center cleft and interacting with the DNA template strand in the immediate vicinity of the RNAP active center, causing it to adopt an A-form helical conformation that facilitates binding of NTPs or NCINs.[63–65] In primer-independent transcription initiation by RNAP II, the same role is performed by the transcription factor IIB (TFIIB) zinc ribbon and B-reader,[66–70] and, in primer-independent transcription initiation by RNAP III, this role is performed by the Brf1 zinc ribbon.[71,72] Surprisingly, although the transcription-factor modules that make these interactions in primer-independent transcription by RNAP I, RNAP II, RNAP III are homologous to each other, the modules show no sequence or structural homology to the bacterial σ finger.[73]

1.3 Initial Transcription

1.3.1 Mechanism: On-pathway Reactions

Initial transcription follows TSS selection.[74,75] Initial transcription starts differently for primer-independent and primer-dependent transcription initiation. Primer-independent initial transcription starts with formation of a first phosphodiester bond between the initiating NTP or NCIN and the extending NTP, bound in the RNAP active-center P and A sites, to yield a nascent RNA of 2 nucleotides length (or $x + 1$ nucleotides for an NCIN comprising x nucleotides), and then proceeds into sequential extension of the nascent RNA to a threshold length of ~11–15 nucleotides (see Figure 1.2). Primer-dependent initial transcription starts with formation of a first phosphodiester bond between the primer 3′ nucleotide and the extending NTP, bound in the RNAP active-center P and A sites, respectively, to yield a nascent RNA of length $y + 1$ nucleotides (where y is the length of the primer), and then proceeds into sequential extension of the nascent RNA to a threshold length of ~11–15 nucleotides.

In both primer-dependent initiation and primer-independent initiation, following formation of the first phosphodiester bond, each subsequent nucleotide-addition step requires translocation of the template strand by one nucleotide step (~3.4 Å). In order to extend the nascent RNA to a threshold length of 11 nucleotides, the DNA template strand must translocate relative to the RNAP active center by 9 nucleotide steps (~31 Å) for primer-independent transcription initiation and by $10 - y$ (where y is the length of the primer) nucleotide steps (*e.g.*, ~27 Å for a 2 nucleotide primer) in primer-dependent transcription initiation.[76]

Strikingly, RNAP remains stationary on promoter DNA, anchored through sequence-specific interactions with promoter elements, during the entire process of initial transcription[76–80] (see Figure 1.2). Accordingly, in order to reconcile the translocation of DNA template strand relative to the RNAP active center by at least ~31 Å without movement of RNAP relative to promoter DNA, there must be a large change of RNAP conformation, a large change of DNA conformation, or both, during initial transcription.[74,76–81]

We now know, definitively, that, following the first nucleotide-addition step, each subsequent nucleotide-addition step involves a large change in DNA conformation: namely, DNA scrunching.[76,80,82] The scrunching that occurs in initial transcription is similar to the scrunching that occurs in TSS selection, but, in this case, takes place concurrently with extension of the nascent RNA (see Figures 1.1 and 1.2). Following the first nucleotide addition, each subsequent nucleotide addition entails scrunching of DNA by 1 bp—unwinding

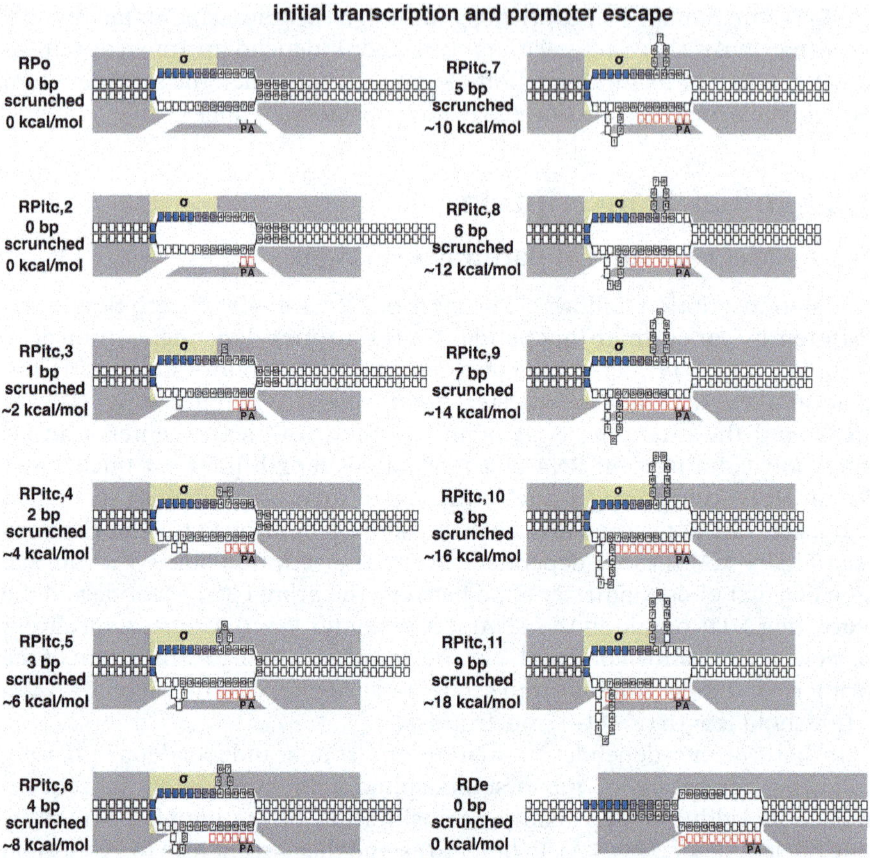

Figure 1.2 **Mechanism of initial transcription and promoter escape.** Each nucleotide-addition step in initial transcription is accompanied by 1 base pair of DNA scrunching (indicated schematically as bulged-out nucleotides), enabling stepwise capture and storage of ~2 kcal mol^{-1} in the form of unwound DNA. Promoter escape occurs upon synthesis of an RNA product of ~11–15 nucleotides in length (shown here upon synthesis of an 11 nucleotide RNA product; compare structure of RPitc,11 to RDe in right column). Promoter escape involves single-step unscrunching, entailing rewinding of the upstream half of the transcription bubble (−10 element through TSS) and extrusion of the rewound double-stranded DNA from the RNAP active-center cleft. This process results in the single-step release of energy stored in the form of unwound DNA (~16 kcal mol^{-1} in RPitc,10), enabling breakage of interactions anchoring RNAP at the promoter (both σ–DNA and σ–RNAP interactions). Positions of scrunched nucleotides are speculative. Red boxes, RNA nucleotides; other colours as in Figure 1.1.

DNA downstream of the RNAP active center, pulling the unwound DNA into and past the RNAP active center, and accommodating the unwound DNA as bulges in the unwound transcription bubble ("bubble expansion")—followed by binding of an extending NTP to the RNAP active-center A site, formation of a phosphodiester bond, and pyrophosphate release.[76,80]

The scrunching that occurs in initial transcription serves as the mechanism to capture and store the free energy required to break the sequence-specific protein–DNA interactions anchoring RNAP on the promoter, thereby enabling promoter escape.[76,80] The free energy anchoring RNAP on promoter DNA is estimated to be ~13–16 kcal mol^{-1}.[83–86] The free energy anchoring RNAP on promoter DNA greatly exceeds the free energy available from the thermal bath (~2 kcal mol^{-1}) and the chemical energy from a single nucleotide addition event (~3 kcal mol^{-1}). The scrunching that occurs in initial transcription enables capture of free energy from multiple nucleotide additions and stepwise storage of the captured free energy in the form of stepwise increases in the amount of DNA unwinding (see Figure 1.2, RPitc,2 to RPitc,11). In each scrunching-dependent nucleotide-addition step, 1 bp of DNA is unwound. This unwinding of 1 bp of DNA has an energetic cost of ~2 kcal mol^{-1} (~1 kcal mol^{-1} for A–T base pairs and ~1 to ~3 kcal mol^{-1} for G–C base pairs[25,87]). Therefore, in each scrunching-dependent-nucleotide addition step, chemical energy available from nucleotide addition (~3 kcal mol^{-1}) is captured and stored by unwinding 1 bp of DNA (~2 kcal mol^{-1}). For example, after 9 scrunching-dependent nucleotide-addition steps, ~18 kcal mol^{-1} has been captured and stored (see Figure 1.2, RPitc,11). This captured and stored free energy is similar to the binding free energy anchoring RNAP on promoter DNA. Upon rewinding of the upstream part of the unwound DNA (−10 element through TSS), the free energy captured and stored in initial transcription is accessed to drive promoter escape (see Figure 1.2; RDe,11).

As the nascent RNA is extended in initial transcription, the RNA 5′ end first approaches, and then collides with, the σ finger inside the RNAP active-centre cleft.[32,34,88,89] The RNA 5′ end then progressively displaces the σ finger from the RNAP active-center cleft during subsequent extension of the RNA to lengths of 6–11 nucleotides.[88] The collision with, and displacement of, the σ finger provides a second mechanism for stepwise capture of chemical energy from multiple nucleotide additions and stepwise storage of captured energy—in this case, in the form of an energetically unfavorable protein conformation (a compressed "protein spring"), and likely assists in driving promoter escape.

Scrunching in initial transcription exhibits the same two defining hallmarks as scrunching in TSS selection: (1) the position of the RNAP leading edge relative to DNA changes, but the position of the RNAP trailing edge relative to DNA does not change; and (2) the size of the transcription bubble changes.[76,80,90] The results of single-molecule fluorescence resonance energy transfer (FRET) experiments and protein–DNA photocrosslinking studies establish that the RNAP leading edge moves, and the RNAP trailing edge does not move, during initial transcription.[80,82,90] The results of magnetic-tweezers studies establish that the size of the transcription bubble changes during initial transcription.[76,82] The results of protein–DNA photocrosslinking studies of complexes engaged in initial transcription indicate that scrunched nontemplate-strand DNA is accommodated between the −10 element and downstream double-stranded DNA, and that scrunched template-strand

DNA is accommodated upstream of the RNA–DNA hybrid.[90] Published crystal structures of complexes engaged in initial transcription show local disorder of DNA template and nontemplate strands at positions corresponding to the positions of scrunched nucleotides indicated by protein–DNA photocrosslinking.[34,89] Unpublished crystal structures define, in detail, the conformation of scrunched nontemplate strand DNA having 1 or 2 nucleotides of scrunched DNA (Y Zhang and RH Ebright unpublished).

1.3.2 Mechanism: Off-pathway Reactions

One important off-pathway reaction during initial transcription is termed "abortive initiation"[74,75,91–94] (see Figure 1.3). Abortive initiation entails the synthesis and release of RNA products shorter than the threshold-length

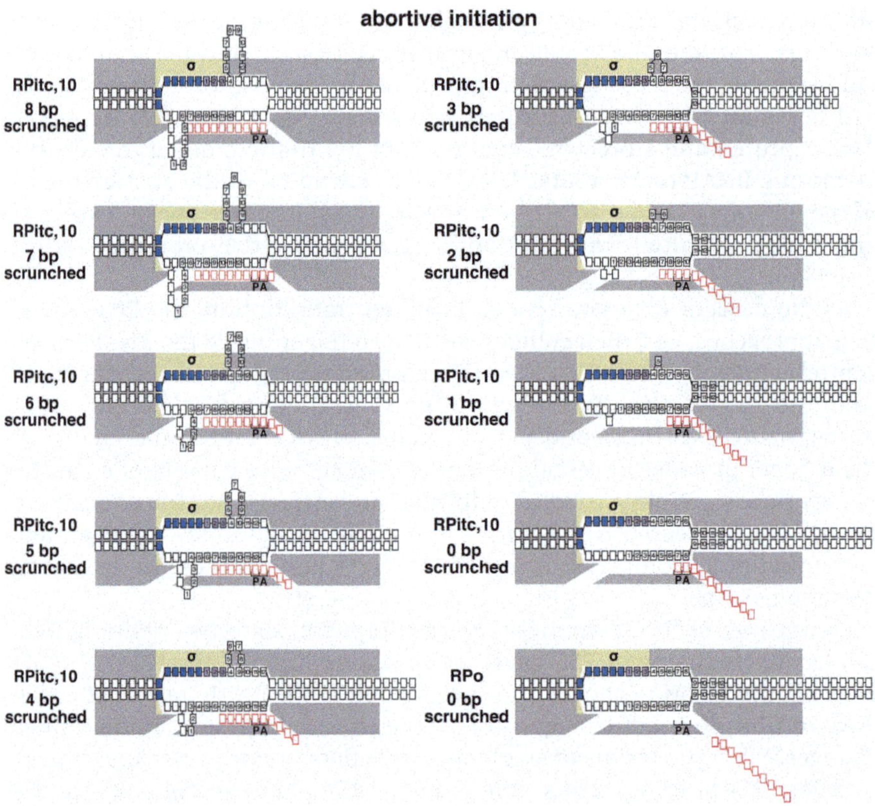

Figure 1.3 **Mechanism of abortive initiation.** The process of abortive release (illustrated for an initial transcribing complex containing a 10 nt RNA product) involves successive steps of reversal of scrunching and extrusion of the 3′ end of the RNA out of the RNAP NTP-entrance channel, followed by release of the RNA to reform an RNAP–promoter open complex. Scrunching is indicated by bulged-out nucleotides. Positions of scrunched and anti-scrunched nucleotides are speculative. Colours are as in Figure 1.2.

RNA product[74,75,91–94] (*i.e.*, RNA <11–15 nucleotides in length). Following each nucleotide-addition step in initial transcription, the RNAP–promoter initial transcribing complex either can proceed to the next nucleotide-addition step or can release the RNA product and reform an RNAP–promoter open complex. Following the first nucleotide-addition step in initial transcription, release of the RNA involves breakage of base-pairing with the template strand and diffusion of the RNA product out of RNAP through the NTP-entrance channel. Following subsequent nucleotide-addition steps in initial transcription, release of the RNA product entails the reversal of scrunching—relaxing the scrunched state to yield an unstressed, "unscrunched" state—followed by breakage of base-pairing to the template strand and diffusion of the RNA product out of RNAP, through the NTP-entrance channel (see Figure 1.3). Relaxing the scrunched state to yield an unscrunched state involves translocation analogous to the "backtracking" that occurs in transcription elongation,[95–97] and results in extrusion of the RNA 3′ end into the NTP-entrance channel and shortening of the RNA–DNA hybrid (to just 2 bp in the case of primer-independent transcription initiation with an NTP as the initiating entity).

Abortive initiation is a significant modulator of promoter activity both *in vitro* and *in vivo*.[94] Abortive initiation is a necessary, unavoidable consequence of the "stressed-intermediate" strategy used by RNAP, involving the stepwise capture and storage of energy during initial transcription and, upon reaching a threshold level, the single-step release and recovery of the stored energy to drive promoter escape.[76,78,80]

Another important off-pathway reaction during initial transcription is "reiterative transcription initiation," also referred to as "transcriptional stuttering," or "pseudo-templated transcription".[98–101] Reiterative transcription initiation occurs both *in vitro* and *in vivo* at promoters that have homopolymeric sequences at, or immediately downstream of, the TSS and results in synthesis of RNA products having 5′ end sequences that contain additional nucleotides not complementary to the DNA template.[14,102–106] RNA products synthesized by reiterative transcription initiation can be released from the RNAP–promoter initial transcribing complex[102–105] ("non-productive reiterative transcription") or can be extended to yield full-length RNA products[14,98] ("productive reiterative transcription). Reiterative transcription initiation provides a means to modulate promoter activity and is a target for transcriptional regulation.[14,18,103,107–114] Classic examples of genes whose expression is regulated by reiterative transcription initiation are *Escherichia coli pyrBI* and *Bacillus subtilis pyrG*.[18,110]

It has been hypothesized that reiterative transcription initiation involves "slippage" of the RNA product relative to the DNA template.[14,98–100,103,105,106,112,114–116] According to this hypothesis, the RNA translocates—slips—relative to both the DNA template strand and RNAP, in a manner that repositions the RNA 3′ end from the RNAP active-center A site into the RNAP active-center P site, enabling binding of an extending NTP, formation of a phosphodiester bond, and pyrophosphate release. An implication of this hypothesis is that RNA extension in reiterative transcription initiation, unlike RNA extension in productive transcription initiation, might not involve scrunching. (Scrunching is required for translocation of DNA and RNA together relative to RNAP, but would not be required for translocation of RNA, but not DNA, relative to RNAP.) Recent work—involving a combination of protein–DNA photocrosslinking, single-molecule DNA nanomanipulation, X-ray crystallography, and cryo-EM—has confirmed the slippage hypothesis, directly demonstrating that RNA, not DNA, translocates relative to RNAP in reiterative transcription initiation, and directly demonstrating that RNA extension in reiterative transcription initiation does not involve DNA scrunching.[106]

1.3.3 Sequence Determinants

The sequence of the initial-transcribed region, between promoter positions +1 and +15, affects the kinetics of initial transcription.[74,75,117–121]

A first example of sequence-dependent effects on initial transcription is provided by dinucleotide sequences associated with inefficient RNAP translocation[122] (*e.g.*, GU, AU, CU, and UU dinucleotide sequences), which exhibit inefficient initial transcription and greater susceptibility to abortive initiation.[120]

A second example of sequence-dependent effects on initial transcription is provided by sequence-dependent reiterative transcription initiation.[98–101] As described above, reiterative transcription initiation is observed primarily for templates that contain homopolymeric sequences at, or immediately downstream of, the TSS. Reiterative transcription initiation also is observed at promoters that contain short repeat sequences at, or immediately downstream of, the TSS.[14,112] According to the slippage hypothesis described above, homopolymeric and repeat sequences enable RNA to translocate—slip—relative to both DNA and RNAP and, after translocation, to re-establish base-pairing interactions with DNA.

A third example of sequence-dependent effects on initial transcription is provided by sequence-dependent initiation pausing.[121,123–126] Nontemplate-strand YG sequences can result in sequence-dependent pausing in transcription elongation.[50,127,128] Nontemplate-strand YG sequences can also result in pausing during initial transcription when present in initial transcribed regions.[121,124] At such sequences, RNAP pauses on seconds to tens-of-seconds time scales (resulting in dwell times ~20 to ~100 times longer than dwell times at other sequences). As with pausing during transcription elongation,[95–97] pausing during initial transcription also can allow entry into "backtracked" states, in which RNAP moves backward relative to DNA and RNA, misaligning the RNAP active centre and the RNA 3′ end.[124,125] The results of a comprehensive analysis of promoter sequences indicate that initial transcription pauses occur in many promoters and can occur at many positions within the promoter initial transcribed region.[121]

1.3.4 Bacterial RNAP *vs.* Eukaryotic RNAP

Available evidence indicates that eukaryotic RNAP I, II, and III perform initial transcription through a DNA scrunching mechanism analogous to that used by bacterial RNAP.[60,69,81] DNA footprinting studies have demonstrated transcription-bubble expansion in RNAP II-dependent initial transcription.[81] Optical-tweezers experiments on RNAP II-dependent initial transcription have provided evidence for movement of the RNAP leading-edge position, but not the RNAP trailing-edge position, characteristic of DNA scrunching.[60] Cryo-EM structures of RNAP II initial transcribing complexes show regions of disorder in the transcription-bubble template and non-template strands[61,69] similar to those seen in crystal structures of bacterial initial transcribing complexes engaged in scrunching.[34,89,106] In the light of the evidence for scrunching in initial transcription by eukaryotic RNAP, it appears likely that a scrunching mechanism mediates the energy capture and storage required for breakage of RNAP–promoter interactions in promoter escape by eukaryotic RNAP.

1.4 Promoter Escape

1.4.1 Mechanism

Promoter escape entails breaking the sequence-specific protein–DNA interactions that anchor RNAP at the promoter and forming a transcription elongation complex.[74,75,77–79] As described above, the primary mechanism to capture, store, and release the energy required to break the interactions that anchor RNAP at the promoter is DNA scrunching.[76,80] DNA scrunching enables stepwise capture of chemical energy from multiple nucleotide additions, stepwise storage of captured energy by unwinding DNA to expand the transcription bubble, followed by single-step release of captured and stored energy by rewinding DNA to contract the transcription bubble[76,80] (see Figure 1.2). An additional mechanism to capture, store, and release energy to break anchoring interactions may involve stepwise RNA extension and concomitant displacement of the σ finger, followed by single-step release of captured and stored energy by displacing the σ finger out of the RNAP active-center cleft and the RNAP RNA-exit channel.[32,34,88]

Promoter escape is thought to involve a series of mechanically coupled, probably concerted, reactions involving: (1) extension of the nascent RNA to a threshold length, which at most promoters is 11–15 nucleotides; (2) entry of the RNA 5′ end into the RNAP RNA-exit channel; (3) displacement of the σ finger from the RNAP RNA-exit channel, driven by steric clash with the RNA 5′ end; (4) disruption of protein–DNA interaction between σR4 and the promoter −35 element; and (5) rewinding of the upstream half of the transcription bubble, from the −10 element through the TSS, a process termed "unscrunching"[32,76,129] (see Figure 1.2). The product of this series of reactions is a transcription elongation complex containing a threshold length of 11–15 nucleotides of RNA and having an altered RNAP-σ interface, in which interactions previously made between RNAP and the σ finger and between RNAP and σR4 are lost.[130,131] (Cryo-EM structures of such early transcription elongation complexes have been determined recently.[132,133]) Because most RNAP-σ interactions are lost upon promoter escape, the affinity of RNAP for σ decreases upon promoter escape, and σ can, and typically does, dissociate from the transcription elongation complex in a time-dependent fashion.[130,134–140]

1.4.2 Sequence Determinants

The sequences of the promoter initial transcribed region, the core promoter, and upstream promoter affect multiple quantitative parameters of promoter escape.[74,75,119,126,141–144] Three key parameters are: (1) the abortive-to-productive ratio (APR; a measure of the efficiency of promoter escape); (2) the maximum size of abortive transcripts (MSAT; a measure of the number of steps of RNA synthesis, and thus the number of steps of scrunching, required before promoter escape; 10–14 nucleotides at most promoters, but as high as 19 at some promoters[118,145,146]), and (3) k_{clear} (also called k_E; the composite rate constant for the set of reactions from formation of a catalytically competent RNAP–promoter open complex through promoter escape[119,143,144]).

The sequence of the initial transcribed region can affect APR, MSAT, and k_{clear} through effects on sequence-dependent translocational bias,[120] as discussed above. The sequence of the initial transcribed region also can affect APR, MSAT, and k_{clear} through effects on the energetics of scrunching.[147] Sequences that require more energy for DNA unwinding in the scrunching process are expected to be associated with higher APR and lower k_{clear}. Sequences that, when unwound during scrunching, store less energy are expected to be associated with higher MSAT.

A large body of evidence indicates that altering the strength of protein–DNA interactions between RNAP and core promoter elements and upstream promoter elements affects promoter escape. One specific instance of a core promoter element that affects promoter escape is the discriminator, the nontemplate-strand segment immediately downstream of the promoter −10 element.[23,148] As described above, different discriminator sequences result in different preferences for TSS selection, due to different free energies of scrunching in TSS selection.[23,148] Promoters with discriminator sequences that result in an unusually downstream shifted TSS—*i.e.*, a TSS 9 nucleotides downstream of the promoter −10 element—exhibit unusually high amounts of scrunching for TSS selection and tend to exhibit unusually low, nearly zero, APR.[23,148] It has been suggested that the unusually high amount of scrunching during TSS selection at these promoters ("pre-scrunching") results in a lower requirement for scrunching in subsequent initial transcription and thus results in a higher efficiency of promoter escape.[23,148]

More generally, the strength of core promoter elements and upstream promoter elements correlates positively with APR and MSAT.[142,143,149–153] This implies that increasing the strength of the interactions anchoring RNAP on promoter DNA tends to decrease the efficiency of escape and to increase the requirement for scrunching and scrunching-dependent energy storage.[142,143,149–153] A striking consequence of this effect is that a fully consensus promoter, paradoxically, can exhibit little, or even no, transcriptional activity, despite having a very high affinity for RNAP and a very high rate of formation of RNAP–promoter open complex, due to having a low, or even zero, rate of promoter escape.[143,154,155] Similarly, protein–protein interactions between sequence-specific DNA binding proteins bound to promoter DNA and RNAP bound to promoter DNA can increase APR and MSAT ("repression through protein–protein contact"), again indicating that increasing the strength of the interactions anchoring RNAP on a promoter tends to decrease the efficiency of escape and to increase the requirement for scrunching and scrunching-dependent energy storage.[156,157]

Efforts have been made to generate mathematical models of promoter escape based on promoter sequence, primarily initial transcribed sequence, but, so far, these have had only limited success.[147]

1.4.3 Bacterial RNAP *vs.* Eukaryotic RNAP

Less information exists about the mechanisms of promoter escape for eukaryotic RNAP I, II, and III. Eukaryotic RNAP I, II, and III all have an active-center cleft that can accommodate an RNA–DNA hybrid of 9–10 bp.[158–160] In all three cases, as in bacterial RNAP, a transcription factor enters the RNAP active-center cleft and interacts with the template strand to facilitate primer-independent transcription initiation, threads through the RNAP RNA-exit channel, and must be displaced out of the RNAP RNA-exit channel upon synthesis of an RNA product of ~11 nucleotide or greater in length.[158–160] Accordingly, it seems possible, even likely, that promoter escape by eukaryotic RNAP I, II, and III, like promoter escape by bacterial RNAP, involves a series of mechanically coupled, probably concerted reactions, involving extension of the RNA to a threshold length of ~11 nucleotides, entry of the RNA 5′ end into the RNAP RNA-exit channel, displacement of an initiation-factor module from the RNAP RNA-exit channel driven by steric clash, disruption of a subset of transcription factor–DNA interactions, and rewinding of the upstream half of the transcription bubble.

1.5 Prospects

A key objective for future work is the determination of high-resolution structures defining the conformations of scrunched nontemplate-strand and template-strand DNA segments at each step of initial transcription. Published structural data for RNAP–promoter initial transcribing complexes show regions of local disorder of nontemplate and template strands that indicate locations of scrunched DNA segments, but do not define conformations of scrunched DNA segments.[34,69,89]

Another key objective for future work is a comprehensive elucidation of the sequence determinants for each quantitative parameter for initial transcription and promoter escape. Based on the information obtained to date, it is expected that the promoter sequence will influence initial transcription and promoter escape through effects on DNA unwinding energies. It also is expected that the promoter sequence will influence initial transcription and promoter escape through effects on protein–DNA interactions with core promoter elements and upstream promoter elements. In principle, promoter sequence also may influence initial transcription and promoter escape through effects on formation of DNA hairpins, or other multi-stranded DNA secondary structures, within the transcription bubble of an RNAP–promoter open complex or within the expanded transcription bubble of an RNAP–promoter initial-transcribing complex.

Another key objective for future work is the analysis of the regulation of initial transcription and promoter escape. Much data indicates that these processes are regulated, but next to nothing is known about how they are regulated.

Finally, elucidation of TSS selection, initial transcription, and promoter escape by eukaryotic RNAPs is a priority. In particular, clarification whether or not RNAP II, together with TFIIH, drives long-range scrunching in long-range TSS selection, as indicated by optical-tweezers and cryo-EM analyses,[60,61] but not by magnetic-tweezers analyses,[62] is an important priority.

Glossary

Transcription start site (TSS)	the promoter position that occupies the RNAP active-center product site (P site) during transcription initiation and, as such, base pairs with the initiating entity in primer-independent transcription initiation and with the primer 3′ end in primer-dependent transcription initiation.
TSS selection	the process of selecting a TSS by placing the TSS template-strand nucleotide in the RNAP active-center P site during transcription initiation.
Initial transcription	the process of synthesizing the first ~11–15 nucleotides of an RNA product by a promoter-anchored RNAP initial transcribing complex.
Promoter escape	the process of breaking interactions anchoring RNAP on promoter DNA upon synthesis of a nascent RNA having a threshold length of ~11–15 nucleotides, enabling transformation of an initial transcribing complex into a transcription elongation complex.
Transcription bubble	the region of unwound DNA within a transcription initiation complex, an initial transcribing complex, or a transcription elongation complex.

Scrunching	expansion of the transcription bubble by unwinding downstream DNA, pulling unwound DNA into the RNAP active-center cleft past the RNAP active center, and accommodating the unwound nontemplate- and template-strand single-stranded DNA as bulges within the transcription bubble. Scrunching occurs in TSS selection when selecting TSS positions downstream of the modal TSS position. Scrunching also occurs in initial transcription during each nucleotide-addition step subsequent to the first. Scrunching during initial transcription enables stepwise capture and storage of free energy for subsequent use in driving promoter escape.
Anti-scrunching	contraction of the transcription bubble by rewinding downstream single-stranded DNA and extrusion of the rewound DNA from the RNAP active-center cleft. Anti-scrunching occurs in TSS selection when selecting TSS positions upstream of the modal position.
Unscrunching	contraction of the transcription bubble by rewinding of upstream single-stranded DNA and extrusion of the rewound DNA from the RNAP active-center cleft. Unscrunching occurs in promoter escape and enables single-step release of captured and stored energy.
Primer-independent transcription initiation	transcription initiation without a primer, using either a nucleoside triphosphate (NTP) or a non-canonical initiating nucleotide (NCIN) as the initiating entity.
Primer-dependent transcription initiation	transcription initiation using an RNA primer.

Acknowledgements

The authors acknowledge support from a Helen Hay Whitney fellowship (JTW) and NIH grants GM041376 (RHE) and GM118059 (BEN).

References

1. D. J. Hoffman and S. K. Niyogi, *Proc. Natl. Acad. Sci. U. S. A.*, 1973, **70**, 574–578.
2. E. G. Minkley and D. Pribnow, *J. Mol. Biol.*, 1973, **77**, 255–277.
3. W. J. Smagowicz and K. H. Scheit, *Nucleic Acids Res.*, 1978, **5**, 1919–1932.

4. P. P. Di Nocera, A. Avitabile and F. Blasi, *J. Biol. Chem.*, 1975, **250**, 8376–8381.
5. J. P. Dausse, A. Sentenac and P. Fromageot, *Eur. J. Biochem.*, 1975, **57**, 569–578.
6. N. Ruetsch and D. Dennis, *J. Biol. Chem.*, 1987, **262**, 1674–1679.
7. W. Smagowicz and K. Scheit, *Nucleic Acids Res.*, 1981, **9**, 2397–2410.
8. J. G. Bird, Y. Zhang, Y. Tian, N. Panova, I. Barvik, L. Greene, M. Liu, B. Buckley, L. Krasny, J. K. Lee, C. D. Kaplan, R. H. Ebright and B. E. Nickels, *Nature*, 2016, **535**, 444–447.
9. I. Barvik, D. Rejman, N. Panova, H. Sanderova and L. Krasny, *FEMS Microbiol. Rev.*, 2017, **41**, 131–138.
10. S. R. Goldman, J. S. Sharp, I. O. Vvedenskaya, J. Livny, S. L. Dove and B. E. Nickels, *Mol. Cell*, 2011, **42**, 817–825.
11. S. Y. Druzhinin, N. T. Tran, K. S. Skalenko, S. R. Goldman, J. G. Knoblauch, S. L. Dove and B. E. Nickels, *PLoS Genet.*, 2015, **11**, e1005348.
12. I. O. Vvedenskaya, J. S. Sharp, S. R. Goldman, P. N. Kanabar, J. Livny, S. L. Dove and B. E. Nickels, *Genes Dev.*, 2012, **26**, 1498–1507.
13. K. S. Skalenko, L. Li, Y. Zhang, I. O. Vvedenskaya, J. Winkelman, A. L. Cope, D. M. Taylor, P. Shah, R. H. Ebright, J. B. Kinney, Y. Zhang and B. E. Nickels, *Proc. Natl. Acad. Sci. U. S. A.*, 2021, **118**, e2106388118.
14. I. O. Vvedenskaya, Y. Zhang, S. R. Goldman, A. Valenti, V. Visone, D. M. Taylor, R. H. Ebright and B. E. Nickels, *Mol. Cell*, 2015, **60**, 953–965.
15. T. Aoyama and M. Takanami, *Nucleic Acids Res.*, 1985, **13**, 4085–4096.
16. K. I. Sorensen, K. E. Baker, R. A. Kelln and J. Neuhard, *J. Bacteriol.*, 1993, **175**, 4137–4144.
17. W. Jeong and C. Kang, *Nucleic Acids Res.*, 1994, **22**, 4667–4672.
18. C. Liu, L. S. Heath and C. L. Turnbough Jr, *Genes Dev.*, 1994, **8**, 2904–2912.
19. J. Liu and C. L. Turnbough Jr, *J. Bacteriol.*, 1994, **176**, 2938–2945.
20. K. A. Walker and R. Osuna, *J. Bacteriol.*, 2002, **184**, 4783–4791.
21. D. E. Lewis and S. Adhya, *Mol. Microbiol.*, 2004, **54**, 692–701.
22. J. T. Winkelman, I. O. Vvedenskaya, Y. Zhang, Y. Zhang, J. G. Bird, D. M. Taylor, R. L. Gourse, R. H. Ebright and B. E. Nickels, *Science*, 2016, **351**, 1090–1093.
23. J. T. Winkelman, P. Chandrangsu, W. Ross and R. L. Gourse, *Proc. Natl. Acad. Sci. U. S. A.*, 2016, **113**, E1787–E1795.
24. I. O. Vvedenskaya, J. G. Bird, Y. Zhang, Y. Zhang, X. Jiao, I. Barvik, L. Krasny, M. Kiledjian, D. M. Taylor, R. H. Ebright and B. E. Nickels, *Mol. Cell*, 2018, **70**, 553–564 e559.
25. L. Yu, J. T. Winkelman, C. Pukhrambam, T. R. Strick, B. E. Nickels and R. H. Ebright, *eLife*, 2017, **6**, e32038.
26. N. C. Robb, T. Cordes, L. C. Hwang, K. Gryte, D. Duchi, T. D. Craggs, Y. Santoso, S. Weiss, R. H. Ebright and A. N. Kapanidis, *J. Mol. Biol.*, 2013, **425**, 875–885.
27. Y. Zhang, D. Degen, M. X. Ho, E. Sineva, K. Y. Ebright, Y. W. Ebright, V. Mekler, H. Vahedian-Movahed, Y. Feng and R. Yin, *eLife*, 2014, **3**, e02450.

28. S. I. Maffioli, Y. Zhang, D. Degen, T. Carzaniga, G. Del Gatto, S. Serina, P. Monciardini, C. Mazzetti, P. Guglierame and G. Candiani, *Cell*, 2017, **169**, 1240–1248.e1223.
29. Y. Zhang, Y. Feng, S. Chatterjee, S. Tuske, M. X. Ho, E. Arnold and R. H. Ebright, *Science*, 2012, **338**, 1076–1080.
30. K. Severinov, D. Fenyö, E. Severinova, A. Mustaev, B. T. Chait, A. Goldfarb and S. A. Darst, *J. Biol. Chem.*, 1994, **269**, 20826–20828.
31. D. Pupov, I. Kuzin, I. Bass and A. Kulbachinskiy, *Nucleic Acids Res.*, 2014, **42**, 4494–4504.
32. K. S. Murakami, S. Masuda and S. A. Darst, *Science*, 2002, **296**, 1280–1284.
33. A. Kulbachinskiy and A. Mustaev, *J. Biol. Chem.*, 2006, **281**, 18273–18276.
34. R. S. Basu, B. A. Warner, V. Molodtsov, D. Pupov, D. Esyunina, C. Fernandez-Tornero, A. Kulbachinskiy and K. S. Murakami, *J. Biol. Chem.*, 2014, **289**, 24549–24559.
35. D. K. Hawley and W. R. McClure, *Nucleic Acids Res.*, 1983, **11**, 2237–2255.
36. S. E. Jorgensen, L. B. Buch and D. P. Nierlich, *Science*, 1969, **164**, 1067–1070.
37. U. Maitra and H. Hurwitz, *Proc. Natl. Acad. Sci. U. S. A.*, 1965, **54**, 815–822.
38. D. D. Anthony, E. Zeszotek and D. A. Goldthwait, *Proc. Natl. Acad. Sci. U. S. A.*, 1966, **56**, 1026.
39. U. Maitra, Y. Nakata and J. Hurwitz, *J. Biol. Chem.*, 1967, **242**, 4908–4918.
40. H. Bremer, M. Konrad, K. Gaines and G. Stent, *J. Mol. Biol.*, 1965, **13**, 540–553.
41. J. S. Krakow and W. J. Horsley, *J. Biol. Chem.*, 1967, **242**, 4796–4800.
42. R. K. Shultzaberger, Z. Chen, K. A. Lewis and T. D. Schneider, *Nucleic Acids Res.*, 2006, **35**, 771–788.
43. C. Wu and D. Goldthwait, *Biochemistry*, 1969, **8**, 4450–4458.
44. T. Naryshkina, A. Mustaev, S. A. Darst and K. Severinov, *J. Biol. Chem.*, 2001, **276**, 13308–13313.
45. S. P. Haugen, M. B. Berkmen, W. Ross, T. Gaal, C. Ward and R. L. Gourse, *Cell*, 2006, **125**, 1069–1082.
46. S. P. Haugen, W. Ross, M. Manrique and R. L. Gourse, *Proc. Natl. Acad. Sci. U. S. A.*, 2008, **105**, 3292–3297.
47. A. Feklistov, N. Barinova, A. Sevostyanova, E. Heyduk, I. Bass, I. Vvedenskaya, K. Kuznedelov, E. Merkiene, E. Stavrovskaya, S. Klimasauskas, V. Nikiforov, T. Heyduk, K. Severinov and A. Kulbachinskiy, *Mol. Cell*, 2006, **23**, 97–107.
48. I. O. Vvedenskaya, H. Vahedian-Movahed, Y. Zhang, D. M. Taylor, R. H. Ebright and B. E. Nickels, *Proc. Natl. Acad. Sci. U. S. A.*, 2016, **113**, E2899–E2905.
49. H. Vahedian Movahed, PhD dissertation, Rutgers University School of Graduate Studies, 2017.
50. I. O. Vvedenskaya, H. Vahedian-Movahed, J. G. Bird, J. G. Knoblauch, S. R. Goldman, Y. Zhang, R. H. Ebright and B. E. Nickels, *Science*, 2014, **344**, 1285–1289.
51. P. Cramer, *Curr. Opin. Struct. Biol.*, 2002, **12**, 89–97.

52. R. Learned and R. Tjian, *J. Mol. Appl. Genet.*, 1982, **1**, 575–584.
53. M. Samuels, A. Fire and P. Sharp, *J. Biol. Chem.*, 1984, **259**, 2517–2525.
54. M. Thomm and G. Wich, *Nucleic Acids Res.*, 1988, **16**, 151–163.
55. W. D. Reiter, U. Hüdepohl and W. Zillig, *Proc. Natl. Acad. Sci. U. S. A.*, 1990, **87**, 9509–9513.
56. P. Fruscoloni, M. Zamboni, G. Panetta, A. De Paolis and G. P. Tocchini-Valentinl, *Nucleic Acids Res.*, 1995, **23**, 2914–2918.
57. G. N. Zecherle, S. Whelen and B. D. Hall, *Mol. Cell. Biol.*, 1996, **16**, 5801–5810.
58. C. Giardina and J. T. Lis, *Science*, 1993, **261**, 759–762.
59. J. N. Kuehner and D. A. Brow, *J. Biol. Chem.*, 2006, **281**, 14119–14128.
60. F. M. Fazal, C. A. Meng, K. Murakami, R. D. Kornberg and S. M. Block, *Nature*, 2015, **525**, 274–277.
61. C. Yang, R. Fujiwara, H. KIm, J. Gorbea Colon, S. Steimle, B. Garcia and K. Murakami, *bioRxiv*, 2021, DOI: 10.1101/2021.05.03.442346.
62. E. J. Tomko, J. Fishburn, S. Hahn and E. A. Galburt, *Nat. Struct. Mol. Biol.*, 2017, **24**, 1139.
63. C. Engel, T. Gubbey, S. Neyer, S. Sainsbury, C. Oberthuer, C. Baejen, C. Bernecky and P. Cramer, *Cell*, 2017, **169**, 120–131.e122.
64. Y. Han, C. Yan, T. H. D. Nguyen, A. J. Jackobel, I. Ivanov, B. A. Knutson and Y. He, *eLife*, 2017, **6**, e27414.
65. Y. Sadian, L. Tafur, J. Kosinski, A. J. Jakobi, R. Wetzel, K. Buczak, W. J. Hagen, M. Beck, C. Sachse and C. W. Müller, *EMBO J.*, 2017, **36**, 2698–2709.
66. D. Kostrewa, M. E. Zeller, K. J. Armache, M. Seizl, K. Leike, M. Thomm and P. Cramer, *Nature*, 2009, **462**, 323.
67. X. Liu, D. A. Bushnell, D. Wang, G. Calero and R. D. Kornberg, *Science*, 2010, **327**, 206–209.
68. S. Sainsbury, C. Bernecky and P. Cramer, *Nat. Rev. Mol. Cell Biol.*, 2015, **16**, 129.
69. Y. He, C. Yan, J. Fang, C. Inouye, R. Tjian, I. Ivanov and E. Nogales, *Nature*, 2016, **533**, 359.
70. C. Plaschka, M. Hantsche, C. Dienemann, C. Burzinski, J. Plitzko and P. Cramer, *Nature*, 2016, **533**, 353.
71. M. K. Vorländer, H. Khatter, R. Wetzel, W. J. Hagen and C. W. Müller, *Nature*, 2018, **553**, 295.
72. G. Abascal-Palacios, E. P. Ramsay, F. Beuron, E. Morris and A. Vannini, *Nature*, 2018, **553**, 301.
73. W. Lin, S. Mandal, D. Degen, M. S. Cho, Y. Feng, K. Das and R. H. Ebright, *Nat. Commun.*, 2019, **10**, 710.
74. L. M. Hsu, *Biochim. Biophys. Acta, Gene Struct. Expression*, 2002, **1577**, 191–207.
75. L. M. Hsu, *EcoSal Plus*, 2008, **3**, DOI: 10.1128/ecosalplus.4.5.2.2.
76. A. Revyakin, C. Liu, R. H. Ebright and T. R. Strick, *Science*, 2006, **314**, 1139–1143.
77. A. J. Carpousis and J. D. Gralla, *J. Mol. Biol.*, 1985, **183**, 165–177.

78. D. C. Straney and D. M. Crothers, *J. Mol. Biol.*, 1987, **193**, 267–278.
79. B. Krummel and M. J. Chamberlin, *Biochemistry*, 1989, **28**, 7829–7842.
80. A. N. Kapanidis, E. Margeat, S. O. Ho, E. Kortkhonjia, S. Weiss and R. H. Ebright, *Science*, 2006, **314**, 1144–1147.
81. M. Pal, A. S. Ponticelli and D. S. Luse, *Mol. Cell*, 2005, **19**, 101–110.
82. E. Margeat, A. N. Kapanidis, P. Tinnefeld, Y. Wang, J. Mukhopadhyay, R. H. Ebright and S. Weiss, *Biophys. J.*, 2006, **90**, 1419–1431.
83. M. Record Jr, W. Reznikoff, M. Craig, K. McQuade and P. Schlax, *Escherichia coli and Salmonella: Cellular and Molecular Biology*, ASM Press, Washington, DC, 1996.
84. S. C. Gill, S. E. Weitzel and P. H. von Hippel, *J. Mol. Biol.*, 1991, **220**, 307–324.
85. K. Henderson, C. E. Evensen, C. Molzahn, L. C. Felth, S. Dyke, G. Liao, I. A. Shkel and M. T. Record, *Biochemistry*, 2019, **58**, 2339–2352.
86. K. L. Henderson, L. C. Felth, C. M. Molzahn, I. Shkel, S. Wang, M. Chhabra, E. F. Ruff, L. Bieter, J. E. Kraft and M. T. Record, *Proc. Natl. Acad. Sci. U. S. A.*, 2017, **114**, E3032–E3040.
87. K. J. Breslauer, R. Frank, H. Blöcker and L. A. Marky, *Proc. Natl. Acad. Sci. U. S. A.*, 1986, **83**, 3746–3750.
88. L. Li, V. Molodtsov, W. Lin, R. H. Ebright and Y. Zhang, *Proc. Natl. Acad. Sci. U. S. A.*, 2020, **117**, 5801–5809.
89. Y. Zuo and T. A. Steitz, *Mol. Cell*, 2015, **58**, 534–540.
90. J. T. Winkelman, B. T. Winkelman, J. Boyce, M. F. Maloney, A. Y. Chen, W. Ross and R. L. Gourse, *Mol. Cell*, 2015, **59**, 768–780.
91. A. J. Carpousis and J. D. Gralla, *Biochemistry*, 1980, **19**, 3245–3253.
92. M. Grachev and E. Zaychikov, *FEBS Lett.*, 1980, **115**, 23–26.
93. L. M. Munson and W. S. Reznikoff, *Biochemistry*, 1981, **20**, 2081–2085.
94. S. R. Goldman, R. H. Ebright and B. E. Nickels, *Science*, 2009, **324**, 927–928.
95. A. Mustaev, J. Roberts and M. Gottesman, *Transcription*, 2017, **8**, 150–161.
96. G. A. Belogurov and I. Artsimovitch, *J. Mol. Biol.*, 2019, **431**, 3975–4006.
97. J. Y. Kang, T. V. Mishanina, R. Landick and S. A. Darst, *J. Mol. Biol.*, 2019, **431**, 4007–4029.
98. C. L. Turnbough Jr and R. L. Switzer, *Microbiol. Mol. Biol. Rev.*, 2008, **72**, 266–300, table of contents.
99. J. P. Jacques and D. Kolakofsky, *Genes Dev.*, 1991, **5**, 707–713.
100. C. L. Turnbough Jr, *Curr. Opin. Microbiol.*, 2011, **14**, 142–147.
101. C. L. Turnbough Jr, *Microbiol. Mol. Biol. Rev.*, 2019, **83**, e00019-19.
102. Y. Cheng, S. M. Dylla and C. L. Turnbough, *J. Bacteriol.*, 2001, **183**, 221–228.
103. H. C. Guo and J. W. Roberts, *Biochemistry*, 1990, **29**, 10702–10709.
104. X. F. Xiong and W. S. Reznikoff, *J. Mol. Biol.*, 1993, **231**, 569–580.
105. K. S. Murakami, Y. Shin, C. L. Turnbough and V. Molodtsov, *Proc. Natl. Acad. Sci. U. S. A.*, 2017, **114**, 8211–8216.
106. Y. Liu, J. T. Winkelman, L. Yu, C. Pukhrambam, E. Firlar, J. T. Kaelber, Y. Zhang, B. E. Nickels and R. H. Ebright, *bioRxiv*, 2021, DOI: 10.1101/2021.05.08.443277.

107. Y. Cheng, S. M. Dylla and C. L. Turnbough Jr, *J. Bacteriol.*, 2001, **183**, 221–228.
108. X. Han and C. L. Turnbough Jr, *J. Bacteriol.*, 1998, **180**, 705–713.
109. I. E. Jensen-MacAllister, Q. Meng and R. L. Switzer, *Mol. Microbiol.*, 2007, **63**, 1440–1452.
110. Q. Meng, C. L. Turnbough Jr and R. L. Switzer, *Proc. Natl. Acad. Sci. U. S. A.*, 2004, **101**, 10943–10948.
111. F. Qi and C. L. Turnbough Jr, *J. Mol. Biol.*, 1995, **254**, 552–565.
112. K. Severinov and A. Goldfarb, *J. Biol. Chem.*, 1994, **269**, 31701–31705.
113. A. H. Tu and C. L. Turnbough Jr, *J. Bacteriol.*, 1997, **179**, 6665–6673.
114. L. A. Wagner, R. B. Weiss, R. Driscoll, D. S. Dunn and R. F. Gesteland, *Nucleic Acids Res.*, 1990, **18**, 3529–3535.
115. L. E. Macdonald, Y. Zhou and W. T. McAllister, *J. Mol. Biol.*, 1993, **232**, 1030–1047.
116. Y. Shin, M. Hedglin and K. S. Murakami, *Nucleic Acids Res.*, 2020, **48**, 2144–2155.
117. L. M. Hsu, N. V. Vo, C. M. Kane and M. J. Chamberlin, *Biochemistry*, 2003, **42**, 3777–3786.
118. L. M. Hsu, I. M. Cobb, J. R. Ozmore, M. Khoo, G. Nahm, L. Xia, Y. Bao and C. Ahn, *Biochemistry*, 2006, **45**, 8841–8854.
119. W. Kammerer, U. Deuschle, R. Gentz and H. Bujard, *EMBO J.*, 1986, **5**, 2995–3000.
120. J. Skancke, N. Bar, M. Kuiper and L. M. Hsu, *Biochemistry*, 2015, **54**, 4267–4275.
121. J. T. Winkelman, C. Pukhrambam, I. O. Vvedenskaya, Y. Zhang, D. M. Taylor, P. Shah, R. H. Ebright and B. E. Nickels, *Mol. Cell*, 2020, **79**, 797–811.e798.
122. P. P. Hein, M. Palangat and R. Landick, *Biochemistry*, 2011, **50**, 7002–7014.
123. D. Duchi, D. L. Bauer, L. Fernandez, G. Evans, N. Robb, L. C. Hwang, K. Gryte, A. Tomescu, P. Zawadzki, Z. Morichaud, K. Brodolin and A. N. Kapanidis, *Mol. Cell*, 2016, **63**, 939–950.
124. D. Dulin, D. L. Bauer, A. M. Malinen, J. J. Bakermans, M. Kaller, Z. Morichaud, I. Petushkov, M. Depken, K. Brodolin and A. Kulbachinskiy, *Nat. Commun.*, 2018, **9**, 1478.
125. E. Lerner, S. Chung, B. L. Allen, S. Wang, J. Lee, S. W. Lu, L. W. Grimaud, A. Ingargiola, X. Michalet, Y. Alhadid, S. Borukhov, T. R. Strick, D. J. Taatjes and S. Weiss, *Proc. Natl. Acad. Sci. U. S. A.*, 2016, **113**, E6562–E6571.
126. E. Heyduk and T. Heyduk, *Nucleic Acids Res.*, 2018, **46**, 4469–4486.
127. M. Imashimizu, H. Takahashi, T. Oshima, C. McIntosh, M. Bubunenko, D. L. Court and M. Kashlev, *Genome Biol.*, 2015, **16**, 98.
128. M. H. Larson, R. A. Mooney, J. M. Peters, T. Windgassen, D. Nayak, C. A. Gross, S. M. Block, W. J. Greenleaf, R. Landick and J. S. Weissman, *Science*, 2014, **344**, 1042–1047.
129. V. Mekler, E. Kortkhonjia, J. Mukhopadhyay, J. Knight, A. Revyakin, A. N. Kapanidis, W. Niu, Y. W. Ebright, R. Levy and R. H. Ebright, *Cell*, 2002, **108**, 599–614.

130. R. A. Mooney, S. A. Darst and R. Landick, *Mol. Cell*, 2005, **20**, 335–345.
131. B. E. Nickels, S. J. Garrity, V. Mekler, L. Minakhin, K. Severinov, R. H. Ebright and A. Hochschild, *Proc. Natl. Acad. Sci. U. S. A.*, 2005, **102**, 4488–4493.
132. J. Shi, X. Gao, T. Tian, Z. Yu, B. Gao, A. Wen, L. You, S. Chang, X. Zhang and Y. Zhang, *Nat. Commun.*, 2019, **10**, 2925.
133. Z. Yin, J. T. Kaelber and R. H. Ebright, *Proc. Natl. Acad. Sci. U. S. A.*, 2019, **116**, 18384–18390.
134. J. Mukhopadhyay, A. N. Kapanidis, V. Mekler, E. Kortkhonjia, Y. W. Ebright and R. H. Ebright, *Cell*, 2001, **106**, 453–463.
135. A. N. Kapanidis, E. Margeat, T. A. Laurence, S. Doose, S. O. Ho, J. Mukhopadhyay, E. Kortkhonjia, V. Mekler, R. H. Ebright and S. Weiss, *Mol. Cell*, 2005, **20**, 347–356.
136. G. Bar-Nahum and E. Nudler, *Cell*, 2001, **106**, 443–451.
137. T. T. Harden, C. D. Wells, L. J. Friedman, R. Landick, A. Hochschild, J. Kondev and J. Gelles, *Proc. Natl. Acad. Sci. U. S. A.*, 2016, **113**, 602–607.
138. M. Raffaelle, E. I. Kanin, J. Vogt, R. R. Burgess and A. Z. Ansari, *Mol. Cell*, 2005, **20**, 357–366.
139. B. E. Nickels, J. Mukhopadhyay, S. J. Garrity, R. H. Ebright and A. Hochschild, *Nat. Struct. Mol. Biol.*, 2004, **11**, 544.
140. N. B. Reppas, J. T. Wade, G. M. Church and K. Struhl, *Mol. Cell*, 2006, **24**, 747–757.
141. L. M. Hsu, in *Meth Enzymol*, Elsevier, 1996, vol. 273, pp. 59–71.
142. T. Ellinger, D. Behnke, R. Knaus, H. Bujard and J. D. Gralla, *J. Mol. Biol.*, 1994, **239**, 466–475.
143. U. Deuschle, W. Kammerer, R. Gentz and H. Bujard, *EMBO J.*, 1986, **5**, 2987–2994.
144. J. Ko and T. Heyduk, *Biochem. J.*, 2014, **463**, 135–144.
145. M. Chander, K. M. Austin, N. N. Aye-Han, P. Sircar and L. M. Hsu, *Biochemistry*, 2007, **46**, 12687–12699.
146. M. Chander, A. Lee, T. K. Vallery, M. Thandar, Y. Jiang and L. M. Hsu, *Biochemistry*, 2015, **54**, 7393–7408.
147. X.-C. Xue, F. Liu and Z.-C. Ou-Yang, *J. Mol. Biol.*, 2008, **378**, 520–529.
148. J. T. Winkelman and R. L. Gourse, *BioEssays*, 2017, **39**, 1600193.
149. A. J. Carpousis, J. E. Stefano and J. D. Gralla, *J. Mol. Biol.*, 1982, **157**, 619–633.
150. T. Ellinger, D. Behnke, H. Bujard and J. D. Gralla, *J. Mol. Biol.*, 1994, **239**, 455–465.
151. P. L. DeHaseth, M. L. Zupancic and M. T. Record, *J. Bacteriol.*, 1998, **180**, 3019–3025.
152. R. Knaus and H. Bujard, in *Nucleic Acids and Molecular Biology 4*, Springer, 1990, pp. 110–122.
153. N. V. Vo, L. M. Hsu, C. M. Kane and M. J. Chamberlin, *Biochemistry*, 2003, **42**, 3798–3811.
154. T. Gaal, W. Ross, S. T. Estrem, L. H. Nguyen, R. R. Burgess and R. L. Gourse, *Mol. Microbiol.*, 2001, **42**, 939–954.

155. M. Brunner and H. Bujard, *EMBO J.*, 1987, **6**, 3139–3144.
156. M. Monsalve, B. Calles, M. Mencía, M. Salas and F. Rojo, *Mol. Cell*, 1997, **1**, 99–107.
157. S. Roy, S. Garges and S. Adhya, *J. Biol. Chem.*, 1998, **273**, 14059–14062.
158. A. L. Gnatt, P. Cramer, J. Fu, D. A. Bushnell and R. D. Kornberg, *Science*, 2001, **292**, 1876–1882.
159. N. A. Hoffmann, A. J. Jakobi, M. Moreno-Morcillo, S. Glatt, J. Kosinski, W. J. Hagen, C. Sachse and C. W. Müller, *Nature*, 2015, **528**, 231.
160. S. Neyer, M. Kunz, C. Geiss, M. Hantsche, V. V. Hodirnau, A. Seybert, C. Engel, M. P. Scheffer, P. Cramer and A. S. Frangakis, *Nature*, 2016, **540**, 607.

CHAPTER 2

Molecular Basis for Transcriptional Fidelity Control by RNA Polymerase II

JUN XU[a], JUNTAEK OH[a], JENNY CHONG[a], LIANG XU*[b] AND DONG WANG*[a]

[a]Division of Pharmaceutical Sciences, Skaggs School of Pharmacy & Pharmaceutical Sciences, Department of Cellular & Molecular Medicine, School of Medicine, University of California, San Diego, La Jolla, California 92093, USA; [b]Department of Chemistry, Sun Yat-Sen University, Guangzhou, China
*E-mail: dongwang@ucsd.edu, xuliang33@mail.sysu.edu.cn

2.1 Introduction

Transcription is the first step of gene expression in all living cells and is mediated by RNA polymerase II (Pol II), which reads the genetic information from DNA sequences and produces pre-mRNA and noncoding RNA transcripts in eukaryotic cells.[1] Accurate gene expression relies on high transcriptional fidelity (error rates: approximately 10^{-4}–10^{-6} *in vitro* and *in vivo*).[2–6] Transcriptional misincorporation into coding RNA generates erroneous nascent mRNAs that alter the codon–anticodon readout during translation and results in miscoded and sometimes misfolded proteins.[7] Misincorporated ribonucleotides can also cause altered secondary or 3D

Chemical Biology No. 19
RNA Polymerases as Molecular Motors: On the Road, 2nd Edition
Edited by Robert Landick, Terence Strick and Jue Wang

Published by the Royal Society of Chemistry, www.rsc.org

structures of noncoding RNA transcripts.[8] These transcription errors can induce transient or heritable phenotypic changes to the cell, which are likely to be involved in tumour biogenesis and development.[7–13] Furthermore, misincorporation during transcription can induce polymerase pausing and lead to conflicts with the trailing or oncoming transcription complexes, replication machinery, or both.[14] Therefore, maintaining high transcriptional fidelity is essential for accurate genetic information transfer.[15,16]

During transcription elongation, Pol II moves along the DNA template and synthesizes a complementary RNA transcript. RNA pol II utilizes the same active centre for both nucleotide addition (polymerization) and error removal by a transcript cleavage reaction (proofreading) (Figure 2.1a). This is in sharp contrast to DNA polymerases, which utilize two distinct active sites for polymerization and proofreading activities, respectively. The RNA chain extension is achieved in a nucleotide addition cycle (NAC).[17] The NAC begins with binding of nucleoside triphosphate (NTP) to the transcription elongation complex (EC) in the post-translocation state (the active site is empty to allow NTP binding). The conserved trigger-loop module undergoes a conformational change from an open state to a closed state upon the binding of complementary NTP. Closure of the trigger loop is important for substrate positioning, catalytic activity, and transcriptional fidelity. Catalytic incorporation of the nucleotide at the growing 3′-end of the RNA leads to the pre-translocation state of the EC, in which the newly added, 3′-terminal RNA nucleotide remains in the +1 site. Finally, the pyrophosphate ion is released and Pol II translocates one base pair downstream in concert with trigger-loop opening. Pol II adopts a new post-translocation state, in which the active site is available for the next round of NTP binding. Multiple repeats of this cycle result in the continuous extension of the RNA transcript. There are at least three fidelity checkpoints for ensuring high transcriptional fidelity:[4,18–25] the insertion step (sequence specific nucleotide selection and incorporation), the extension step (from a matched *versus* a mismatched 3′-RNA terminus), and the proofreading step (removal of mis-incorporated nucleotides from the 3′-RNA terminus) (Figure 2.1b)

In the past two decades, a combination of genetic, biochemical, structural and computational studies has provided extensive details on the molecular mechanisms of RNA polymerase transcription and its fidelity control.[2,19,20,26–35] Several excellent reviews focusing on Pol II enzymatic catalysis and transcriptional regulation have been published.[17,36–39] In this chapter we will discuss recent progress toward understanding of the mechanisms of these three key fidelity checkpoint steps in controlling Pol II transcriptional fidelity.[34] We will mainly focus on the use of chemical biology approaches to study the molecular basis of Pol II transcriptional fidelity control during transcript elongation.

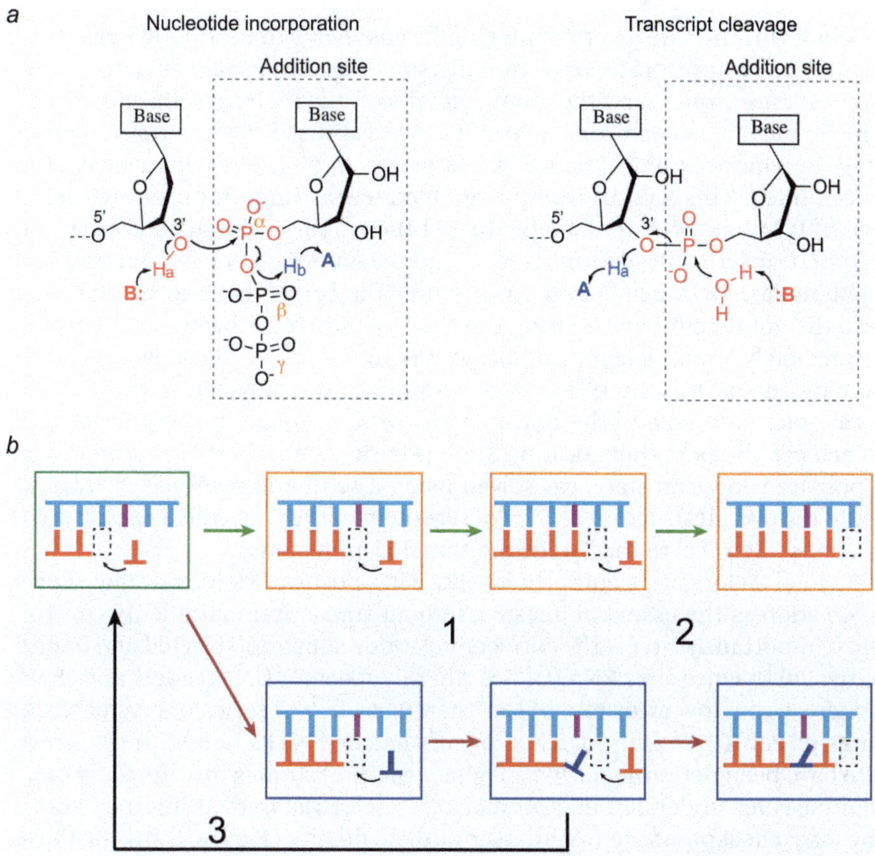

Figure 2.1 **RNA polymerase II reaction mechanism and transcriptional fidelity control.** (a) The active site of Pol II can catalyse both polymerization and cleavage reactions. Nucleotide incorporation (left) and RNA hydrolysis (right) reactions occur *via* a SN2 nucleophilic attack mechanism in the same Pol II active site (addition site). Black arrows indicate the electron path. Mg^{2+} ions are not shown for clarity. A, general acid. B, general base. The addition site is indicated by a dashed box. (b) A schematic representation of the three fidelity checkpoints of Pol II transcription elongation. During the incorporation step (step 1), the matched nucleotide (orange) is selected over the incorrect substrate (blue) and the RNA transcript can be efficiently extended from a matched pair (step 2). In step 3, cleavage of the mismatched end with misincorporated nucleotides is significantly more efficient than that for the matched end. The template DNA and transcript RNA are shown in cyan and orange, respectively. The active site of Pol II is indicated by the boxed area.

2.2 Structural Insights into Transcriptional Fidelity Control by RNA Polymerase II

Advances in structure determination of the eukaryotic transcription machinery provided a remarkable increase in understanding the mechanism and regulation of transcription and the atomic details of this molecular machinery.[30,40] The first crystal structure of a Pol II EC was captured in a pre-translocation state in which the Pol II active site is occupied by the newly incorporated nucleotide.[30] This crystal structure revealed several important features of the EC: (i) RNA and DNA are held by the Pol II clamp in a closed conformation; (ii) the transcript RNA forms a short upstream RNA–DNA hybrid; and (iii) the template DNA bends and crosses over the bridge helix to be delivered into the Pol II active site. During the NAC, the pre-translocation state forms immediately after nucleotide addition. In order to make the active site available for the next round of nucleotide addition, the polymerase must translocate one base pair of the downstream DNA template into the active site to achieve the post-translocation state (Figure 2.2a). The Pol II complex in a post-translocation state was solved using a synthetic scaffold.[14,41] Intriguingly, the overall structure of the Pol II protein in the post-translocation state is almost identical to that in the pre-translocation state.

The Pol II elongation complex structure in a post-translocation state allows us to address the essential questions about transcriptional fidelity control. Most importantly, how is the correct nucleotide substrate selected and bound to the Pol II-active site? RNA Pol II is a highly specific "DNA reader" and "RNA writer" with a low error rate of less than 0.001%.[5,6] A series of crystal structures of Pol II ECs with matched or mismatched NTPs bound in the active site have been determined.[14,28,41] These structural studies provide the molecular basis for understanding of nucleotide selection and addition by Pol II, the first checkpoint step of transcriptional fidelity (Figure 2.2b). In these structures, only the matched NTP binds to the canonical "addition site" (A site), in which the nucleobase of substrate forms a Watson–Crick base pair with the DNA template base (Figure 2.3a), whereas a mismatched NTP binds in an inverted conformation with its Watson–Crick face oriented away from the DNA template, in a location referred to as the "entry site" or E site.[19,42–44] However, the base pair hydrogen-bonding between substrate and DNA template alone would not provide sufficient energetic discrimination against non-complementary NTPs or against complementary 2′-deoxy NTPs[14,19] to account for the $\geq 10^5$ fidelity. Additional factors are required to explain substrate discrimination. Indeed, the discovery of substrate-dependent conformational changes of the conserved trigger loop provides additional contacts that can explain substrate discrimination. Binding of correct NTP in the active site induces folding of the trigger loop to seal the active site and form an extensive interaction network with the cognate NTP, which in turn promotes the nucleoside monophosphate (NMP) incorporation into the transcripts.[2,19,45] In the cognate NTP binding state, all three moieties of the substrate (nucleobase, sugar, and triphosphate) are recognized by Pol II through this extensive substrate recognition network.[19] The nucleobase of

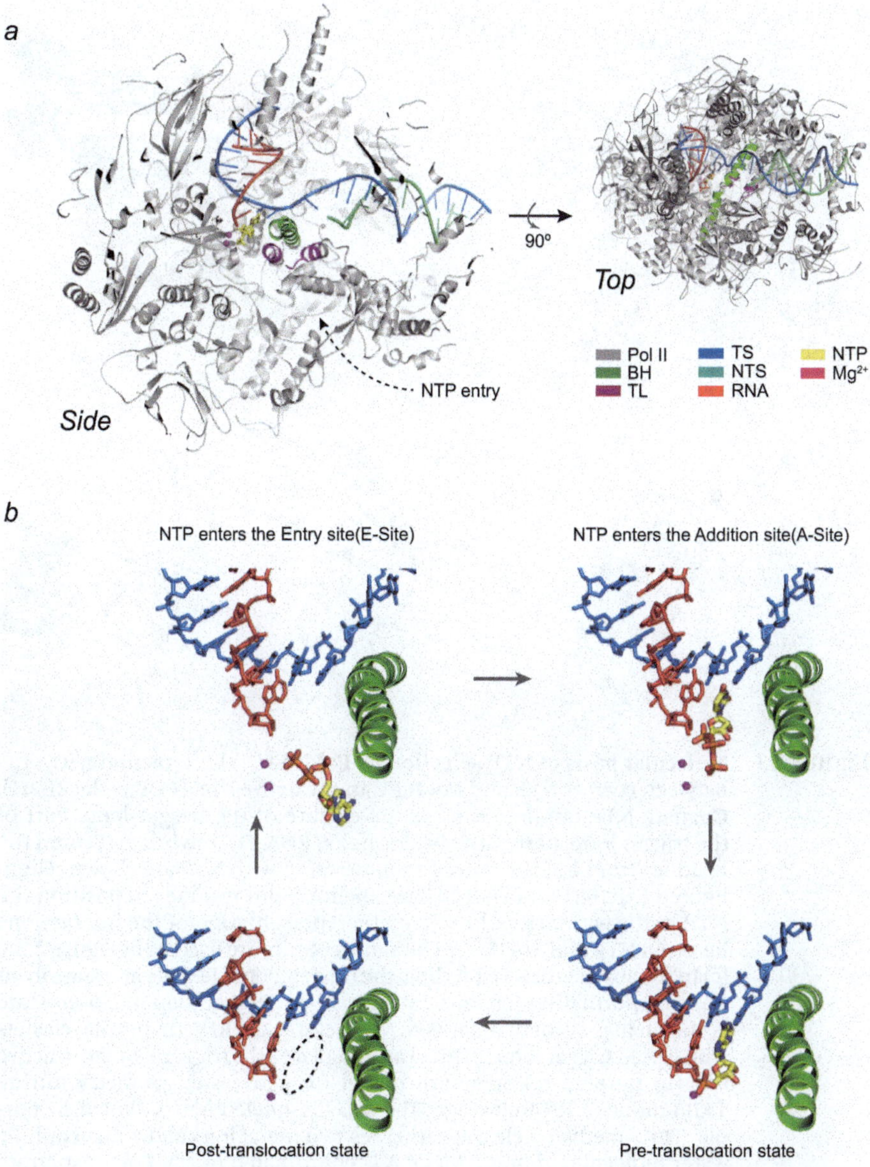

Figure 2.2 **Structures of Pol II elongation complexes (EC) and critical steps of the nucleotide addition cycle.** (a) Overall architecture of Pol II EC (PDB code: 2E2H). The side view is shown on the left and the top view is shown on the right. Pol II is shown in grey, the bridge helix (BH) is shown in green, and the trigger loop is shown in purple. RNA, the template DNA strand (TS), the non-template DNA strand (NTS), nucleotide substrate (NTP), and magnesium ions are shown in red, blue, green, yellow, and magenta, respectively. (b) Transcriptional nucleotide addition cycle of Pol II. NTP first binds to the "entry site", followed by rotating into the "addition site" by base pairing with the matched DNA template. After the nucleotide incorporation step, Pol II translocates one base forward from the pre-translocation state to the post-translocation state, creating a new empty active site for the next round of nucleotide addition.

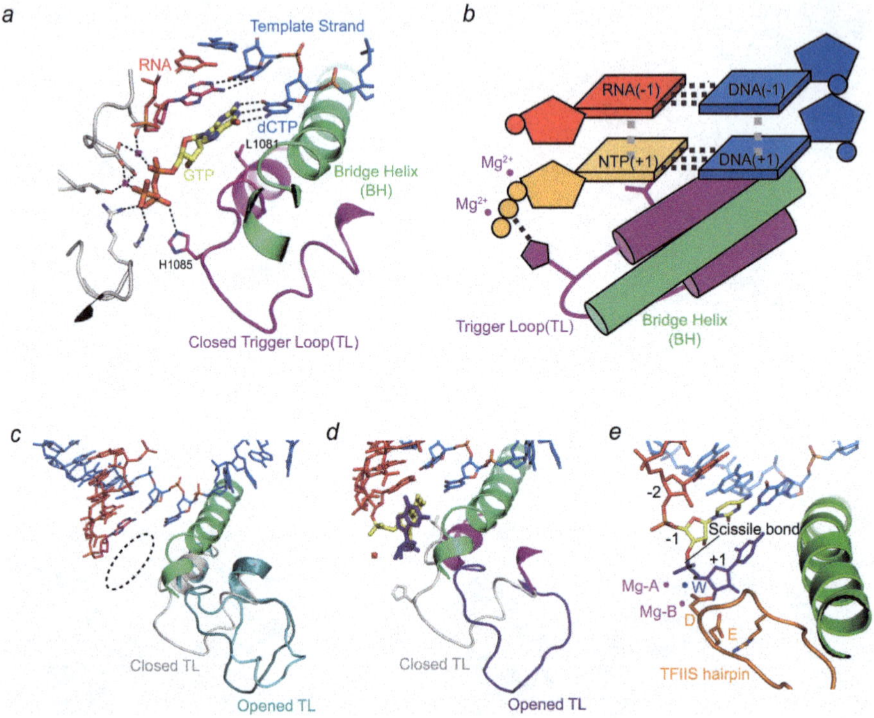

Figure 2.3 **Molecular basis of NTP selection by Pol II EC.** (a) Interaction networks between correct NTP and Pol II EC in the active site. (PDB code: 2E2H). Cognate NTP binding induces the closure of the trigger loop. Part of the trigger loop, part of the bridge helix, RNA–DNA hybrid, NTP and the residues from Pol II involved in interacting with NTP are shown. Highlighted Leu1081 and His1085 are anchor point residues to position the NTP in the active site. (b) Schematic representation of the interactions between NTP and Pol II EC. The same colours are used as in Figure 2.3a. (c) In the absence of NTP binding, the trigger loop is flexible and stays in an open conformation (coloured in cyan). The trigger loop in closed state is shown in grey for comparison. (d) Backtracked RNA (in purple) clashes with closed trigger loop (in grey) and displaces the trigger loop from active site (in purple). (e) Structure model for active site geometry during TFIIS-induced RNA cleavage, the scissile nucleotide is shown in purple. TFIIS mediated cleavage involves two metal ions and a nucleophilic water molecule (in blue). Metal A is coordinated by the Pol II aspartate loop (not shown) and the RNA phosphate. Metal B is coordinated by TFIIS hairpin residues D290 and E291 (in orange).

the NTP substrate is paired with the DNA template and sandwiched between the RNA 3′ nucleotide and the Pol II trigger loop. These interactions ensure correct nucleobase positioning in both lateral and vertical directions.[19,33] In sharp contrast, when the mismatched nucleotide binds to the "E site", the trigger loop is in the inactive open conformation. Therefore, the rate of mismatched nucleotide incorporation (trigger loop-independent nucleotide addition) is several orders of magnitude slower than the rate of correct substrate incorporation (trigger loop-dependent nucleotide addition).[4,45] This key induced-fit mechanistic difference provides additional discrimination power to achieve high transcriptional fidelity in the insertion step. As is also observed for many other nucleic-acid enzymes, the closure of the active site induced by correct substrates is a common strategy to facilitate enzymatic efficiency and selectivity.[46–48] Indeed, side-chain substitutions in the trigger loop and nearby active site, or blockage of trigger-loop closure by alpha-amanitin, significantly affects Pol II transcription fidelity and catalytic efficiency.[15,19,34,49,50]

Structural studies also provide important insights into the molecular basis of the second fidelity checkpoint, the extension step. In this step, the fully matched upstream RNA–DNA hybrid and nearby active site residues play essential roles for the proper alignment of the 3′-RNA terminus with active site metal ions and the incoming NTP. In the Pol II structures containing a mismatched RNA terminus, the mismatched 3′-RNA terminal nucleotide either adopts a frayed conformation or causes Pol II to move in a reverse direction into backtracked states, extruding from the active site into the "secondary channel" of Pol II, which is also the channel for NTP entry by diffusion.[32,51,52] Consequently, the mismatched 3′-RNA terminus is positioned in an inactive conformation and the RNA chain partially overlaps with the canonical substrate binding site, greatly compromising the next round of nucleotide binding and addition. Therefore, RNA transcript extension from a mismatched 3′-RNA terminus becomes much slower than the extension from a fully matched end.

Proofreading is the third fidelity checkpoint. Pol II proofreading activity depends on cleavage of the transcripts containing a mismatched 3′-RNA terminus through a backtracking-dependent proofreading mechanism to regenerate a new post-translocation state, allowing Pol II to have a "second chance" to select the correct nucleotide for incorporation and elongation. In fact, slow forward extension from the mismatched 3′-RNA terminus also provides a time window for Pol II backtracking and proofreading.[20,27,32,51] Structures of backtracked Pol II provide a detailed understanding of these backtracked states and proofreading mechanisms. The mismatched single-stranded 3′-RNA terminus moves toward the secondary channel (Figure 2.3d). Dinucleotides or short oligomers containing the mis-incorporated nucleotide are preferentially cleaved.[32,43,51] There are two different proofreading mechanisms: intrinsic cleavage and transcription factor IIS (TFIIS)-mediated cleavage.[24,32,51,53–55] In the intrinsic cleavage mode (Figure 2.3d), Pol II removes the one or two backtracked nucleotides from the 3′-RNA terminus without the recruitment of specific proofreading accessory protein. A metal ion activated water molecule (Figure 2.3d) is thought to attack the phosphodiester backbone of RNA transcript for intrinsic cleavage.[56,57] In the case of more extensive backtracking, transcription factor TFIIS is recruited to the backtracked complex to stimulate the transcript cleavage.[28,32,43,51] The domain III of TFIIS inserts into the Pol II active site and provides acidic residues for chelating a magnesium ion to activate the attacking water molecule[51] (Figure 2.3e). Meanwhile, the inserted domain III of TFIIS pushes the Pol II trigger loop into an inactive open conformation, changing the catalytic properties of the polymerase from the slower intrinsic cleavage to the highly efficient factor-mediated cleavage. To re-enter the productive Pol II elongation mode, the inserted domain III of TFIIS must be displaced from the secondary channel, allowing the trigger loop to access its closed conformation during nucleotide addition cycles. Multiple rounds of backtracking, cleavage, and reactivation can also happen when Pol II encounters translocation barriers or DNA damage. Multiple rounds of proofreading provide a mechanism for increasing overall transcriptional bypass by either suppressing DNA lesion-induced transcriptional mutagenesis or providing enough time for the recruitment of DNA damage repair machinery.

2.3 Chemical Biology Approaches to Study the Transcription Fidelity Mechanism

Structural studies have revealed the sophisticated interaction network between Pol II, NTP nucleotide, and the DNA–RNA hybrid to achieve the high transcription fidelity. This interaction network is composed of various types of chemical interactions: base stacking, hydrogen bonding, hydrophobic interactions, and salt bridge interactions.[19,33] However, the individual extents of the contributions of these interactions to transcriptional fidelity control have been unclear. Dissecting individual contributions of these chemical interactions, in particular the interactions of the nucleic-acid scaffold and substrate, will provide insights into the molecular mechanism by which Pol II reads the DNA template and maintains high transcriptional fidelity. Moreover, this knowledge also helps us to understand how DNA lesions and modifications affect transcription fidelity, since lesions or modifications typically change the functional groups and structural features of nucleic acids. Isolating and dissecting the contribution of each individual chemical interaction to transcription fidelity is difficult by conventional biological approaches. Hence, synthetic nucleic-acid chemistry has been employed to provide the opportunity to investigate the functional interplay between nucleic-acid substrates and nucleic-acid enzymes. These valuable chemical tools have allowed and facilitated the study of how individual chemical interactions and intrinsic structural features of nucleic acids are specifically recognized by Pol II and utilized for fidelity control.

2.3.1 Distinct Contributions of Base-pair Hydrogen Bonding in Transcriptional Fidelity Control Checkpoints

In collaboration with the Kool group, our lab used a series of synthetic "hydrogen-bond-deficient" nucleobase analogues as chemical probes to dissect hydrogen-bonding interactions governing RNA Pol II transcriptional fidelity.[4] A series of nonpolar thymidine analogues in the template strand were synthesized to mimic the shape of the thymine nucleoside but lacking the capability to form hydrogen bonds. The overall size increment is only 1.0 Å from the smallest (dH) to the largest (dI) nucleoside (Figure 2.4a).[58,59]

This system allowed us to evaluate the effects of the loss of hydrogen bonding as well as the size effect of a nucleobase on the three individual "checkpoint" steps.[4] The results revealed that base-pair hydrogen-bonding makes distinct contributions in the three fidelity control checkpoints. In the insertion step, Pol II completely loses the ability to discriminate between cognate ATP and non-cognate UTP for the hydrogen-deficient DNA templates (dF, dL, dB, and dI) with approximately 1.0 selectivity for ATP over UTP, whereas there is an approximately 10^5 fold selectivity of ATP over UTP for canonical dT template.[4] This result highlights the critical role of

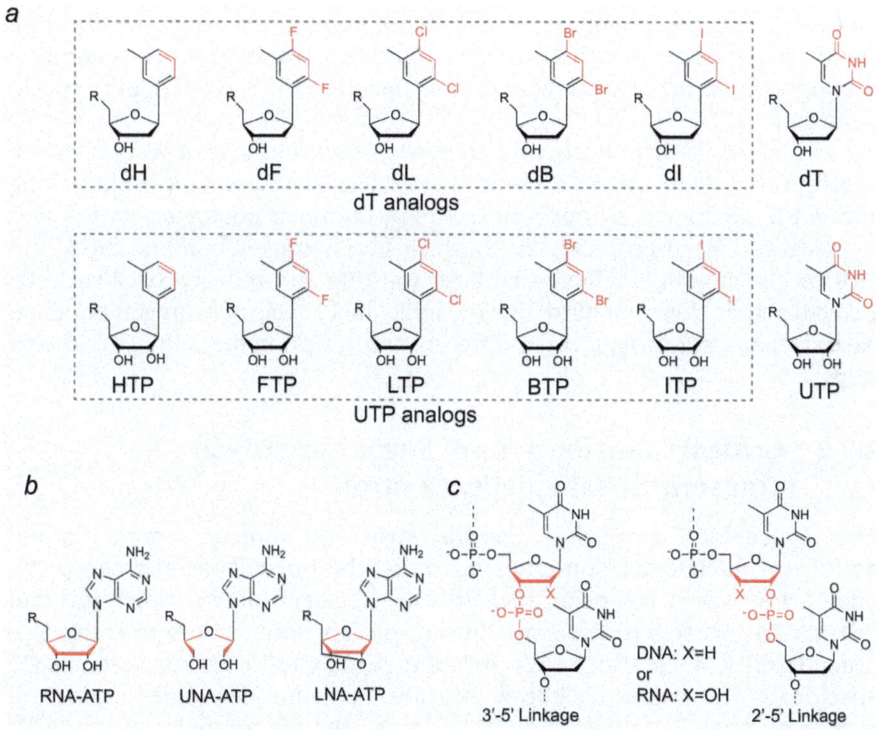

Figure 2.4 Nucleotide analogues as chemical probes to dissect individual contributions of chemical interactions and structural motifs to transcriptional fidelity. (a) Structures of "hydrogen bonding-deficient" dT analogues with sub-angstrom increments in size to investigate the effect of hydrogen bonding base pairing. The analogues are shown in dashed boxes. R, indicates the triphosphate group. (b) Unlocked nucleotide (UNA) was designed to evaluate the functions of the sugar backbone in Pol II transcriptional fidelity. (c) Two different phosphodiester linkages were used to probe the contribution of phosphodiester linkage orientation.

Watson–Crick hydrogen bonding in ensuring correct substrate incorporation in the first fidelity checkpoint step. The results of further kinetic studies demonstrated that the loss of substrate selection ability is mainly attributed to decreased incorporation rates (k_{pol}) rather than lower substrate binding affinities ($K_{d,\ app}$). This is in sharp contrast to many high-fidelity DNA polymerases where dNTP incorporation is relatively less affected by hydrogen bond deficiency.

Unexpectedly, hydrogen-bonding plays a different role during the extension step. The loss of hydrogen bonds between the matched 3′-terminus base pair has little effect on CTP incorporation specificity. More strikingly, we observed significant increases (100–1000-fold) in extension from a U–T mismatched pair upon the removal of the wobble hydrogen bonding. These results indicate that an unexpected contribution of wobble-pair hydrogen bonding is crucial to prevent extension of mismatched transcripts by Pol II. These two different effects caused by the loss of hydrogen bonds indicate that while canonical hydrogen bonding assures correct nucleotide binding and insertion in the first fidelity checkpoint step, wobble hydrogen bonding is important in preventing mismatch extension in the second fidelity checkpoint step.

In the third fidelity checkpoint step, we found that loss of Watson–Crick hydrogen bonds results in a significant increase in transcript backtracking and Pol II cleavage activities. Interestingly, nonpolar analogues with larger sizes (dB and dI) can compensate for the loss of hydrogen bonds at the 3′-end of RNA, presumably by increased base stacking and reduced backtracking. Taken together, these results define the individual effects of hydrogen bonding, base size, and stacking at three different transcriptional fidelity checkpoint steps.

2.3.2 Critical Contributions of Sugar Integrity in Transcriptional Fidelity Control

The nucleoside sugar is the central structural moiety connecting the peripheral nucleic acid functional groups: the nucleobase and the phosphate. However, it has been unclear how the sugar affects transcriptional fidelity. To investigate the contribution of this core moiety to transcriptional fidelity, nucleotides with different sugar modifications were used.[60] Specifically, a "sugar backbone mutant" (a ribonucleotide analogue), termed an unlocked nucleoside (UNA), was used for comparative studies[61] (Figure 2.4b). Compared with a "wild-type" ribonucleotide, the UNA analogue contains all of the same peripheral functional groups except for a bond that connects the C2′ and C3′ atoms of the ribose ring. UNA residues still maintain the capability to form Watson–Crick base pairing within DNA strands and form a similar duplex structure in comparison with a RNA–DNA hybrid.[62] Intriguingly, we found that the sugar backbone is a dominant but underappreciated factor in controlling all three checkpoint steps of Pol II transcriptional fidelity.[60] Nucleobase discrimination is almost completely abolished in every single fidelity checkpoint step for Pol II EC containing UNA analogues. Indeed, the sugar backbone integrity is a prerequisite for correct nucleotide selection in Pol II transcription. Pol II has an unexpectedly strong selectivity for interrogating substrate ribose integrity, which is about 100-fold stronger than its ability to discriminate against a mismatched base and 10^3–10^4-fold stronger than its ability to discriminate against dNTPs. Therefore, the correct peripheral functional groups of the nucleotide *per se* are insufficient for efficient RNA extension. Rather, the correct spatial arrangement of all peripheral functional groups (enforced by the sugar backbone) is essential in maintaining high Pol II catalytic activity and fidelity. Indeed, the Pol II active site is not fully preassembled (*e.g.*, the trigger loop remains in an open state) and only binding of correct nucleotide substrate with the correct spatial arrangement of its functional groups can induce trigger-loop closure and full assembly of the Pol II active site poised for catalysis.[19,63,64]

2.3.3 Asymmetric Discrimination of Phosphodiester Linkage During Pol II Transcription

The phosphodiester linkage is another critical structural moiety for maintaining RNA Pol II transcriptional fidelity. The non-enzymatic polymerization of ribonucleotides introduces backbone heterogeneity with a mixture of 2′–5′ and 3′–5′ linkages[65–67] (Figure 2.4c). Pol II is a key enzyme evolved specifically for synthesizing 3′–5′ linked RNA. It is not clear how Pol II selectively recognizes the 3′–5′ linkage over the 2′–5′ linkage. Our lab systematically investigated how phosphodiester linkages of nucleic acids control Pol II transcriptional efficiency and fidelity.[18] By using site-specific alteration of 3′–5′ to 2′–5′ linkages in either RNA or DNA, we quantitatively dissected the influence of phosphodiester linkage on Pol II substrate incorporation, elongation, and proofreading. Intriguingly, we found that Pol II recognizes phosphodiester linkages in an asymmetric (strand-specific) manner. The presence of 2′–5′ linkage in the RNA strand only modestly reduces Pol II transcriptional efficiency without affecting Pol II transcriptional fidelity. In sharp contrast, the presence of 2′–5′ linkage in the DNA template strand leads to a dramatic decrease in both transcriptional efficiency and fidelity. These results reveal essential contributions of the template phosphodiester linkages to Pol II transcription and provide new understanding of nucleic acid recognition and genetic information transfer during molecular evolution. Alteration of template phosphodiester backbone by 2′–5′ phosphodiester linkage has similar effect to bulky DNA lesions, both of which often change the configuration of phosphodiester backbone and affect transcriptional fidelity. Indeed, we found similarities in how Pol II processes several bulky CPD and CydA lesions and a 2′–5′ linkage template.[34,68,69]

Why does Pol II recognize the phosphodiester linkage in an asymmetric (strand-specific) manner? We are tempted to speculate that the reason for this asymmetric pattern of linkage discrimination may reflect a universal principle of template-dependent genetic information transfer: the DNA template strand and RNA product strand have distinct roles in supporting template-dependent genetic information transfer. The DNA template strand plays a vital role in genetic information transfer. It provides both the genetic coding instructions and the structural scaffold that supports selection and positioning of the correct incoming NTP through hydrogen bonding, size, and shape recognition as well as by base stacking. In contrast, the RNA product strand plays a lesser role. It stabilizes the incoming NTP mainly by base stacking and provides the terminal hydroxyl group poised for attacking the alpha-phosphate of the substrate NTP. Thus, the penalty of 2′–5′ phosphodiester linkage in the template strand may be significantly higher than that in the product strand because the altered DNA linkage distorts important structural contacts with Pol II, whereas the RNA makes relatively few side-chain contacts with Pol II.[19] This structural role may have provided high evolutionary pressure to evolve an enzyme to recognize the desired phosphodiester linkage in the DNA template strand.

2.3.4 Using Chemical Biology Tools to Understand Trigger-loop Recognition and Conformational Changes

Our studies using synthetic nucleic-acid analogues highlight the importance of the intrinsic structural features of nucleic acids in controlling Pol II transcriptional efficiency and fidelity that could not have been defined easily using conventional approaches.[4,70–72] The nucleic-acid analogues are also useful probes to investigate specific interactions between Pol II functional domains and nucleic-acid moieties. The trigger loop (TL) is a conserved structural motif of Pol II that is crucial for both catalytic activity and transcriptional fidelity. Our lab recently utilized a series of synthetic nucleotide analogues and alpha-amanitin to reveal the chemical interactions and substrate structural signatures governing Pol II trigger loop closure.[73]

The trigger loop is a flexible module of Pol II that is localized close to the nucleotide addition site. It remains in an inactive open conformation in the absence of nucleotide binding or in the presence of mismatched substrate binding. In contrast, binding of cognate substrate induces movement of TL from an inactive open state to an active closed state to facilitate nucleotide addition. However, a comprehensive understanding of the specific chemical interactions and substrate structural signatures that are essential to this TL conformational switch has been elusive. We employed synthetic nucleotide analogues as ‘chemical mutation’ tools in an alpha-amanitin transcription inhibition assay to systematically dissect the key chemical interactions and structural signatures governing the substrate-coupled TL closure in Pol II. Alpha-amanitin binds near the Pol II active site, traps the TL in an open inactive conformation and greatly reduces the rate of TL-dependent nucleotide addition of cognate substrate by several hundred-fold.[19,74,75]

In contrast, alpha-amanitin treatment does not significantly affect the incorporation rate of a mismatched substrate, since nucleotide misincorporation proceeds through a TL-independent mechanism (slow and low fidelity) in which the TL is already in an open conformation. Therefore, the comparison of substrate incorporation rate with *vs.* without alpha-amanitin treatment provides valuable information about the TL conformational profile upon the binding of any given substrate (Figure 2.5). If the TL is able to switch to a closed conformation upon the binding of a given substrate (such as a cognate substrate), one would expect a significant decrease in the substrate incorporation rate with alpha-amanitin treatment. On the other hand, if the TL remains in an open conformation upon the binding of a given substrate (such as a mismatched substrate), one would expect limited change in the substrate incorporation rate (TL-independent transcription) upon alpha-amanitin treatment. Therefore, sensitivity to alpha-amanitin treatment can be used as a measure of the extent of TL-dependent substrate incorporation for any given substrate.

Three classes of nucleotide analogues (with specific mutations in nucleobase, sugar, and phosphate leaving group) were used as 'chemical mutation' tools, coupled with the alpha-amanitin transcription inhibition assay.[73,76] These chemical tools allow us to systematically dissect the key

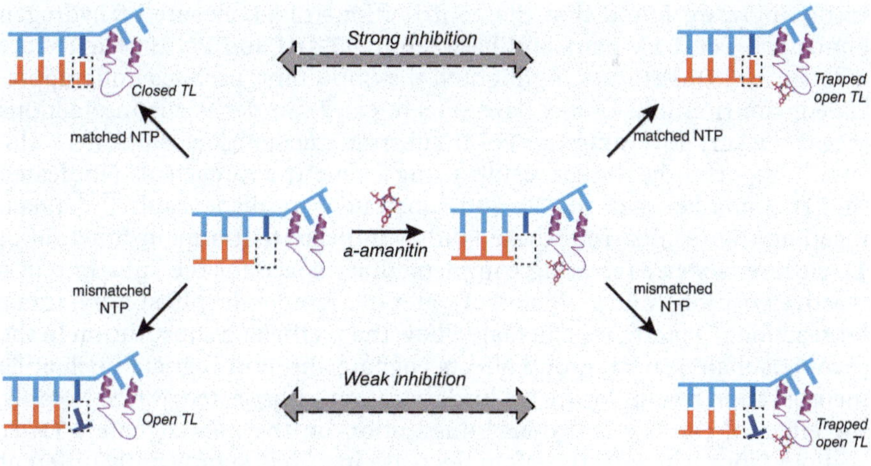

Figure 2.5 α-Amanitin as a probe to dissect the functions of trigger loop in Pol II transcriptional fidelity control. Cognate NTP binding in the A-site (indicated by dashed boxes) induces folding of the TL into the closed state, which, in turn, participates in the catalysis of the NTP incorporation. Binding of non-cognate NTP in the A-site cannot induce productive folding of the TL (left). α-Amanitin binds near the Pol II active site and traps the TL in an open inactive conformation (middle). α-Amanitin exhibits distinct inhibitory behaviours with different nucleotide incorporation (right).

chemical interactions and structural signatures governing the substrate-coupled TL conformation closure. First, our data indicate the formation of Watson–Crick hydrogen bonds is dispensable in TL closure, and the loss of hydrogen bonds can be fully compensated by increasing the size (base stacking) of a nucleobase. The substitution of natural UTP by hydrogen bond-deficient LTP or BTP (Figure 2.4a) does not alter the TL closure behaviour, as evidenced by similar effect of alpha-amanitin treatment. Second, the sugar ring conformation plays a more important role in facilitating nucleotide incorporation and TL closure than does hydrogen bonding of the 2′-OH. Results from previous studies have indicated that the main cause of observed difference in discriminating NTP over dNTP is the loss of hydrogen bonds between 2′-OH and Pol II residues, including the TL. Here we revealed a dominant role, which is underappreciated, of the sugar ring conformation in nucleotide incorporation and TL recognition. Comparing UNA-ATP (with all of the functional groups but without a constrained sugar ring) and LNA-ATP (with correct sugar pucker ring conformation but lacking 2′-OH) with canonical ATP (Figure 2.4b), we conclude that correct sugar ring conformation, rather than hydrogen bonds, is the more crucial determinant of RNA Pol II substrate incorporation and TL closure. In summary, these studies revealed that closure of the TL upon substrate binding is controlled by multiple chemical interactions including hydrogen bonding, base stacking, and van der Waals interactions (Figures 2.3 and 2.5). Intriguingly, we found that the contributions to TL closure of hydrogen bonds between base pairs and between the 2′-OH and Pol II residues are dispensable. Our studies highlighted the important, perhaps underappreciated, contributions of π–π base stacking and van der Waals interactions (sugar pucker) governing the Pol II substrate recognition and full TL closure.[77] Importantly, similar results using a genetic approach also indicated that H-bonds between the TL and substrate were dispensable.[77,78] Taken together, the results from these studies indicate that a main function of TL closure is steric (creating complementary space for the substrate in a reactive position), and exclusion of water to promote the phosphoryl transfer reaction.[79] Finally, we elucidated how the methylene substitution in the pyrophosphate leaving group affects cognate and non-cognate nucleotide incorporation by employing β,γ-bridging atom-‘substituted’ NTPs. Notably, the effect of the β,γ-methylene substitution on the non-cognate UTP/dT scaffold (approximately threefold decrease in k_{pol}) is significantly different from that of the cognate ATP/dT scaffold (approximately 130-fold decrease in k_{pol}). Removal of the wobble hydrogen bonds in U:dT recovers a strong response to methylene substitution of UTP. The results from our kinetic and modelling studies are consistent with a unique altered transition state for bond formation and cleavage for UTP/dT incorporation compared with ATP/dT incorporation. Collectively, our data reveal the functional interplay between the NTP triphosphate moiety and base pair hydrogen bonding recognition during nucleotide incorporation.

2.4 Conclusions

In summary, results from previous studies using structural and biochemical approaches have provided a structural framework for understanding the fidelity control checkpoints during RNA synthesis. Synthetic nucleic acid analogues are powerful tools to probe interactions between nucleic acids and Pol II by 'substituting' individual nucleic acid moieties or ablating specific interactions. These nucleic acid analogues enabled us to dissect the individual contributions of the nucleobase, sugar, and triphosphate moieties to accurate Pol II transcription. This synthetic nucleic acid chemistry approach is highly complementary with classic genetic and biochemical approaches that focus on mutations in Pol II residues to elucidate the roles of key protein residues. These studies have greatly advanced our understanding of how intrinsic structural features of nucleic acid moieties are recognized and how specific interactions are involved in substrate selection and incorporation during Pol II transcription.

Through studies from many labs in the field, we have now learned the atomic details of how Pol II recognizes DNA template and selects substrates, how the trigger loop is folded upon the cognate substrate binding, and how a mismatch prevents subsequent substrate addition and leads to a backtracked state that is subject to proofreading cleavage. As shown in Figure 2.6, there are at least three important "checkpoint steps" contributing to RNA Pol II high transcriptional fidelity during transcription elongation. At the insertion step, Pol II is able to binds cognate NTP and trigger loop is folded to form a closed state. The nucleobase of the cognate NTP substrate is paired with the DNA template and sandwiched between the RNA 3′ nt and the Pol II trigger loop. The substrate is well positioned, the metal ions and Pol II

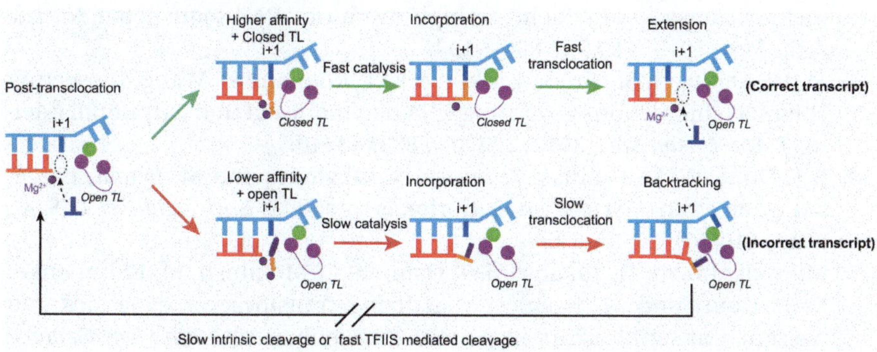

Figure 2.6 **Molecular mechanisms of transcriptional fidelity.** Different transcription intermediates of the Pol II active site are shown in schematic representations. Template DNA is shown in blue and RNA is shown in red. The bridge helix and trigger loop are shown in green and purple, respectively.

active site residues are reconfigured poised for catalytic reaction. The Pol II bound with a mismatched substrate, on the other hand, fails to achieve into a fully-active closed state and is not optimal for reaction. Therefore, the rate of mismatched nucleotide incorporation (trigger loop-independent nucleotide addition) is several orders of magnitude slower than the rate of correct substrate incorporation (trigger loop-dependent nucleotide addition). The subsequent forward Pol II translocation after cognate substrate incorporation is also faster than that of Pol II after misincorporation. For the extension step, extension from a mismatched 3′ end is substantially slow. The mismatched 3′ end base pair is not stable and the RNA 3′ nucleotide is often misaligned (frayed), preventing binding of next cognate NTP and formation of fully-active closed state for the next round of nucleotide addition. The third step, backtracking-dependent proofreading, also contributes to fidelity. Indeed, Pol II containing a mismatched 3′-RNA terminus favoured backtracking. The transcript containing a mismatched 3′-RNA terminus is subject to intrinsic cleavage or TFIIS-stimulated cleavage. As a result, Pol II is reactivated and has a "second chance" to select the correct nucleotide for incorporation and elongation.

References

1. R. Kornberg, The molecular basis of eukaryotic transcription (Nobel Lecture), *Angew. Chem., Int. Ed.*, 2007, **46**(37), 6956–6965.
2. M. L. Kireeva, Y. A. Nedialkov, G. H. Cremona, Y. A. Purtov, L. Lubkowska, F. Malagon, Z. F. Burton, J. N. Strathern and M. Kashlev, Transient reversal of RNA polymerase II active site closing controls fidelity of transcription elongation, *Mol. Cell*, 2008, **30**(5), 557–566.
3. M. Imashimizu, T. Oshima, L. Lubkowska and M. Kashlev, Direct assessment of transcription fidelity by high-resolution RNA sequencing, *Nucleic Acids Res.*, 2013, **41**(19), 9090–9104.
4. M. W. Kellinger, S. Ulrich, J. Chong, E. T. Kool and D. Wang, Dissecting chemical interactions governing RNA polymerase II transcriptional fidelity, *J. Am. Chem. Soc.*, 2012, **134**(19), 8231–8240.
5. J. F. Gout, W. K. Thomas, Z. Smith, K. Okamoto and M. Lynch, Large-scale detection of *in vivo* transcription errors, *Proc. Natl. Acad. Sci. U. S. A.*, 2013, **110**(46), 18584–18589.
6. M. Imashimizu, H. Takahashi, T. Oshima, C. McIntosh, M. Bubunenko, D. L. Court and M. Kashlev, Visualizing translocation dynamics and nascent transcript errors in paused RNA polymerases *in vivo*, *Genome Biol.*, 2015, **16**, 98.
7. T. T. Saxowsky and P. W. Doetsch, RNA polymerase encounters with DNA damage: transcription-coupled repair or transcriptional mutagenesis?, *Chem. Rev.*, 2006, **106**(2), 474–488.
8. D. Bregeon and P. W. Doetsch, Transcriptional mutagenesis: causes and involvement in tumour development, *Nat. Rev. Cancer*, 2011, **11**(3), 218–227.

9. A. J. Gordon, J. A. Halliday, M. D. Blankschien, P. A. Burns, F. Yatagai and C. Herman, Transcriptional infidelity promotes heritable phenotypic change in a bistable gene network, *PLoS Biol.*, 2009, **7**(2), e44.
10. A. J. Gordon, D. Satory, J. A. Halliday and C. Herman, Heritable change caused by transient transcription errors, *PLoS Genet.*, 2013, **9**(6), e1003595.
11. A. Viswanathan, H. J. You and P. W. Doetsch, Phenotypic change caused by transcriptional bypass of uracil in nondividing cells, *Science*, 1999, **284**(5411), 159–162.
12. L. B. Carey, RNA polymerase errors cause splicing defects and can be regulated by differential expression of RNA polymerase subunits, *eLife*, 2015, **4**, e09945.
13. K. James, P. Gamba, S. J. Cockell and N. Zenkin, Misincorporation by RNA polymerase is a major source of transcription pausing in vivo, *Nucleic Acids Res.*, 2017, **45**(3), 1105–1113.
14. P. Gamba and N. Zenkin, Transcription fidelity and its roles in the cell, *Curr. Opin. Microbiol.*, 2018, **42**, 13–18.
15. R. T. Libby and J. A. Gallant, The role of RNA polymerase in transcriptional fidelity, *Mol. Microbiol.*, 1991, **5**(5), 999–1004.
16. C. L. Zheng, N. J. Wang, J. Chung, H. Moslehi, J. Z. Sanborn, J. S. Hur, E. A. Collisson, S. S. Vemula, A. Naujokas, K. E. Chiotti, J. B. Cheng, H. Fassihi, A. J. Blumberg, C. V. Bailey, G. M. Fudem, F. G. Mihm, B. B. Cunningham, I. M. Neuhaus, W. Liao, D. H. Oh, J. E. Cleaver, P. E. LeBoit, J. F. Costello, A. R. Lehmann, J. W. Gray, P. T. Spellman, S. T. Arron, N. Huh, E. Purdom and R. J. Cho, Transcription restores DNA repair to heterochromatin, determining regional mutation rates in cancer genomes, *Cell Rep.*, 2014, **9**(4), 1228–1234.
17. F. W. Martinez-Rucobo and P. Cramer, Structural basis of transcription elongation, *Biochim. Biophys. Acta*, 2013, **1829**(1), 9–19.
18. L. Xu, L. Da, S. W. Plouffe, J. Chong, E. Kool and D. Wang, Molecular basis of transcriptional fidelity and DNA lesion-induced transcriptional mutagenesis, *DNA Repair*, 2014, **19**, 71–83.
19. D. Wang, D. A. Bushnell, K. D. Westover, C. D. Kaplan and R. D. Kornberg, Structural basis of transcription: role of the trigger loop in substrate specificity and catalysis, *Cell*, 2006, **127**(5), 941–954.
20. M. J. Thomas, A. A. Platas and D. K. Hawley, Transcriptional fidelity and proofreading by RNA polymerase II, *Cell*, 1998, **93**(4), 627–637.
21. J. D. Kahn and J. E. Hearst, Reversibility of nucleotide incorporation by Escherichia coli RNA polymerase, and its effect on fidelity, *J. Mol. Biol.*, 1989, **205**(2), 291–314.
22. M. G. Izban and D. S. Luse, The RNA polymerase II ternary complex cleaves the nascent transcript in a 3'----5' direction in the presence of elongation factor SII, *Genes Dev.*, 1992, **6**(7), 1342–1356.
23. D. A. Erie, O. Hajiseyedjavadi, M. C. Young and P. H. von Hippel, Multiple RNA polymerase conformations and GreA: control of the fidelity of transcription, *Science*, 1993, **262**(5135), 867–873.

24. S. Sigurdsson, A. B. Dirac-Svejstrup and J. Q. Svejstrup, Evidence that transcript cleavage is essential for RNA polymerase II transcription and cell viability, *Mol. Cell*, 2010, **38**(2), 202–210.
25. T. L. Johnson and M. J. Chamberlin, Complexes of yeast RNA polymerase II and RNA are substrates for TFIIS-induced RNA cleavage, *Cell*, 1994, **77**(2), 217–224.
26. D. A. Erie, T. D. Yager and P. H. von Hippel, The single-nucleotide addition cycle in transcription: a biophysical and biochemical perspective, *Annu. Rev. Biophys. Biomol. Struct.*, 1992, **21**, 379–415.
27. C. Jeon and K. Agarwal, Fidelity of RNA polymerase II transcription controlled by elongation factor TFIIS, *Proc. Natl. Acad. Sci. U. S. A.*, 1996, **93**(24), 13677–13682.
28. H. Kettenberger, K. J. Armache and P. Cramer, Complete RNA polymerase II elongation complex structure and its interactions with NTP and TFIIS, *Mol. Cell*, 2004, **16**(6), 955–965.
29. D. Reines, R. C. Conaway and J. W. Conaway, Mechanism and regulation of transcriptional elongation by RNA polymerase II, *Curr. Opin. Cell Biol.*, 1999, **11**(3), 342–346.
30. A. L. Gnatt, P. Cramer, J. Fu, D. A. Bushnell and R. D. Kornberg, Structural basis of transcription: an RNA polymerase II elongation complex at 3.3 A resolution, *Science*, 2001, **292**(5523), 1876–1882.
31. C. D. Kaplan and R. D. Kornberg, A bridge to transcription by RNA polymerase, *J. Biol.*, 2008, **7**(10), 39.
32. D. Wang, D. A. Bushnell, X. Huang, K. D. Westover, M. Levitt and R. D. Kornberg, Structural basis of transcription: backtracked RNA polymerase II at 3.4 angstrom resolution, *Science*, 2009, **324**(5931), 1203–1206.
33. X. Huang, D. Wang, D. R. Weiss, D. A. Bushnell, R. D. Kornberg and M. Levitt, RNA polymerase II trigger loop residues stabilize and position the incoming nucleotide triphosphate in transcription, *Proc. Natl. Acad. Sci. U. S. A.*, 2010, **107**(36), 15745–15750.
34. C. Walmacq, A. C. Cheung, M. L. Kireeva, L. Lubkowska, C. Ye, D. Gotte, J. N. Strathern, T. Carell, P. Cramer and M. Kashlev, Mechanism of translesion transcription by RNA polymerase II and its role in cellular resistance to DNA damage, *Mol. Cell*, 2012, **46**(1), 18–29.
35. L. Zhang, F. Pardo-Avila, I. C. Unarta, P. P. Cheung, G. Wang, D. Wang and X. Huang, Elucidation of the Dynamics of Transcription Elongation by RNA Polymerase II using Kinetic Network Models, *Acc. Chem. Res.*, 2016, **49**(4), 687–694.
36. E. Nogales, R. K. Louder and Y. He, Structural Insights into the Eukaryotic Transcription Initiation Machinery, *Annu. Rev. Biophys.*, 2017, **46**, 59–83.
37. M. Hantsche and P. Cramer, The Structural Basis of Transcription: 10Years After the Nobel Prize in Chemistry, *Angew. Chem., Int. Ed.*, 2016, **55**(52), 15972–15981.
38. R. D. Kornberg, The molecular basis of eucaryotic transcription, *Cell Death Differ.*, 2007, **14**(12), 1989–1997.

39. J. Q. Svejstrup, RNA polymerase II transcript elongation, *Biochim. Biophys. Acta*, 2013, **1829**(1), 1.
40. P. Cramer, D. A. Bushnell and R. D. Kornberg, Structural basis of transcription: RNA polymerase II at 2.8 angstrom resolution, *Science*, 2001, **292**(5523), 1863–1876.
41. K. D. Westover, D. A. Bushnell and R. D. Kornberg, Structural basis of transcription: separation of RNA from DNA by RNA polymerase II, *Science*, 2004, **303**(5660), 1014–1016.
42. P. C. Hanawalt and G. Spivak, Transcription-coupled DNA repair: two decades of progress and surprises, *Nat. Rev. Mol. Cell Biol.*, 2008, **9**(12), 958–970.
43. H. Kettenberger, K. J. Armache and P. Cramer, Architecture of the RNA polymerase II-TFIIS complex and implications for mRNA cleavage, *Cell*, 2003, **114**(3), 347–357.
44. K. D. Westover, D. A. Bushnell and R. D. Kornberg, Structural basis of transcription: nucleotide selection by rotation in the RNA polymerase II active center, *Cell*, 2004, **119**(4), 481–489.
45. Y. Yuzenkova, A. Bochkareva, V. R. Tadigotla, M. Roghanian, S. Zorov, K. Severinov and N. Zenkin, Stepwise mechanism for transcription fidelity, *BMC Biol.*, 2010, **8**, 54.
46. S. Doublie, M. R. Sawaya and T. Ellenberger, An open and closed case for all polymerases, *Structure*, 1999, **7**(2), R31–R35.
47. P. Cramer, Common structural features of nucleic acid polymerases, *BioEssays*, 2002, **24**(8), 724–729.
48. W. A. Beard and S. H. Wilson, Structure and mechanism of DNA polymerase Beta, *Chem. Rev.*, 2006, **106**(2), 361–382.
49. I. Toulokhonov, J. Zhang, M. Palangat and R. Landick, A central role of the RNA polymerase trigger loop in active-site rearrangement during transcriptional pausing, *Mol. Cell*, 2007, **27**(3), 406–419.
50. H. Braberg, H. Jin, E. A. Moehle, Y. A. Chan, S. Wang, M. Shales, J. J. Benschop, J. H. Morris, C. Qiu, F. Hu, L. K. Tang, J. S. Fraser, F. C. Holstege, P. Hieter, C. Guthrie, C. D. Kaplan and N. J. Krogan, From structure to systems: high-resolution, quantitative genetic analysis of RNA polymerase II, *Cell*, 2013, **154**(4), 775–788.
51. A. C. Cheung and P. Cramer, Structural basis of RNA polymerase II backtracking, arrest and reactivation, *Nature*, 2011, **471**(7337), 249–253.
52. J. F. Sydow, F. Brueckner, A. C. Cheung, G. E. Damsma, S. Dengl, E. Lehmann, D. Vassylyev and P. Cramer, Structural basis of transcription: mismatch-specific fidelity mechanisms and paused RNA polymerase II with frayed RNA, *Mol. Cell*, 2009, **34**(6), 710–721.
53. M. G. Izban and D. S. Luse, The increment of SII-facilitated transcript cleavage varies dramatically between elongation competent and incompetent RNA polymerase II ternary complexes,*J. Biol. Chem.*, 1993, **268**(17), 12874–12885.
54. P. H. von Hippel, An integrated model of the transcription complex in elongation, termination, and editing, *Science*, 1998, **281**(5377), 660–665.

55. N. Zenkin, Y. Yuzenkova and K. Severinov, Transcript-assisted transcriptional proofreading, *Science*, 2006, **313**(5786), 518–520.
56. M. E. Gottesman and A. Mustaev, Inorganic phosphate, arsenate, and vanadate enhance exonuclease transcript cleavage by RNA polymerase by 2000-fold, *Proc. Natl. Acad. Sci. U. S. A.*, 2018, **115**(11), 2746–2751.
57. C. K. M. Tse, J. Xu, L. Xu, F. K. Sheong, S. Wang, H. Y. Chow, X. Gao, X. Li, P. P.-H. Cheung and D. J. N. C. Wang, Intrinsic cleavage of RNA polymerase II adopts a nucleobase-independent mechanism assisted by transcript phosphate, *Nat. Catal.*, 2019, 228–235.
58. T. W. Kim, J. C. Delaney, J. M. Essigmann and E. T. Kool, Probing the active site tightness of DNA polymerase in subangstrom increments, *Proc. Natl. Acad. Sci. U. S. A.*, 2005, **102**(44), 15803–15808.
59. T. W. Kim and E. T. Kool, A series of nonpolar thymidine analogues of increasing size: DNA base pairing and stacking properties, *J. Org. Chem.*, 2005, **70**(6), 2048–2053.
60. L. Xu, S. W. Plouffe, J. Chong, J. Wengel and D. Wang, A chemical perspective on transcriptional fidelity: dominant contributions of sugar integrity revealed by unlocked nucleic acids, *Angew. Chem., Int. Ed.*, 2013, **52**(47), 12341–12345.
61. N. Langkjaer, A. Pasternak and J. Wengel, UNA (unlocked nucleic acid): a flexible RNA mimic that allows engineering of nucleic acid duplex stability, *Bioorg. Med. Chem.*, 2009, **17**(15), 5420–5425.
62. A. Pasternak and J. Wengel, Thermodynamics of RNA duplexes modified with unlocked nucleic acid nucleotides, *Nucleic Acids Res.*, 2010, **38**(19), 6697–6706.
63. C. D. Kaplan, K. M. Larsson and R. D. Kornberg, The RNA polymerase II trigger loop functions in substrate selection and is directly targeted by alpha-amanitin, *Mol. Cell*, 2008, **30**(5), 547–556.
64. A. M. Malinen, M. Turtola, M. Parthiban, L. Vainonen, M. S. Johnson and G. A. Belogurov, Active site opening and closure control translocation of multisubunit RNA polymerase, *Nucleic Acids Res.*, 2012, **40**(15), 7442–7451.
65. E. H. Ekland and D. P. Bartel, RNA-catalysed RNA polymerization using nucleoside triphosphates, *Nature*, 1996, **383**(6596), 192.
66. J. P. Ferris and G. Ertem, Oligomerization of ribonucleotides on montmorillonite: reaction of the 5'-phosphorimidazolide of adenosine, *Science*, 1992, **257**(5075), 1387–1389.
67. F. R. Bowler, C. K. Chan, C. D. Duffy, B. Gerland, S. Islam, M. W. Powner, J. D. Sutherland and J. Xu, Prebiotically plausible oligoribonucleotide ligation facilitated by chemoselective acetylation, *Nat. Chem.*, 2013, **5**(5), 383–389.
68. C. Walmacq, L. F. Wang, J. Chong, K. Scibelli, L. Lubkowska, A. Gnatt, P. J. Brooks, D. Wang and M. Kashlev, Mechanism of RNA polymerase II bypass of oxidative cyclopurine DNA lesions, *Proc. Natl. Acad. Sci. U. S. A.*, 2015, **112**(5), E410–E419.

69. L. Xu, W. Wang, L. Zhang, J. Chong, X. Huang and D. Wang, Impact of template backbone heterogeneity on RNA polymerase II transcription, *Nucleic Acids Res.*, 2015, **43**(4), 2232–2241.
70. S. A. Benner, Understanding nucleic acids using synthetic chemistry, *Acc. Chem. Res.*, 2004, **37**(10), 784–797.
71. A. M. Sismour and S. A. Benner, Synthetic biology, *Expert Opin. Biol. Ther.*, 2005, **5**(11), 1409–1414.
72. O. Khakshoor, S. E. Wheeler, K. N. Houk and E. T. Kool, Measurement and theory of hydrogen bonding contribution to isosteric DNA base pairs, *J. Am. Chem. Soc.*, 2012, **134**(6), 3154–3163.
73. L. Xu, K. V. Butler, J. Chong, J. Wengel, E. T. Kool and D. Wang, Dissecting the chemical interactions and substrate structural signatures governing RNA polymerase II trigger loop closure by synthetic nucleic acid analogues, *Nucleic Acids Res.*, 2014, **42**(9), 5863–5870.
74. D. A. Bushnell, P. Cramer and R. D. Kornberg, Structural basis of transcription: alpha-amanitin-RNA polymerase II cocrystal at 2.8 A resolution, *Proc. Natl. Acad. Sci. U. S. A.*, 2002, **99**(3), 1218–1222.
75. F. Brueckner and P. Cramer, Structural basis of transcription inhibition by alpha-amanitin and implications for RNA polymerase II translocation, *Nat. Struct. Mol. Biol.*, 2008, **15**(8), 811–818.
76. L. Xu, W. Wang, J. Chong, J. H. Shin, J. Xu and D. Wang, RNA polymerase II transcriptional fidelity control and its functional interplay with DNA modifications, *Crit. Rev. Biochem. Mol. Biol.*, 2015, **50**(6), 503–519.
77. C. Qiu, O. C. Erinne, J. M. Dave, P. Cui, H. Jin, N. Muthukrishnan, L. K. Tang, S. G. Babu, K. C. Lam, P. J. Vandeventer, R. Strohner, J. Van den Brulle, S. H. Sze and C. D. Kaplan, High-Resolution Phenotypic Landscape of the RNA Polymerase II Trigger Loop, *PLoS Genet.*, 2016, **12**(11), e1006321.
78. C. D. Kaplan, H. Jin, I. L. Zhang and A. Belyanin, Dissection of Pol II trigger loop function and Pol II activity-dependent control of start site selection in vivo, *PLoS Genet.*, 2012, **8**(4), e1002627.
79. S. A. Seibold, B. N. Singh, C. Zhang, M. Kireeva, C. Domecq, A. Bouchard, A. M. Nazione, M. Feig, R. I. Cukier, B. Coulombe, M. Kashlev, M. Hampsey and Z. F. Burton, Conformational coupling, bridge helix dynamics and active site dehydration in catalysis by RNA polymerase, *Biochim. Biophys. Acta*, 2010, **1799**(8), 575–587.

CHAPTER 3

RNA Polymerase as a Torsional Motor

JESSICA L. KILLIAN[†a,b], JIE MA[†c,d] AND MICHELLE D. WANG*[a,b]

[a]Howard Hughes Medical Institute, Ithaca, NY 14853, USA; [b]Department of Physics – LASSP, Cornell University, Ithaca, NY 14853, USA; [c]School of Physics, Sun Yat-sen University, Guangzhou, 510275, People's Republic of China; [d]State Key Laboratory of Optoelectronic Materials and Technologies, Sun Yat-sen University, Guangzhou, 510275, People's Republic of China
*E-mail: mwang@physics.cornell.edu

3.1 Introduction

3.1.1 Transcription Generates Torsion

RNA polymerase (RNAP) is an intrinsic torsional motor.[1–4] During transcription, RNAP tracks the major helical groove of double-stranded DNA (dsDNA) as the DNA moves through the enzyme (Figure 3.1), necessarily producing a relative rotation between RNAP and its DNA substrate. *In vivo*, RNAP's ability to rotate freely about its DNA substrate is thought to be significantly restricted; RNAP and its transcript are potentially tethered to cell membranes and other cellular structures, and the RNA transcript is known to be associated with large, multi-subunit processing machineries (*e.g.* ribosomes in prokaryotes and spliceosomes in eukaryotes)[5,6] that impose substantial viscous drag on RNAP rotation.

[†]These authors contributed equally.

Chemical Biology No. 19
RNA Polymerases as Molecular Motors: On the Road, 2nd Edition
Edited by Robert Landick, Terence Strick and Jue Wang

Published by the Royal Society of Chemistry, www.rsc.org

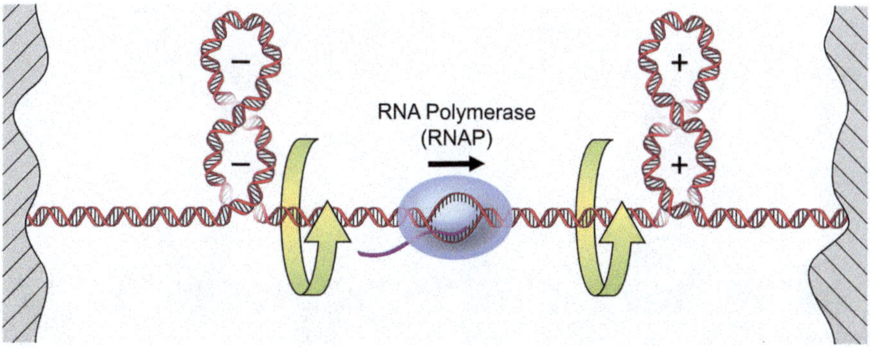

Figure 3.1 The "twin-supercoiled-domain model".[5] Because RNAP tracks the major helical groove of DNA, the translocation of DNA through the RNAP requires the rotation of the DNA relative to the enzyme, generating supercoiling. Here and in all figures, for the sake of simplicity and convention, we designate the relative motion of RNAP and substrate *via* an arrow indicating RNAP motion. The reader should note that RNAP is not free to rotate, and that relative rotation between RNAP and DNA occurs primarily *via* rotation of the DNA. Thus, when both the free rotation of RNAP and DNA are restricted, transcription will drive downstream DNA to form (+) supercoiling and upstream DNA to form (−) supercoiling.

Thus, it is believed that a transcribing RNAP remains rather stationary with respect to the cell cytoplasm, and the DNA twists as it passes through the enzyme instead. This action generates torsion, adding one supercoil to the DNA for each helical pitch of DNA transcribed. Because DNA is confined within cellular structures or topological domains,[7,8] the accumulated torsion in the DNA cannot be efficiently dissipated *via* DNA end rotation. Consequently, RNAP will over-twist DNA downstream, creating (+) supercoiling, and under-twist DNA upstream, creating (−) supercoiling, as described in the "twin-supercoiled-domain" model proposed by Liu and Wang (Figure 3.1).[5]

3.1.2 The Biological Implications of Torsion

The effects of transcription-generated torsion have proven to be far-reaching, inducing genome-wide supercoiling that can transmit thousands of base pairs from a transcribing RNAP.[9,10] A recent study revealed that dynamic (−) DNA supercoiling can transmit over 1.5 kbp upstream from a transcription start site,[9] and results from other recent genome-wide studies have indicated that transcription-generated torsion is broadly present both upstream of and within actively transcribing genes.[9–13] For example, Matsumoto and Hirose[11] found that transcription coupled (−) DNA supercoiling spread over 150 loci on the polytene chromosomes of *Drosophila melanogaster* even in the presence of active topoisomerases.

Torsion has been shown to serve a broad range of regulatory roles in transcription (Figure 3.2), such as facilitating the binding of regulatory proteins,[14–16] activating neighboring genes,[17–19] encouraging R-loop

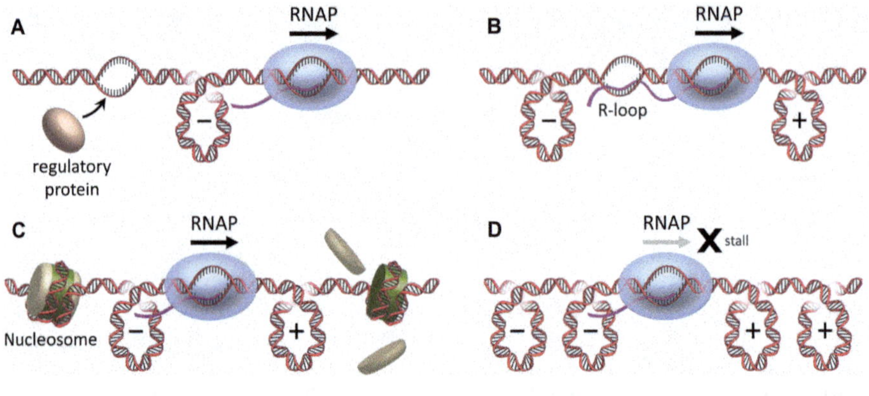

Figure 3.2 RNAP-generated torsion during transcription has diverse biological consequences: (A) Transcription generated (−) supercoiling upstream of an RNAP encourages DNA to form non-B-DNA structures which can help recruit regulatory proteins like transcription factors; (B) transcription generated (−) supercoiling upstream of an RNAP promotes R-loop formation; (C) transcription generated (+) supercoiling ahead of an RNAP can destabilize nucleosomes in downstream DNA, while generated (−) supercoiling can facilitate nucleosome formation in upstream DNA; (D) transcription generated DNA (−) supercoiling upstream and (+) supercoiling downstream can stall an elongating RNAP.

formation,[20–22] promoting protein dissociation,[23] increasing RNAP pausing,[24] inducing transcription bursting,[25] and slowing down or even stalling transcription.[24,26] For example, Guo *et al.*[16] recently reported that the protein GapR preferentially binds to overtwisted DNA and stimulates gyrase and topo IV to relax positive supercoils. Chen *et al.*[17] showed that the *leu*-500 promotor of *S. typhimurium* can be stimulated by an adjacent gene *via* transcription-coupled DNA supercoiling in *topA* strains. Zhi *et al.*[19] demonstrated that even transient (−) DNA supercoiling generated by transcription at a divergent promoter is capable of activating the *leu*-500 promoter without the need of topoisomerase depletion. Kouzine *et al.*[14,15] demonstrated that the torsional stress generated by RNAP could amplify the transient melting of a far upstream DNA element (FUSE), in this case the human c-*myc* gene, which may facilitate transcription initiation. The results of these studies confirmed that RNAP's torque generation capacity is large enough to enhance spontaneous DNA melting events even far from active transcription sites.

Although transcription-generated torsion can promote transcription in certain contexts, torsion accumulated to excess can have deleterious consequences, indicating that torsion itself should be regulated in the cell. Recently, Ma *et al.* demonstrated that (+) torsional stress ahead of RNAP can slow down elongation, increase RNAP's pause density and duration, and eventually cause RNAP to stall.[24] *In vivo*, it has been shown that accumulated (+) torsional stress greatly diminishes mRNA synthesis in yeast.[27] Chong *et al.* also found that torsional stress can slow down transcription elongation and additionally demonstrated that transient DNA supercoiling generated locally during active transcription is the primary origin of transcriptional bursting in bacteria.[25] It therefore appears that transcription-induced torsion and DNA supercoiling may underlie many of the mechanisms influencing gene expression noise and consequent variations in cell-to-cell viability.

Transcription-generated torsion has also been shown to mediate interactions between multiple transcribing RNAPs. Using live cell imaging, Tantale *et al.*[28] found that HIV-1 RNA was synthesized by "convoys" of closely spaced RNAPs traveling at the same speed. Physical modeling indicates that torsional coupling between adjacent polymerases is sufficient to maintain convoy speed and that they may transcribe more efficiently than isolated RNAPs. Indeed, Kim *et al.*[29] recently demonstrated that, both *in vivo* and *in vitro*, multiple co-transcribing RNAPs can display either collaborative or antagonistic dynamics over long distances through transcription-coupled DNA supercoiling. As a result, it was found that a group of co-transcribing RNAPs can transcribe at a faster rate than a single RNAP alone. It was further shown that promoter repression can dramatically slow transcription by inhibiting the collaborative torsional coupling between RNAPs, which may lead to slowed elongation rates and premature RNAP dissociation. Taken together, these discoveries demonstrate the importance of understanding the torsional features of RNAP and their regulation.

3.1.3 Understanding Biological Torsion

To fully understand biological torsion in the context of RNAP and DNA, we must characterize two parameters: torque and DNA supercoiling. We begin with a discussion of torque. RNAP, because it both translocates on DNA and rotates DNA, is both a linear and a torsional motor. A linear motor converts chemical energy into translational motion and exerts force along its direction of translocation. A torsional motor converts chemical energy into rotational motion, and therefore generates torque. Torque is defined as the product of the magnitude of the force driving the rotation and the distance to the axis of rotation (Figure 3.3). Distinct from the translocational force in a linear motor, the force involved in generating the torque of a torsional motor operates perpendicular to the axis of rotation. Force and torque are thus both key but different physical parameters of biological machines and play important roles in the elucidation of their mechanisms of action.

In addition to generating torque, RNAP also experiences torques from external sources. RNAP experiences a substantial viscous drag torque from the rotation of the RNAP, RNA transcript, and associated protein complexes through the cell cytoplasm. This torque tends to restrict RNAP rotation, and causes the DNA to rotate instead as it is pulled through the motor. Additionally, as RNAP transcribes, the supercoiling it generates results in a buildup of torsional stress in the DNA when the system is torsionally constrained. This torsional stress is experienced by the RNAP as a resistive torque to continued transcription. The torque RNAP generates as it transcribes and the opposing torque from the DNA that RNAP must work against are thus analogous to the forces between two objects in Newton's Third Law and are always exactly equal in magnitude and opposite in direction.

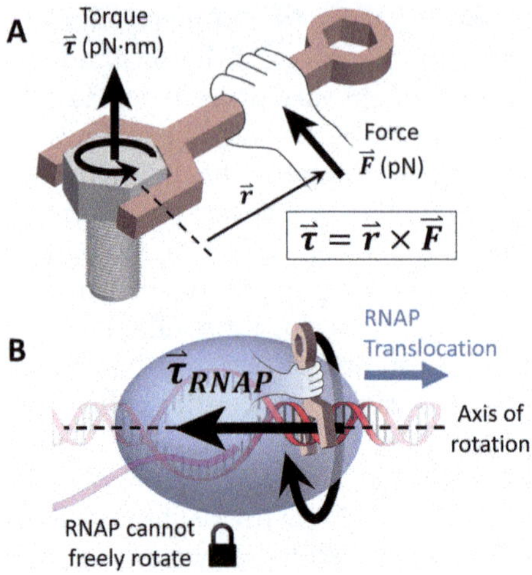

Figure 3.3 (A) Torque is a mechanical quantity that tends to rotate objects, occurring when a force $\vec{F}$ acts on an object at a distance $\vec{r}$ from its axis of rotation. The torque is then defined as the vector cross product between $\vec{r}$ (sometimes called a lever arm) and $\vec{F}$. Here, torque is applied to a bolt, causing it to rotate. The direction of the torque vector and the resulting object rotation are given by the right-hand rule for the vector cross product. (B) Because RNAP tracks the major helical groove of dsDNA but cannot freely rotate through the cytoplasm, RNAP instead applies torque to rotate the DNA as RNAP translocates.

Though torque drives DNA into supercoiled forms (Box 3.1), torque and DNA supercoiling are not equivalent. Transcription-generated DNA supercoiling indicates the number of extra turns RNAP adds to DNA, whereas transcription-generated torque reflects how difficult it is for RNAP to add those turns. The relationship between torque and supercoiling can be very complex, depending on the specific nature of the DNA substrate (*e.g.* with or without bound proteins or intercalating reagents). In general, it is the value of torque rather than the amount of supercoiling that more directly determines outcomes such as DNA structural changes, whether RNAP stalls, and whether RNAP can dissociate a bound protein from DNA. Therefore, the maximum torque that RNAP can generate before stalling against the resistive torque buildup in the DNA (the stall torque) is an important property that should be characterized in order to understand the nature and consequences of transcription-generated torque.

Although a great deal has been learned about transcription-generated DNA supercoiling *via* biochemical tools, such as two-dimensional gel electrophoresis,[14,21,27,30] much less is known about transcription-generated torque. This is primarily because direct, real-time measurement of torque in an active biological system poses significant technical challenges. However, recent advances in single-molecule techniques are beginning to enable measurements of this important physical parameter. In this chapter, we focus our discussion on how transcription-generated torque has been measured, how the torque-generation capacity of RNAP may be regulated, and the implications of these measurements for understanding cellular functions.

Box 3.1 Torque and DNA Supercoiling

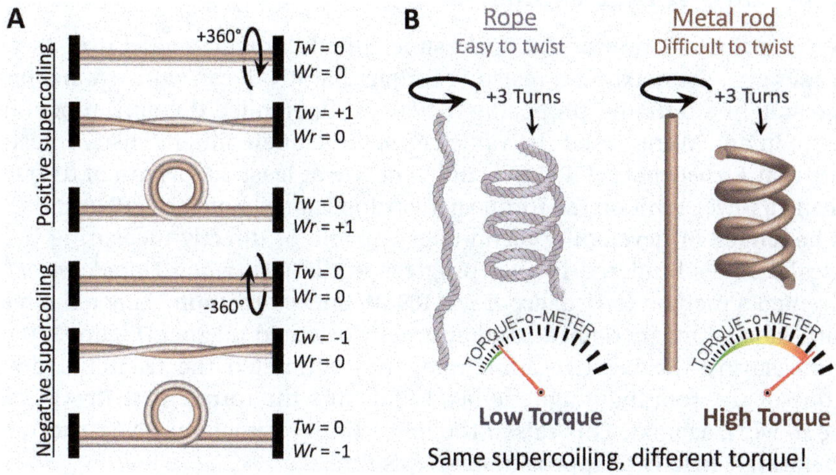

(A) Torque drives DNA into supercoiled forms. DNA supercoiling may be described by the parameters of twist (Tw), writhe (Wr), and linking number (Lk).[106] Simply stated, twist counts the number of helical turns in the DNA while writhe counts how many times the double helix crosses itself (*e.g.* plectonemes and toroids). The linking number is equal to the sum of twist and writhe, *i.e.* Lk=Tw + Wr, according to White's theorem.[106] For torsionally relaxed B-DNA, Wr = 0 and Tw will be equal to the number of helical turns (relaxed DNA has one helical turn per 10.5 bp). Winding DNA in either the (+) or (−) direction, which requires torque, will change Lk *via* a change in Tw, Wr, or both. The partition of Lk into Tw and Wr in naked DNA is dependent on both the force on the DNA and its degree of supercoiling. For example, when winding DNA under a constant force from an initially relaxed state, Tw normally increases first. The torque under this condition will increase proportionally with the change in Tw. Eventually, with continued winding, a plectoneme is formed and Wr starts to rise while Tw remains constant. The torque under this condition will remain constant and only depend on force, which has been estimated to be about 0.4 pN *in vivo*.[107]

(B) The degree of supercoiling (proportional to turns added) in DNA is often mistakenly assumed to correspond to its torque; however, the relationship between supercoiling and torque is complex. In this example, the uncoiled rope and metal rod are both initially in a relaxed state with zero torque. To add an equal amount of supercoiling to these two materials, however, requires a very different amount of torque and leaves the two substrates in very different torsional states! In general, the torque required to twist a substrate is determined by its torsional stiffness, a physical property which not only changes across different biological substrates (*e.g.* naked DNA *versus* protein-bound DNA), but can vary within a given substrate as a nonlinear function of supercoiling. For example, the relationship between torque and supercoiling is dramatically changed by the formation of plectonemes as well as the conversion of naked DNA into chromatin. Thus, for each type of substrate, torque must be measured over a wide range of supercoiling values in order to fully establish the torsional properties of a biological material.

3.2 Measuring the Torque Generated by RNAP

3.2.1 Early Torque Estimates

Since the introduction of the "twin-supercoiled-domain" model over three decades ago, the maximum resisting torque that RNAP can work against has been sought after. After proposing the "twin-supercoiled-domain" model in 1987, Liu and Wang[5] used the values of *in vivo* supercoiling density to estimate that *Escherichia coli* RNAP could generate at least 2.8 pN nm of torque, placing a lower limit on the torque generation capacity of *E. coli* RNAP.

Challenges in developing techniques capable of directly measuring biological torques hindered further progress until 2001, when Harada *et al.*[31] presented a method to visualize *E. coli* RNAP-induced rotation using a fluorescently labeled magnetic bead and an axially oriented magnet (Figure 3.4D). By measuring the bead rotation rate, they estimated the frictional drag torque of the rotating magnetic bead and thus the torque that RNAP was able to work against. This raised the lower limit estimate of RNAP's torque generation capacity to approximately 5 pN nm.

3.2.2 The Path to Direct Torque Measurements

Measuring the stall torque that RNAP generates during transcription requires a methodology or instrument with the capacity of torque detection. Ideally, the motion of RNAP should be monitored concurrently, as the RNAP translocates along DNA, causes an accumulation of resistive torque in its substrate, and eventually becomes stalled. These real-time data contain information not only on the torque required to stall the RNAP but also on how torque may alter transcription speed and pausing at the single-molecule level. Although this feat was eventually achieved using an angular optical trap,[24,26] as described later, several earlier studies laid down important foundations to enable these measurements.[32–42]

One important achievement was the ability to track the motion of RNAP *via* a reporter particle using the "tethered particle motion" technique.[32,33,35] In the 1990's, Schafer *et al.*[32] and Yin *et al.*[33,35] tethered an RNAP to the surface of a microscope coverslip and attached a microsphere to the distal end of its DNA substrate (Figure 3.4A). In this assay, the volume of space the bead is able to explore through Brownian motion is directly related to the length of the DNA tether. Thus, by measuring the spread of the Brownian motion of the tethered microsphere, they were able to follow the progression of the RNAP as it translocated along DNA. Although future studies would improve the resolution of RNAP translocation measurements, the experimental configuration established in this study has been instrumental to many subsequent single-molecule studies of transcription requiring a tethered particle.

Meanwhile, another important milestone came from single-molecule studies of the force-generation capacity of *E. coli* RNAP where the translocation of RNAP was monitored in real time as the RNAP worked against an external force.[34,36] These experiments adopted a similar configuration to that established by Schafer *et al.* and Yin *et al.* except that the microsphere was held in

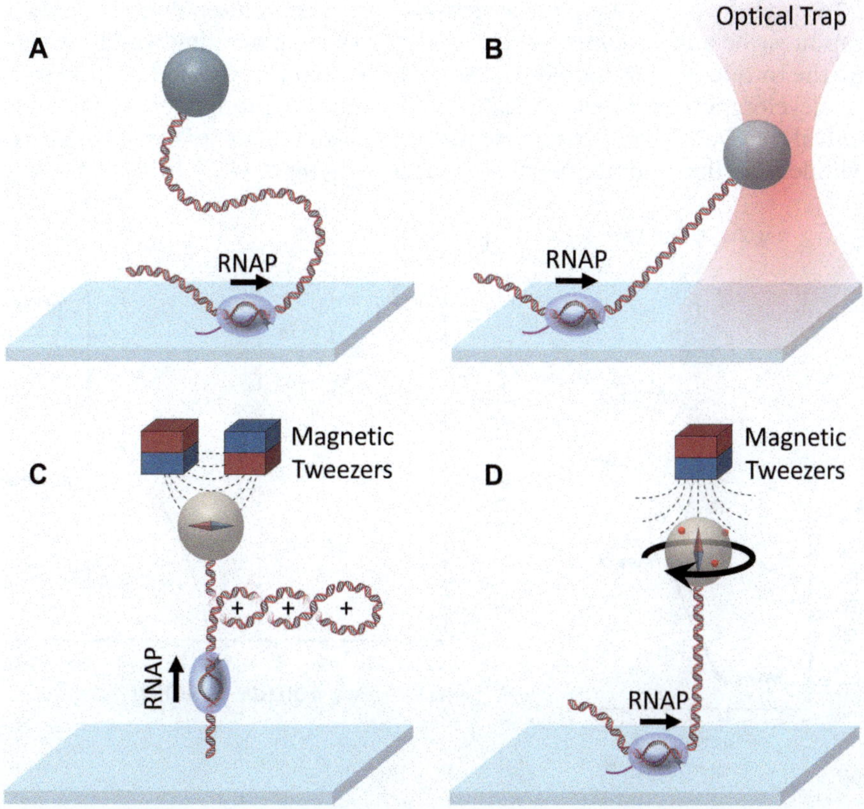

Figure 3.4 Techniques and configurations used in early single molecule studies of transcription in real time. (A) Tethered particle motion technique to monitor transcription elongation.[32,33,35] RNAP of a transcription elongation complex is tethered to the surface of a microscope coverslip with one end of the DNA template attached to a microsphere. The extent of the Brownian motion of the tethered microsphere provides a measure of the progression of the RNAP. (B) An optical trap to measure RNAP elongation and force-generation capacity.[34,36] An optical trap enables precise, real-time measurements of RNAP position as RNAP translocates along DNA and stalls under an increasing external force. (C) Magnetic tweezers to detect transcription initiation on a supercoiled DNA template.[41,42] Formation of a transcription initiation complex creates (+) supercoiling in the rest of the DNA template and is detected as an amplified change in the extension of the plectonemic DNA. (D) Axial magnetic tweezers to visualize RNAP rotation of the DNA.[31] A fluorescently labeled magnetic bead in an axial magnetic tweezers was used to visualize *E. coli* RNAP-induced rotation of its DNA substrate.

an optical trap, allowing precise, real-time measurements of the force and displacement of the trapped microsphere affected by the transcribing RNAP (Figure 3.4B). Subsequently, the position of RNAP on the DNA template was monitored as RNAP worked against an increasing force before being stalled. This also established the force generation capacity of *E. coli* RNAP as a linear motor to be approximately 25 pN.

To extend optical trapping assays for torque measurements, an optical trap must be able to control the degree of DNA supercoiling while measuring the torque in the supercoiled DNA. These requirements led ultimately to the development of a new optical trapping instrument in 2004, the angular optical trap (AOT),[37] and the subsequent realization of nanofabricated quartz cylinders as the trapping particle in 2007 (Figure 3.5A).[38] In contrast to a

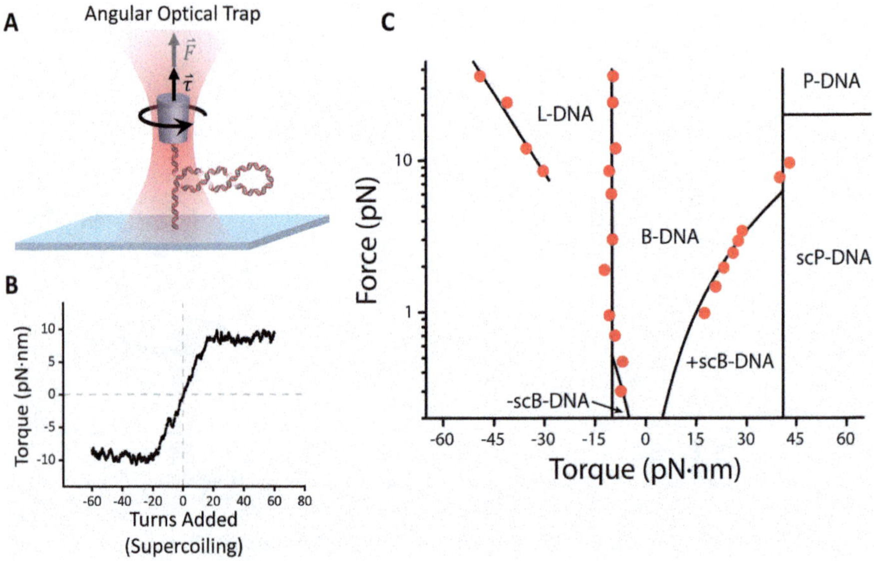

Figure 3.5 (A) An angular optical trap (AOT).[37,38] Like a conventional optical trap, an AOT can apply and measure force (grey arrow). In contrast to a convention optical trap, an AOT additionally enables both the direct application and the direct measurement of torque. Torque is applied *via* the rotation of a trapped cylinder (black arrow), accomplished by rotating the trapping beam's linear polarization. The torque exerted on the cylinder is measured by the change in the angular momentum of the transmitted trapping beam. An AOT can thus exert and measure torque, control rotation and supercoiling, and measure the displacement and force of the trapped cylinder at the same time. (B) The torque measured by an angular optical trap as a function of the twist added into naked DNA, under a constant force of 0.6 pN. The linear region corresponds to DNA before buckling. The (+) and (−) torque plateau regions correspond to DNA after buckling or melting. Data were acquired by Dr Xiang Gao and are shown here with permission. (C) DNA force and torque measurements obtained by an AOT as summarized in a force–torque phase diagram. This diagram describes the force and torque conditions under which DNA will exist in one or more distinct structures or "phases," including B-DNA (the standard relaxed right-handed double helix), −scB-DNA (negative-supercoiled B-DNA), +scB-DNA (positive-supercoiled B-DNA), L-DNA (denatured left-handed DNA), P-DNA (denatured right-handed DNA), and scP-DNA (supercoiled P-DNA). Phase transitions between specific states of DNA are represented by solid black lines.[45,103,104] Red points indicate torque values measured during phase transitions using an AOT.[38–40,75] Adapted from ref. 105 with permission from Annual Reviews Inc, Copyright 2008.

conventional optical trap, an AOT can also rotate the trapped birefringent cylinder about its cylindrical axis *via* the rotation of the trapping beam's linear polarization (Box 3.2), and the torque exerted on the cylinder can be measured *via* the change in the angular momentum of the transmitted beam.[37,43] Thus an AOT allows simultaneous control and measurement of rotation, torque, displacement and force of the trapped cylinder.

Box 3.2 The Angular Optical Trap

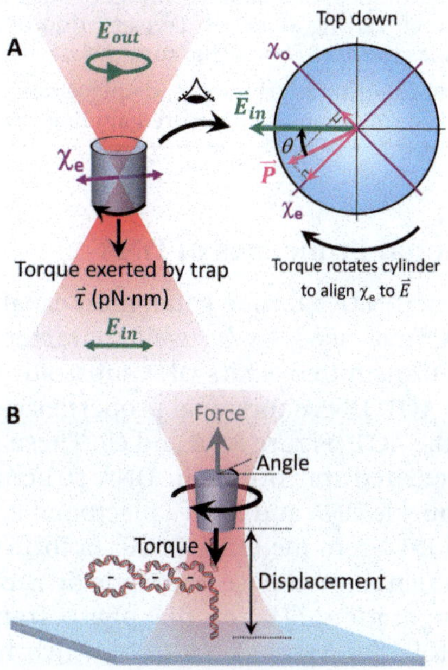

(A) The angular optical trap (AOT) has enabled direct control and detection of torque and rotation in individual biomolecules. The trapping particle is a nanofabricated quartz cylinder that, when trapped, aligns with its cylinder axis along the direction of light propagation.[38] When the trap is linearly polarized with electric field $\vec{E}_{in}$, the quartz cylinder develops a polarization $\vec{P}$ depending on the electric susceptibility of the cylinder χ, which is anisotropic (χ_e along the extraordinary axis is greater than χ_o, the ordinary axis). The cylinder is fabricated so that the extraordinary axis is perpendicular to the cylinder axis. Once polarized, the cylinder will experience a torque analogous to an electric dipole $\vec{p}$ in an electric field $\vec{E}$. This torque tends to align $\vec{p}$ to $\vec{E}$ and has a magnitude given by $\tau = |\vec{p} \times \vec{E}| = pE\sin\theta$ where θ is the angle between $\vec{E}$ and $\vec{p}$.[4] Though the equation for the torque on a cylinder is more complex due to its geometry, the essential idea is the same. Note that the relationship between the direction of the net torque vector applied to an object and its resulting rotation is given by the right-hand-rule. This torque will act until the extraordinary axis and $\vec{E}_{in}$ are aligned. The cylinder can thus be rotated about its axis by rotation of the laser polarization. Attaching a biological molecule specifically to one end of the cylinder allows the independent application of force and torque to the molecule. The torque that

(*continued*)

Box 3.2 (*continued*)

rotates the cylinder to align the extraordinary axis to the trapping laser's polarization is provided by the angular momentum intrinsically carried by light. Because angular momentum must be conserved, the torque imparted to the cylinder by the trapping laser can be detected *via* a change in the angular momentum of the trapping laser after it passes through the trapped cylinder.[37] This is illustrated here by the change in the laser polarization from linear ($\vec{E}_{in}$) to elliptical after it passes through the cylinder ($\vec{E}_{out}$). The same trapping beam is thus used for torque application as well as the detection of the trapped particle's angle and torque, without the need for a secondary detection beam or imaging method. Such a detection method is exceedingly direct, relying solely on conservation of angular momentum, and distinguishes the AOT from other torque detection methods.[37]

(B) During a typical experiment, the force, displacement, torque, and angle of the cylinder are simultaneously measured at kilohertz frequencies.[38–40]

3.2.3 The Torsional Properties of DNA

In order to determine RNAP's torque generation capabilities, the torsional properties of DNA first need to be well-characterized, as the RNAP-generated torque is transmitted *via* its DNA substrate. Following the initial development of the AOT, DNA's torsional properties were comprehensively investigated using the AOT (Figure 3.5B and C). These measurements characterized torque-mediated transitions in DNA structure, establishing the torque required to buckle DNA and form a plectoneme,[39] for the phase transition from normal B-DNA to melted L-DNA[40] (a form of underwound and stretched DNA in which normal base pairing is disrupted and DNA adopts a left-handed configuration[44,45]), for the phase transition from B-DNA to supercoiled P-DNA[38] (an exotic DNA state which has a smaller helical repeat than B-DNA and unpaired bases extruding to the exterior of the molecule[44,46,47]), and to migrate a Holliday junction.[48] Some of these experimental results are concisely summarized in a phase diagram as shown in Figure 3.5C. Can a torsional motor protein like RNAP drive these structural transitions *in vivo*?

The maximum torque RNAP can generate is exhibited when RNAP stalls due to the presence of resistive torque in a supercoiled DNA template that is equal and opposite in value to the motor's own maximum torque output. However, applied force that opposes an RNAP's forward translocation can also stall the motor. Thus, to accurately measure RNAP's stall torque, the applied force on the DNA substrate must be kept low to minimize the contribution of stalling due to force. The unfortunate result of this requirement is a substantial increase in the Brownian noise in the measured DNA length, limiting the precision with which one may resolve the translocation of the DNA through the RNAP. The inspiration for a solution to this problem came from studies by Revyakin *et al.*[41,42] who, using magnetic tweezers, studied transcription initiation on a supercoiled DNA template (Figure 3.4C). By first forming a plectoneme by supercoiling the DNA substrate, the researchers showed that changes in DNA extension due to the formation of a transcription initiation complex could be amplified by an order of magnitude relative to the case where a plectoneme was not present. This same principle may be adapted to amplify changes in DNA extension resulting from RNAP transcription on a torsionally constrained template.

With these many technical hurdles out of the way, torque measurements during transcription using the AOT became possible.

3.3 RNAP's Torsional Properties

3.3.1 The Torque Generation Capacity of RNAP

Using an AOT, Ma *et al.*[24] developed a method to monitor RNAP movement as it worked against an increasing opposing torque and eventually became stalled. An *E. coli* RNAP was torsionally anchored to a surface, and either the downstream or upstream end of the DNA template was torsionally anchored to one end of a nanofabricated quartz cylinder held in an AOT. As the RNAP translocated the DNA, it also rotated the DNA, forming a (+) plectoneme downstream (Figure 3.6A) or a (−) plectoneme upstream (Figure 3.6B). In both cases, the researchers allowed torque to build up and eventually stall transcription.

The results of these experiments indicated that accumulated torsional stress, whether in the (+) supercoiled DNA ahead of the RNAP or in the (−) supercoiled DNA behind the RNAP, could cause an elongating RNAP to stall. The measured average stall torque value in both cases was approximately 11 pN nm, which is significantly greater than the previously estimated lower bound of RNAP-generated torque. In this and following AOT stall torque assays, stalling occurs at many different sequence locations and thus stall torque measurements reflect an average over many different local sequence configurations. The measured stall torque showed a spread of approximately 3 pN nm around the average value, probably as a result of sequence-dependence of stalling.

Intriguingly, under (−) DNA supercoiling behind a transcribing RNAP, the mean stall torque value of 11 pN nm coincides with the torque required for DNA melting, where DNA undergoes a phase transition from B-DNA to melted, denatured DNA.[40,49] This transition is known to play critical roles in gene regulation, such as facilitating transcription initiation[18,41,42] and the binding of transcription factors.[14,15] Meanwhile, ahead of the RNAP, (+) DNA supercoiling has the potential to disrupt protein roadblocks to facilitate the passage of an elongating RNAP.[23] Thus, perturbations to RNAP's torque generation capacity could lead to sensitive tuning of the level of gene expression. The measured stall torque values, in establishing a physiologically relevant torque scale, represent a baseline from which other factors may influence the regulation of transcription through RNAP-generated torque.

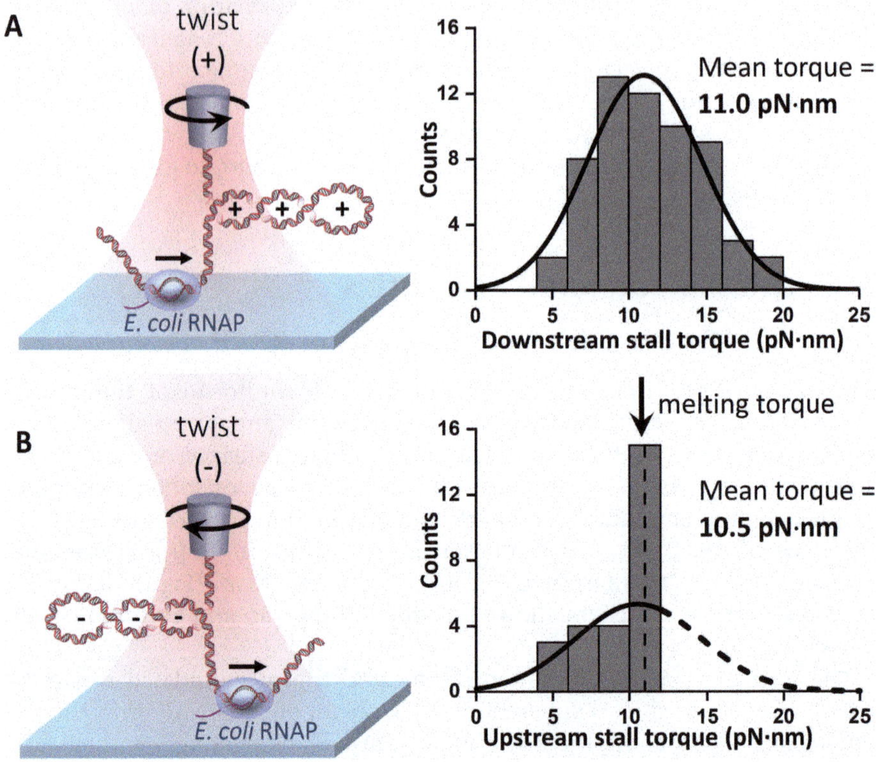

Figure 3.6 Stall torque of a transcribing RNAP measured using an AOT. In these experiments, an *E. coli* RNAP was torsionally anchored to a surface with either the downstream end (A) or the upstream end (B) of the DNA template torsionally constrained to a quartz cylinder held in an AOT. The translocation of the DNA through the RNAP and the corresponding torque generated by RNAP were monitored in real-time. RNAP elongation accumulated (+) or (−) DNA supercoiling, for the downstream and upstream configurations, respectively. As torque increased, RNAP was eventually stalled and the stall torque was recorded. Note the asymmetry of the upstream stall torque distribution is a result of DNA structural changes, *i.e.* DNA under (−) torque undergoes a melting transition at 10.5 pN nm. Adapted from ref. 24 with permission from The American Association for the Advancement of Science, Copyright 2013.

3.3.2 Other Torsional Characteristics of RNAP

Ma *et al.*[24] also found that, upon torque release, a significant fraction of stalled RNAPs resumed transcription within a short time. This result indicates that stalled RNAPs *in vivo* can be rescued if torsional stress is released in a timely fashion, either by topoisomerases or *via* the individual or combined processes of DNA and RNAP rotation when possible. The prompt recovery also prevents a stalled RNAP from becoming a roadblock and hindering other vital cellular processes, such as replication.[50,51]

In addition to slowly-varying torque loads, RNAP may encounter dynamic torque fluctuations during transcription. *In vivo*, the torsional stress on DNA can change quickly due to the spontaneous loss of torsional constraints in DNA, action of topoisomerases, or the action of other torsional motors. For example, the replication fork is known to move at a speed several times faster than transcription[52] and can thus potentially build up torsional stress more quickly. The authors investigated the response of RNAP to transient torque pulses, and found that RNAP is relatively immune to fast torsional fluctuations that may occur in the cell. The insensitivity of RNAP to torsional fluctuations may reduce unnecessary pausing during transcription, making the process more robust.

Finally, the authors also showed that torque can directly affect both transcription speed and pausing, which are known to be important regulators for transcription and are frequently targeted by transcription factors. For example, the binding of the transcription factor NusA to RNAP can slow elongation and increases the rates of RNAP entry into both short- and long-pausing states,[53] whereas NusG can accelerate transcription speed and dramatically suppress the rate of RNAP entry into long pauses.[54] Ma *et al.*'s experiments thus provided direct evidence for DNA torque as an effective alternative regulator of transcription elongation and pausing in addition to transcription factors.

3.4 Regulation of RNAP's Torque Generation Capacity

3.4.1 RNAP's Torque Generation Can be Regulated by a Transcription Factor

In vivo, RNAP interacts closely with many transcription factors, which regulate RNAP activities, and therefore it is possible that these transcription factors might also regulate the torsional properties of RNAP. Although under (−) supercoiling, the torque RNAP can generate is limited to approximately 11 pN nm due to the DNA melting transition, under (+) supercoiling, DNA is able to sustain a much greater torque without undergoing any structural denaturation. Therefore, it is possible that RNAP may further increase its torque generation capacity under (+) supercoiling with the help of transcription factors, promoting its ability to work against situations where excessive torsional stress might be encountered.

In a follow up study, Ma *et al.*[26] found that when RNAP worked against a resisting torque, it was highly prone to extensive backtracking, during which RNAP reverse translocates along DNA with its catalytic site disengaged from the 3′-end of the RNA, inactivating transcription.[55,56] RNAP was observed to backtrack up to 110 bp over a stalling time of 60 s (Figure 3.7A). These results demonstrate that backtracking is a primary mechanism limiting RNAP's torsional capacity.

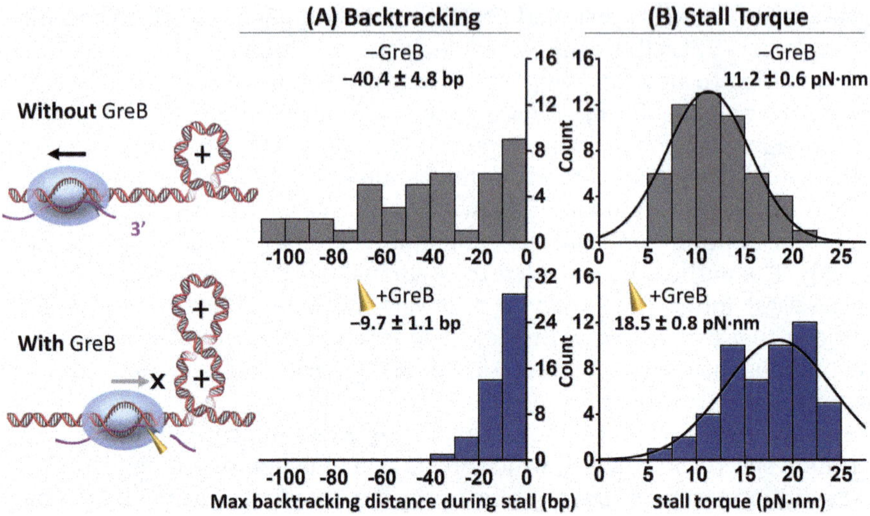

Figure 3.7 RNAP stalling under torque in the absence and presence of GreB. (A) The distributions of the maximum backtracking distance during stalling in the absence and presence of GreB. (B) The distributions of the stall torque in the absence and presence of GreB. Remarkably, the presence of GreB dramatically reduced RNAP backtracking and increased the maximum torque generation capacity of RNAP by 65%. These results demonstrate that GreB significantly increases the stall torque of RNAP. Reproduced from ref. 26 with permission from PNAS, Copyright 2019.

A universal class of transcription elongation factors, including TFIIS in eukaryotes[57,58] and GreB in prokaryotes,[59–61] rescues backtracked RNAP complexes and promotes transcription through obstructive regions of DNA, such as regulatory transcription pause sites. These proteins act either as auxiliary factors or, in the case of RNA polymerases I and III, as intrinsic parts of the core RNAP enzyme.[62,63] They stimulate the cleavage activities of RNAP, leading to the removal of the 3′-end of the RNA. The newly generated RNA 3′-end then becomes aligned with the catalytic site, reactivating transcription.

Ma *et al.* investigated transcription in the presence of GreB and found that GreB greatly suppressed backtracking and remarkably increased the torque that RNAP was able to generate by 65%, from 11.2 to 18.5 pN nm (Figure 3.7B). Thus, the limitation to RNAP's torsional capacity due to induced backtracking can be mitigated in part by GreB. These results demonstrate that the torsional capacity of RNAP can be regulated in a transcription factor-dependent manner *via* backtracking suppression and rescue.

3.4.2 Kinetics of RNAP Backtracking and Rescue by a Transcription Factor

Interestingly, the researchers also observed that upon stalling in the presence of GreB, RNAP positional trajectories exhibited minute fluctuations that should reflect rapid cycles of RNAP backtracking and rescue by GreB. The small amplitude of these fluctuations, however, made it difficult to clearly discern transition events. Fortuitously, if backtracking and rescue events are stochastic in nature, the positional variance of RNAP should increase with time and thus contain valuable information on the kinetics of backtracking and GreB rescue.

Ma *et al.*[26] developed a statistical method to determine the rate of GreB rescue, obtaining analytical solutions for both the variance and mean velocity over time of RNAP after a stall by solving a general two-state transition problem using rigorous statistical mechanics and Eigenvector analysis. In their derivations, the processes of torque induced stalling and recovery in the presence of GreB are modeled as reversible first-order transitions between backtracking and forward elongation. The mean RNAP position $\bar{x}$ and the variance of the RNAP position σ_x^2 were found to evolve with time t as: $\bar{x} = \frac{k_- v_- + k_+ v_+}{k_- + k_+} t = \bar{v} t$ and $\sigma_x^2 = \frac{2k_- k_+}{(k_- + k_+)^3} (v_- - v_+)^2 t$, where v_- is the backtracking velocity, v_+ is the active elongation velocity, k_+ is the GreB rescue rate, and k_- is the rate of entering backtracking.

The authors showed that, while stalled at maximum torque generation in the presence of GreB, RNAP backtracks by approximately 3 bp over approximately 4 s before being rescued by GreB. RNAP then actively forward elongates for approximately 3 bp over approximately 1 s before backtracking again (Figure 3.8A). In addition, the 4 s time to rescue indicates that a single rescue event requires multiple rounds (approximately 10) of GreB binding and unbinding events, in agreement with observations from a recent study of GreB performed under zero torsion and tension.[64] As before, these measurements represent averages taken from stalling events over many different sequences, however the specific effects of sequence on these kinetics has not yet been studied.

This theoretical model and the use of variance analysis provide a framework for obtaining kinetic parameters from complex translocation trajectories even when individual events are not discernable. This framework is general and may be applied to many other two-state transition problems. We highly recommend that readers take advantage of these approaches and theoretical solutions when applicable.

The experimental results by Ma *et al.* support the following torque enhancement model for transcription under torsion in the presence of GreB (Figure 3.8B). Transcription-generated torsional stress hinders RNAP forward translocation and is highly effective in inducing RNAP backtracking, which may lead to transcription arrest. GreB binding to RNAP stimulates 3′-RNA cleavage and thus prevents RNAP from extensive backtracking. Consequently, the stalled RNAP can be rescued and transcription can be resumed. As this cycle repeats, RNAP has the opportunity to work against an increasingly greater torque which further hinders forward transcription elongation. RNAP ultimately stalls when forward elongation is exactly countered by backtracking. Upon stalling, RNAP repeats the cycle of forward elongation and backtracking, with no or minimal net motion.

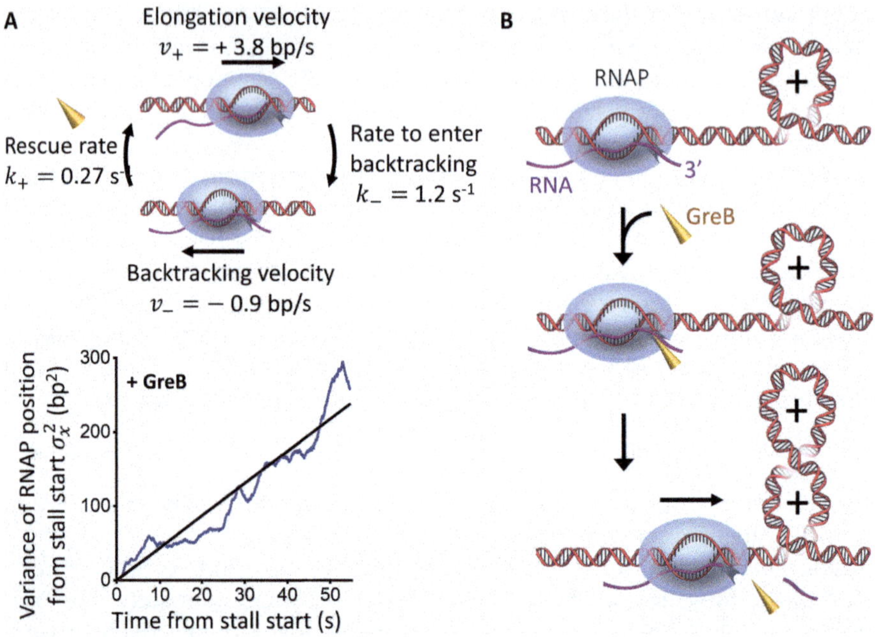

Figure 3.8 (A) GreB rescue during torque-induced stalling. Top: Kinetics of backtracking and GreB rescue with the measured kinetic rates indicated. Bottom: Measured variance of RNAP position from the start of a stalling event *versus* time in the presence of GreB (blue). The black solid line is a linear fit passing through the origin, yielding a slope of 4.4 bp^2 s^{-1}. (B) Torque enhancement model. During transcription, the accumulation of torsional stress encourages RNAP to backtrack and ultimately stall. GreB can bind to the backtracked RNAP and stimulate its intrinsic RNA cleavage activity. RNAP can thus be rescued and elongation can be resumed. Once transcription is resumed, the RNAP will be able to work against a stronger torsional barrier. Reproduced from ref. 26 with permission from PNAS, Copyright 2019.

3.4.3 Other Factors that May Regulate RNAP Torque Generation Capacity

In the cell, in addition to GreB, RNAP also works closely with many other transcription factors. These factors play important roles in regulating RNAP properties including encouraging or suppressing backtracking as well as rescuing a backtracked RNAP,[53,54,65] thereby potentially regulating RNAP's torsional capacity as well. For example, NusA increases RNAP pausing[53] and, thus, potentially decreases RNAP's stall torque, whereas factors like GreA,[59] DksA,[66] and NusG[54] prevent backtracking and may increase RNAP's stall torque.

In contrast to these "passive" transcription factors, Mfd, an ATP-dependent motor protein that is essential for transcription-coupled repair,[67–70] is an active transcription factor. Recently, it has been shown that Mfd actively assists RNAP in forward translocation, preventing RNAP backtracking and enhancing RNAP's stall force.[71] It is thus very possible that Mfd may also increase RNAP's stall torque. Though interacting with RNAP differently, these factors may work in coordination to downregulate or upregulate RNAP's torsional capacity *via* encouraging or suppressing its backtracking properties to achieve a broad range of regulation. Therefore, backtracking may serve as a general mechanism for transcription factors to regulate RNAP's torsional properties.

In the cell, where RNAP may work in conjunction with other transcription factors, an enhanced torsional capacity of RNAP may be necessary for overcoming obstacles, such as removing roadblock proteins or transcribing a long gene under high torsional stress. For example, *E. coli* RNAP must transcribe through various bound proteins such as histone like factor U (HU) and histone-like nucleoid structuring protein (H-NS) that create and maintain specific DNA conformations including DNA bridges and loops. Recently, Kotlajich *et al.*[72] showed that bridged, but not linear, H-NS filaments slow down transcription and strongly increase pausing by *E. coli* RNA polymerase at backtracking-prone pausing sites. These results favor a model in which the bridged H-NS filaments create topologically fixed domains that prevent dissipation of transcription-generated DNA supercoiling, increasing torsional stress. Interestingly, the presence of anti-backtracking factors like GreA and GreB greatly facilitates transcription and reduces pausing in these regions. These observations are entirely consonant with the recent findings that GreB can promote transcription through torsional stress, exemplifying how torque may provide the cell with additional modes to regulate transcription.

3.5 RNAP Torque Generation in Context

3.5.1 A Physiologically Relevant Torque Scale

It is important to place the measured values of RNAP's torque generation capacity in perspective. Studies using the rotary bead method,[45,49,73] magnetic tweezers,[74] and the AOT[39,40,75,108] have characterized DNA's torsional properties and measured the torque values driving structural phase transitions in DNA (Figure 3.4C). In particular, the critical torques have been measured to be approximately +34 pN nm for the B-DNA to P-DNA transition, approximately −3 pN nm for the B-DNA to Z-DNA (an alternative DNA structure with a left-handed, zig-zag double helix for sequences with GC repeats[76]) transition, and approximately −10 pN nm for the B-DNA to melted DNA transition.[39,40,45,73] The measured torque that RNAP alone can generate[24] coincides with the DNA melting torque[39,45,73] and is much larger than the torque required for the B-DNA to Z-DNA transition.[73] This indicates that transcription induced torsional stress can significantly modify DNA and encourage DNA to form alternative structures (*e.g.* melted DNA, Z-DNA, R-loops, *etc.*). Such non-B-DNA structures can serve as distinct targets for regulatory proteins that influence gene regulation.[14,15] The torque generated by RNAP is also sufficient to convert DNA into a plectonemic state,[39] which by coiling and condensing DNA may allow distant segments of DNA to interact with each other. These results establish RNAP as a powerful torsional motor in a biological context[3,24] and indicate that the torque-generating capacity of RNAP may have been tuned to important transitions in DNA.

3.5.2 Torque- *Versus* Force-generation

As we have discussed, the translocation of RNAP during transcription requires concurrent rotation relative to the dsDNA helical axis. Consequently, RNAP is able to generate not only translocational force as a linear motor but also rotational torque as a torsional motor, both of which regulate transcription. How do these two mechanical parameters compare to each other in the context of a molecular motor? If we assume RNAP converts chemical energy to the mechanical work of translocation and rotation with the same thermodynamic efficiency, then the maximum force F_{max} and torque τ_{max} are expected to be closely related: $F_{max}d = \tau_{max}\theta$, where $d = 0.34$ nm is the translocation distance of a single base pair and $\theta = 0.59$ radians is the rotation angle over a single base pair.[34]

In contrast to the scarcity of RNAP torque-generation capacity measurements, RNAP's force-generation capacity has been much better characterized for both *E. coli* RNAP[34,36] and RNA polymerase II (Pol II).[77] Previous studies have revealed that *E. coli* RNAP alone can generate approximately 25 pN of force.[36] Using the relationship above to convert F_{max} to τ_{max}, this stall force predicts a stall torque of 14 pN nm, which is close to but somewhat greater than the measured stall torque of 11 pN nm.[24,26] The difference could come from multiple causes that are not mutually exclusive. It is possible that RNAP does not generate force and torque with the same thermodynamic efficiency when converting chemical energy to mechanical energy. It is also possible that torque and force may induce different RNAP conformational changes that differentially affect RNAP's mechanical functions.

Both RNAP-generated force and torque have been shown to induce RNAP backtracking, whereas anti-backtracking factors have been found to enhance both RNAP force- and torque- generation capacities. For example, Galburt *et al.*[77] found that the resistive force under which Pol II stalled due to backtracking increased from 7.5 pN to 16.9 pN in the presence of TFIIS, analogous to the findings of Ma *et al.*[24,26] showing torque-induced stalling and backtracking of *E. coli* RNAP and torque enhancement by GreB. Taken together, these results demonstrate that force and torque share some mechanisms in regulating transcription.

Despite these commonalities, evidence indicates that transcription-generated torque may be a more potent regulator of transcription. Consider the displacement of a bound protein that is a roadblock to an advancing RNAP. In order for RNAP to use force to displace the protein, RNAP may need to come into direct contact with the protein, making force a local or short-distance regulator of transcription. In contrast, torsional stress can rapidly transmit to even a distant region of DNA *via* "action-at-a-distance"[78] such that RNAP can remotely perturb and displace a protein without the need for direct contact. This ability makes torque a potential global or long-distance regulator of transcription.

In addition, since torque alters DNA topology, torque can create DNA structures that are difficult to attain using force. For example, the force required to melt DNA or form DNA bubbles was found to be larger than 45 pN,[79,80] far beyond what RNAP can generate. The melting torque of DNA, however, is approximately 11 pN nm, which is well attainable by RNAP. Torque, not force, has also been found to encourage formation of other nucleic acid structures, such as Z-DNA and R-loops.[81,82] Therefore, torque may play a more diverse role in regulating transcription and could potentially be more important than force.

3.5.3 Comparison of Torque Generation Capacity with Other Motor Proteins

Although there are limited studies of torque generated by DNA-based motors, with the only examples discussed above, torque generated by other molecular motors has been measured. Among the most studied torsional motors are those responsible for the rotation of bacterial flagella, the long filaments that enable bacterial cell movement. The *E. coli* flagellar motor has been studied with several techniques, including viscous drag on cells[83] or beads,[84,85] optical trapping,[86] and electrorotation.[87,88] The maximum torque reported for this motor is approximately 1300 pN nm,[85] while the flagellar motor of *Vibrio alginolyticus* was shown to generate torques of up to 4000 pN nm during its rotation.[89] In contrast to DNA translocases like RNAP, which are driven directly by individual nucleotide hydrolysis events, the remarkable torque-generation capacity of flagellar motors is driven by ion-motive forces across the cell membrane.[90]

The torque generation capacity of the F_oF_1-ATPase rotational motors has also been studied. F_oF_1-ATPase, also called ATP synthase and responsible for ATP synthesis in the cell, consists of two rotary motors: the transmembrane F_o and the soluble F_1, coupled by a central rotary shaft. The transmembrane motor F_o is powered by the ion motive force, whereas the soluble motor F_1 is driven by ATP hydrolysis. The rotary motor F_1-ATPase has been studied using single-molecule methods. By attaching a fluorescent actin filament to a surface-immobilized motor and observing the resulting rotation rate, Noji *et al.*[91] estimated that a single rotary unit can generate torques of up to approximately 40 pN nm. In comparison to F_1-ATPase, RNAP generates significantly less torque. However, F_1-ATPase is a purely rotary motor, whereas RNAP has a linear motor and rotary motor embodied in a single protein.

Although there have been no other torque-related studies of DNA-based motors, the torque that RNAP can generate is likely to be representative of the relevant torque scale for torsional motors that translocate on DNA *in vivo*. Measuring these small torque values remains an experimentally challenging undertaking, requiring instruments with exceptional torque resolution, bandwidth, and stability.

3.5.4 Future Outlook

These recent studies have demonstrated that *E. coli* RNAP is an intrinsic torsional motor, capable of generating torques sufficient to significantly alter DNA topology and melt DNA, and that its torsional capacity can be regulated in a manner dependent on transcription factors. These experiments were enabled by the development of AOT-based assays. It is worth noting that besides the AOT, other techniques, such as electrorotation,[87] rotary bead assays,[45,49,73] and magnetic tweezers[92–95] have also enabled torque detection. Although measurements of DNA-based biological torque are technically challenging, advancements in torque detection techniques are paving the path towards more complex measurements.

The AOT-based approach for *E. coli* RNAP discussed above[24,26] may also be used to investigate torque generation by other polymerases (*e.g.* eukaryotic RNAP I, RNAP II, and RNAP III). As the eukaryotic analog of RNAP, Pol II's torque generation capacity is likely to be regulated by other transcription factors, such as TFIIS, while Pol I and Pol III may afford a stronger torque capacity even in the absence of transcription factors because their intrinsic cleavage activity is stimulated by one of their core subunits.[62,63] The pointed lack of this ability in *E. coli* RNAP and Pol II indicates that the opportunity to regulate torque-generation capacity *via* the binding of external factors is valuable.

Indeed, the enhancement of torque generated by eukaryotic RNAPs may be essential for proper cellular function, as these RNAPs must transcribe through nucleosomes. Notably, a recent genomic study revealed that transcription generated (+) torsional stress increases nucleosome turnover within gene bodies.[13] It is possible that with the help of transcription factors, the enhanced torque-generation capacity of RNAP could allow RNAP to not only destabilize but also disrupt nucleosome structures, and make the DNA within or beyond a nucleosome more accessible to transcription.

The chirality of the nucleosome may also serve as an important mediator of torsion with implications for the regulation of transcription (Figure 3.2C). In a nucleosome, DNA wraps around the core histones in a left-handed fashion. Thus (+) torsion generated in front of an RNAP should faciliate disassembly of nucleosomes and (−) torsion generated behind an RNAP should faciliate nucleosome assembly. Indeed, results from studies have indicated that (+) torsion destabilizes the nucleosome[13,23,96] and (−) torsion facilitates nucleosome formation.[97] In particular, Sheinin *et al.*[23] showed that a (+) torque of approximately 19 pN nm or larger can significantly facilitate displacement of the H2A–H2B dimers in a nucleosome.

Transcription-generated torsion may also affect chromatin topology and structure.[98,99] Bancaud *et al.*[100,101] found that a nucleosome array can reversibly accommodate a large amount of added supercoiling and that under high torsional stress, the array's nucleosomes undergo a chiral transition from a left-handed to a right-handed configuration. Recently, Le *et al.*[102] provided direct measurements of the torsional properties of chromatin fibers. Although the work focused on replication, it has implications for transcription. They found that the torsional stiffness of a single chromatin substrate under (+) supercoiling is substantially smaller than that of naked DNA or a single chromatin substrate under (−) supercoiling. This indicates that during transcription, the softer (+) torsional stiffness of a chromatin fiber in front of the RNAP may facilitate transcription elongation, whereas the relatively stiffer (−) torsional stiffness behind the RNAP may facilitate nucleosome assembly. These single-molecule studies have provided significant quantitative information concerning the effects of torsion on chromatin topology and nucleosome structure, laying the groundwork for understanding the interplay of DNA supercoiling and transcription.

Ultimately RNAP's torque-generation capacity should be examined *in vivo*. Thus far, genomic analyses of microarrays have been able to detect (−) supercoiling using DNA intercalators, such as psoralen or its derivatives, followed by DNA photo-cross-linking. However, this approach currently only provides a snapshot of (−) supercoiling in the genome. A full picture of the dynamics of torsion *in vivo* would require methods that can quantitatively assay both torque and DNA supercoiling in real time.

Acknowledgements

This work was supported by the Howard Hughes Medical Institute (to M.D.W.), the National Science Foundation (MCB-1517764 to M.D.W.), the National Institutes of Health (R01GM136894 to M.D.W), the National Natural Science Foundation of China (NSFC-12074445 to J.M.), and the Fundamental Research Funds for the Central Universities (18lgzd16 to J.M.).

References

1. J. Gelles and R. Landick, *Cell*, 1998, **93**, 13–16.
2. J. Ma and M. D. Wang, *Transcription*, 2014, **5**, e28636.
3. J. Ma and M. D. Wang, *Cell Cycle*, 2014, **13**, 337–338.
4. J. Ma and M. D. Wang, *Biophys. Rev.*, 2016, **8**, 75–87.
5. L. F. Liu and J. C. Wang, *Proc. Natl. Acad. Sci. U. S. A.*, 1987, **84**, 7024–7027.
6. D. A. Koster, C. Aurélien, S. Stewart, B. Mary-Ann and N. H. Dekker, *Cell*, 2010, **142**, 519–530.
7. J. R. Dixon, S. Selvaraj, F. Yue, A. Kim, Y. Li, Y. Shen, M. Hu, J. S. Liu and B. Ren, *Nature*, 2012, **485**, 376–380.
8. A. Pombo and N. Dillon, *Nat. Rev. Mol. Cell Biol.*, 2015, **16**, 245–257.
9. F. Kouzine, A. Gupta, L. Baranello, D. Wojtowicz, K. Ben-Aissa, J. Liu, T. M. Przytycka and D. Levens, *Nat. Struct. Mol. Biol.*, 2013, **20**, 396–403.
10. C. Naughton, N. Avlonitis, S. Corless, J. G. Prendergast, I. K. Mati, P. P. Eijk, S. L. Cockroft, M. Bradley, B. Ylstra and N. Gilbert, *Nat. Struct. Mol. Biol.*, 2013, **20**, 387–395.
11. K. Matsumoto and S. Hirose, *J. Cell Sci.*, 2004, **117**, 3797–3805.
12. I. Bermúdez, J. García-Martínez, J. E. Pérez-Ortín and J. Roca, *Nucleic Acids Res.*, 2010, **38**, e182.

13. S. S. Teves and S. Henikoff, *Nat. Struct. Mol. Biol.*, 2014, **21**, 88–94.
14. F. Kouzine, J. Liu, S. Sanford, H.-J. Chung and D. Levens, *Nat. Struct. Mol. Biol.*, 2004, **11**, 1092–1100.
15. F. Kouzine, S. Sanford, Z. Elisha-Feil and D. Levens, *Nat. Struct. Mol. Biol.*, 2008, **15**, 146–154.
16. M. S. Guo, D. L. Haakonsen, W. Zeng, M. A. Schumacher and M. T. Laub, *Cell*, 2018, **175**, 583–597 e523.
17. D. Chen, R. Bowater, C. J. Dorman and D. M. Lilley, *Proc. Natl. Acad. Sci. U. S. A.*, 1992, **89**, 8784–8788.
18. D. M. J. Lilley and C. F. Higgins, *Mol. Microbiol.*, 1991, **5**, 779–783.
19. X. Zhi, S. Dages, K. Dages, Y. Liu, Z.-C. Hua, J. Makemson and F. Leng, *J. Biol. Chem.*, 2017, **292**, 14566–14575.
20. M. Drolet, S. Broccoli, F. Rallu, C. Hraiky, C. Fortin, E. Massé and I. Baaklini, *Front. Biosci.*, 2003, **8**, d210–221.
21. F. Leng, L. Amado and R. McMacken, *J. Biol. Chem.*, 2004, **279**, 47564–47571.
22. J. M. Santos-Pereira and A. Aguilera, *Nat. Rev. Genet.*, 2015, **16**, 583–597.
23. M. Y. Sheinin, M. Li, M. Soltani, K. Luger and M. D. Wang, *Nat. Commun.*, 2013, **4**, 2579.
24. J. Ma, L. Bai and M. D. Wang, *Science*, 2013, **340**, 1580–1583.
25. S. Chong, C. Chen, H. Ge and X. S. Xie, *Cell*, 2014, **158**, 314–326.
26. J. Ma, C. Tan, X. Gao, R. Fulbright, J. Roberts and M. D. Wang, *Proc. Natl. Acad. Sci. U. S. A.*, 2019, **116**, 2583–2588.
27. M. R. Gartenberg and J. C. Wang, *Proc. Natl. Acad. Sci. U. S. A.*, 1992, **89**, 11461–11465.
28. K. Tantale, F. Mueller, A. Kozulic-Pirher, A. Lesne, J. M. Victor, M. C. Robert, S. Capozi, R. Chouaib, V. Backer, J. Mateos-Langerak, X. Darzacq, C. Zimmer, E. Basyuk and E. Bertrand, *Nat. Commun.*, 2016, **7**, 12248.
29. S. Kim, B. Beltran, I. Irnov and C. Jacobs-Wagner, *Cell*, 2019, **179**, 106–119 e116.
30. H.-Y. Wu, S. Shyy, J. C. Wang and L. F. Liu, *Cell*, 1988, **53**, 433–440.
31. Y. Harada, O. Ohara, A. Takatsuki, H. Itoh, N. Shimamoto and K. Kinosita, *Nature*, 2001, **409**, 113–115.
32. D. A. Schafer, J. Gelles, M. P. Sheetz and R. Landick, *Nature*, 1991, **352**, 444–448.
33. H. Yin, R. Landick and J. Gelles, *Biophys. J.*, 1994, **67**, 2468–2478.
34. H. Yin, M. D. Wang, K. Svoboda, R. Landick, S. M. Block and J. Gelles, *Science*, 1995, **270**, 1653–1657.
35. H. Yin, I. Artsimovitch, R. Landick and J. Gelles, *Proc. Natl. Acad. Sci. U. S. A.*, 1999, **96**, 13124–13129.
36. M. D. Wang, M. J. Schnitzer, H. Yin, R. Landick, J. Gelles and S. M. Block, *Science*, 1998, **282**, 902–907.
37. A. La Porta and M. D. Wang, *Phys. Rev. Lett.*, 2004, **92**, 190801.
38. C. Deufel, S. Forth, C. R. Simmons, S. Dejgosha and M. D. Wang, *Nat. Methods*, 2007, **4**, 223–225.
39. S. Forth, C. Deufel, M. Y. Sheinin, B. Daniels, J. P. Sethna and M. D. Wang, *Phys. Rev. Lett.*, 2008, **100**, 148301.

40. M. Y. Sheinin, S. Forth, J. F. Marko and M. D. Wang, *Phys. Rev. Lett.*, 2011, **107**, 108102.
41. A. Revyakin, R. H. Ebright and T. R. Strick, *Proc. Natl. Acad. Sci. U. S. A.*, 2004, **101**, 4776–4780.
42. A. Revyakin, C. Liu, R. H. Ebright and T. R. Strick, *Science*, 2006, **314**, 1139–1143.
43. J. Inman, S. Forth and M. D. Wang, *Opt. Lett.*, 2010, **35**, 2949–2951.
44. J. F. Allemand, D. Bensimon, R. Lavery and V. Croquette, *Proc. Natl. Acad. Sci. U. S. A.*, 1998, **95**, 14152–14157.
45. Z. Bryant, M. D. Stone, J. Gore, S. B. Smith, N. R. Cozzarelli and C. Bustamante, *Nature*, 2003, **424**, 338–341.
46. J. F. Leger, G. Romano, A. Sarkar, J. Robert, L. Bourdieu, D. Chatenay and J. F. Marko, *Phys. Rev. Lett.*, 1999, **83**, 1066–1069.
47. L. Pauling and R. B. Corey, *Proc. Natl. Acad. Sci. U. S. A.*, 1953, **39**, 84–97.
48. S. Forth, C. Deufel, S. S. Patel and M. D. Wang, *Biophys. J.*, 2011, **101**, L5–L7.
49. F. C. Oberstrass, L. E. Fernandes and B. Zev, *Proc. Natl. Acad. Sci. U. S. A.*, 2012, **109**, 6106–6111.
50. D. Dutta, K. Shatalin, V. Epshtein, M. E. Gottesman and E. Nudler, *Cell*, 2011, **146**, 533–543.
51. T. T. Le and M. D. Wang, *J. Mol. Biol.*, 2018, **430**, 4513–4524.
52. G. P. Rédei, in *Encyclopedia of Genetics, Genomics, Proteomics and Informatics*, ed. G. P. Rédei, Springer Netherlands, Dordrecht, 2008, pp. 535–536.
53. J. Zhou, K. S. Ha, A. La Porta, R. Landick and S. M. Block, *Mol. Cell*, 2011, **44**, 635–646.
54. K. M. Herbert, J. Zhou, R. A. Mooney, A. L. Porta, R. Landick and S. M. Block, *J. Mol. Biol.*, 2010, **399**, 17–30.
55. N. Komissarova and M. Kashlev, *Proc. Natl. Acad. Sci. U. S. A.*, 1997, **94**, 1755–1760.
56. E. Nudler, A. Mustaev, E. Lukhtanov and A. Goldfarb, *Cell*, 1997, **89**, 33–41.
57. M. G. Izban and D. S. Luse, *J. Biol. Chem.*, 1993, **268**, 12864–12873.
58. D. Reinberg and R. G. Roeder, *J. Biol. Chem.*, 1987, **262**, 3310–3321.
59. M. T. Marr and J. W. Roberts, *Mol. Cell*, 2000, **6**, 1275–1285.
60. E. Stepanova and M. Wang, *J. Biol. Chem.*, 2009, **284**, 35702–35713.
61. E. J. Strobel and J. W. Roberts, *Nucleic Acids Res.*, 2014, **42**, 5097–5108.
62. A. Lisica, C. Engel, M. Jahnel, E. Roldan, E. A. Galburt, P. Cramer and S. W. Grill, *Proc. Natl. Acad. Sci. U. S. A.*, 2016, **113**, 2946–2951.
63. S. Chedin, M. Riva, P. Schultz, A. Sentenac and C. Carles, *Genes Dev.*, 1998, **12**, 3857–3871.
64. L. E. Tetone, L. J. Friedman, M. L. Osborne, H. Ravi, S. Kyzer, S. K. Stumper, R. A. Mooney, R. Landick and J. Gelles, *Proc. Natl. Acad. Sci. U. S. A.*, 2017, **114**, E1081–E1090.
65. S. Borukhov, J. Lee and O. Laptenko, *Mol. Microbiol.*, 2005, **55**, 1315–1324.
66. Y. Zhang, R. Mooney, J. Grass, P. Sivaramakrishnan, C. Herman, R. Landick and J. Wang, *Mol. Cell*, 2014, **53**, 766–778.

67. C. P. Selby and A. Sancar, *Science*, 1993, **260**, 53–58.
68. J. Roberts and J. S. Park, *Curr. Opin. Microbiol.*, 2004, **7**, 120–125.
69. A. J. Smith, M. D. Szczelkun and N. J. Savery, *Nucleic Acids Res.*, 2007, **35**, 1802–1811.
70. A. J. Smith, C. Pernstich and N. J. Savery, *Nucleic Acids Res.*, 2012, **40**, 10408–10416.
71. T. T. Le, Y. Yang, C. Tan, M. M. Suhanovsky, R. M. Fulbright Jr, J. T. Inman, M. Li, J. Lee, S. Perelman, J. W. Roberts, A. M. Deaconescu and M. D. Wang, *Cell*, 2018, **172**, 344–357.
72. M. V. Kotlajich, D. R. Hron, B. A. Boudreau, Z. Sun, Y. L. Lyubchenko and R. Landick, *eLife*, 2015, **4**, e04970.
73. F. C. Oberstrass, L. E. Fernandes, P. Lebel and Z. Bryant, *Phys. Rev. Lett.*, 2013, **110**, 178103.
74. T. R. Strick, J.-F. Allemand, D. Bensimon, A. Bensimon and V. Croquette, *Science*, 1996, **271**, 1835–1837.
75. M. Y. Sheinin and M. D. Wang, *Phys. Chem. Chem. Phys.*, 2009, **11**, 4800–4803.
76. Y. Mitsui, R. Langridge, B. E. Shortle, C. R. Cantor, R. C. Grant, M. Kodama and R. D. Wells, *Nature*, 1970, **228**, 1166–1169.
77. E. A. Galburt, S. W. Grill, W. Anna, L. Lucyna, C. Jason, N. Eva, K. Mikhail and B. Carlos, *Nature*, 2007, **446**, 820–823.
78. J. C. Wang and G. N. Giaever, *Science*, 1988, **240**, 300–304.
79. S. B. Smith, Y. Cui and C. Bustamante, *Science*, 1996, **271**, 795–799.
80. X. Zhang, H. Chen, S. Le, I. Rouzina, P. S. Doyle and J. Yan, *Proc. Natl. Acad. Sci. U. S. A.*, 2013, **110**, 3865–3870.
81. F. Kouzine, D. Wojtowicz, L. Baranello, A. Yamane, S. Nelson, W. Resch, K. R. Kieffer-Kwon, C. J. Benham, R. Casellas, T. M. Przytycka and D. Levens, *Cell Syst.*, 2017, **4**, 344–356 e347.
82. S. Scott, Z. M. Xu, F. Kouzine, D. J. Berard, C. Shaheen, B. Gravel, L. Saunders, A. Hofkirchner, C. Leroux, J. Laurin, D. Levens, C. J. Benham and S. R. Leslie, *Nucleic Acids Res.*, 2018, **46**, 4622–4631.
83. G. Lowe, M. Meister and H. C. Berg, *Nature*, 1987, **325**, 637–640.
84. X. Chen and H. C. Berg, *Biophys. J.*, 2000, **78**, 1036–1041.
85. S. W. Reid, M. C. Leake, J. H. Chandler, C.-J. Lo, J. P. Armitage and R. M. Berry, *Proc. Natl. Acad. Sci. U. S. A.*, 2006, **103**, 8066–8071.
86. R. M. Berry and H. C. Berg, *Proc. Natl. Acad. Sci. U. S. A.*, 1997, **94**, 14433–14437.
87. H. C. Berg and L. Turner, *Biophys. J.*, 1993, **65**, 2201–2216.
88. M. Washizu, Y. Kurahashi, H. Iochi, O. Kurosawa, S. I. Aizawa, S. Kudo, Y. Magariyama and H. Hotani, *IEEE Trans. Ind. Appl.*, 1993, **29**, 286–294.
89. Y. Sowa, H. Hotta, M. Homma and A. Ishijima, *J. Mol. Biol.*, 2003, **327**, 1043–1051.
90. K. M. Thormann and A. Paulick, *Microbiology*, 2010, **156**, 1275–1283.
91. H. Noji, R. Yasuda, M. Yoshida and K. Kinosita Jr, *Nature*, 1997, **386**, 299.
92. J. Lipfert, J. W. J. Kerssemakers, T. Jager and N. H. Dekker, *Nat. Methods*, 2010, **7**, 977–980.

93. X. J. Janssen, J. Lipfert, T. Jager, R. Daudey, J. Beekman and N. H. Dekker, *Nano Lett.*, 2012, **12**, 3634–3639.
94. A. Celedon, I. M. Nodelman, B. Wildt, R. Dewan, P. Searson, D. Wirtz, G. D. Bowman and S. X. Sun, *Nano Lett.*, 2009, **9**, 1720–1725.
95. F. Mosconi, J. F. Allemand and V. Croquette, *Rev. Sci. Instrum.*, 2011, **82**, 034302.
96. P. Gupta, J. Zlatanova and M. Tomschik, *Biophys. J.*, 2009, **97**, 3150–3157.
97. H.-G. Patterton and C. von Holt, *J. Mol. Biol.*, 1993, **229**, 623–636.
98. B. A. Shen and R. Landick, *J. Mol. Biol.*, 2019, **431**, 4040–4066.
99. P. B. Talbert, M. P. Meers and S. Henikoff, *Nat. Rev. Genet.*, 2019, **20**, 283–297.
100. A. Bancaud, N. Conde e Silva, M. Barbi, G. Wagner, J.-F. Allemand, J. Mozziconacci, C. Lavelle, V. Croquette, J.-M. Victor, A. Prunell and J.-L. Viovy, *Nat. Struct. Mol. Biol.*, 2006, **13**, 444–450.
101. A. Bancaud, G. Wagner, N. Conde e Silva, C. Lavelle, H. Wong, J. Mozziconacci, M. Barbi, A. Sivolob, E. Le Cam, L. Mouawad, J.-L. Viovy, J.-M. Victor and A. Prunell, *Mol. Cell*, 2007, **27**, 135–147.
102. T. T. Le, X. Gao, S. H. Park, J. Lee, J. T. Inman, J. H. Lee, J. L. Killian, R. P. Badman, J. M. Berger and M. D. Wang, *Cell*, 2019, **179**, 619–631 e615.
103. J. F. Marko, *Phys. Rev. E: Stat., Nonlinear, Soft Matter Phys.*, 2007, **76**, 021926.
104. A. Sarkar, J. F. Léger, D. Chatenay and J. F. Marko, *Phys. Rev. E*, 2001, **63**, 051903.
105. S. Forth, M. Y. Sheinin, J. Inman and M. D. Wang, *Annu. Rev. Biophys.*, 2013, **42**, 583–604.
106. A. V. Vologodskii and N. R. Cozzarelli, Conformational and thermodynamic properties of supercoiled DNA, *Annu. Rev. Biophys. Biomol. Struct.*, 1994, **23**, 609–643.
107. G. Charvin, J. F. Allemand, T. R. Strick, D. Bensimon and V. Croquette, Twisting DNA: single molecule studies, *Contemp. Phys.*, 2004, **45**, 383–403.
108. X. Gao, Y. F. Hong, F. Ye, J. T. Inman and M. D. Wang, *Phys. Rev. Lett.*, 2021, **127**, 028101.

CHAPTER 4

RNA Polymerase-associated Transcription Elongation Factors

SHUN-ICHI SEKINE*[a], TAMAMI UEJIMA[a] AND HARUHIKO EHARA[a]

[a]Laboratory for Transcription Structural Biology, RIKEN Center for Biosystems Dynamics Research, 1-7-22 Suehiro-cho, Tsurumi-ku, Yokohama 230-0045, Japan
*E-mail: shunichi.sekine@riken.jp

4.1 Introduction

DNA-dependent RNA polymerase (RNAP) governs gene transcription. Transcription is mediated through three major stages: initiation, elongation, and termination. In each stage, different factors associate with RNAP to form huge complexes specialized for each stage-specific function. After RNAP has initiated transcription by synthesizing an approximately 10 or more nucleotide RNA, the RNAP transforms into the elongation complex, and binds several transcription elongation factors. These factors support RNAP processivity and allow extensive RNA elongation to complete gene transcription. In eukaryotes, RNAPs transcribe DNA within the highly organized chromatin structure, and transcription is coupled with multiple functions, such as chromatin remodeling, chromatin modification, DNA repair, and RNA processing. Therefore, eukaryotes have developed many

Chemical Biology No. 19
RNA Polymerases as Molecular Motors: On the Road, 2nd Edition
Edited by Robert Landick, Terence Strick and Jue Wang

Published by the Royal Society of Chemistry, www.rsc.org

proteins that support these functions.[1] In this chapter, we focus on the "basal" elongation factors that tightly associate with the transcribing RNAP and support essential functions, especially those accompanied by eukaryotic transcription.

4.2 Factors Associated with RNAP During Transcription Elongation

While RNAP alone can catalyze RNA elongation *in vitro*, transcription *in vivo* is highly efficient and processive. For example, RNA polymerase II (RNAP II) is capable of transcribing a 2.3 Mbp gene in humans.[2] Early studies identified many protein factors that support transcription elongation. Transcription elongation factor SII (TFIIS) is the first eukaryotic elongation factor to be identified, and has the ability to support the production of long transcripts in an *in vitro* transcription system.[3] Genetic studies in the yeast *Saccharomyces cerevisiae* revealed a number of *trans*-acting factors that suppress the transcription defect caused by the insertion of the Ty transposable element.[4] These suppressor of Ty (*spt*) gene products include Spt4, Spt5, Spt6, and Spt16, which differently affect gene expression and are phenotypically related to mutations in histones H2A (Spt11) and H2B (Spt12).[4] Similarly, a genetic screen identified Rtf1 (restores TBP functions), a component of the RNAP II-associated factor (PAF), as a factor that regulates DNA binding by the TATA-box binding protein (TBP).[5] Mutations in Rtf1 were later shown to cause synthetic lethality in combination with mutations in transcription regulators, including Spt4 and Spt5.[6] A proteomic study with a tandem affinity purification scheme revealed PAF to be an elongation factor that directly associates with RNAP II.[7] A subsequent yeast genetic screen for mutations that cause synthetic lethality together with mutations in TFIIS and Spt6 identified elongation factor 1 (Elf1).[8] In higher eukaryotes, transcription elongation is regulated immediately downstream of gene promoters. Negative transcription elongation factor (NELF) was discovered as a factor implicated in the maintenance of this promoter-proximal pausing, in cooperation with 5,6-dichloro-1-β-D-ribofuranosylbenzimidazole (DRB)-sensitivity inducing facto (DSIF; Spt4–5).[9]

TFIIS, Elf1, Spt4–5, Spt6, and PAF are conserved in eukaryotes. TFIIS, Elf1, and Spt4–5 are also distributed in archaea. Bacteria conserve the Spt5 counterpart, N-utilization substance G (NusG). The results of many chromatin immunoprecipitation (ChIP) studies and genome-wide analyses have indicated that these proteins are associated with transcribing RNAP II on gene bodies.[10] Exceptionally, NELF is only found in limited animal classes, and is associated with promoter-proximal regions. These proteins directly bind to RNAP II, and have been biochemically and structurally characterized, as described in the following sections.

4.3 Architecture of the RNAP II Elongation Complex – an Overview

The crystal structure of the yeast (*S. cerevisiae*) RNAP II bound to TFIIS was the first solved structure of RNAP in complex with a transcription elongation factor.[11] TFIIS is the best-characterized elongation factor, and is conserved in eukaryotes and archaea.[11–14] It enhances the RNAP II processivity during transcription elongation by reactivating a backtracked or arrested RNAP II by its transcript cleavage stimulating activity, similar to its bacterial functional counterparts growth regulator A and B (GreA and GreB).[15] TFIIS inserts its C-terminal zinc-ribbon domain into the secondary channel of RNAP II, to remodel the active site for the hydrolytic cleavage of the RNA 3′-end.[12,13] (details about TFIIS can be found in Chapter 3).

The structure of the RNAP II elongation complex bound with Elf1, Spt4/5, and TFIIS from the yeast *Komagataella pastoris* has also been solved.[16] Elf1 and Spt4/5 cover a wide area of the RNAP II surface (Figure 4.1). They bind near the nucleic-acid binding sites, reinforcing or establishing the paths for DNA and RNA. Elf1, the NusG N-terminal (NGN) domain of Spt5, and Spt4 are aligned along the DNA within the RNAP II main channel, establishing a continuous tunnel covering approximately 30 bp of the DNA from the downstream part, through the transcription bubble, to the upstream part. The Kyrpides Ouzounis and Woese 4 and 5 (KOW4 and KOW5) domains of Spt5 bind near the RNA exit site, directing the nascent RNA. The structure of the mammalian RNAP II elongation complex bound with DSIF (Spt4–5) revealed an array of KOW2-3-X-4 domains between the RNAP II clamp and stalk.[17] The Spt5 KOW1 domain intervenes between the exiting DNA and RNA, preventing their possible interaction. The elongation complex architecture is quite different from those of the initiation complexes, and satisfies the requirement

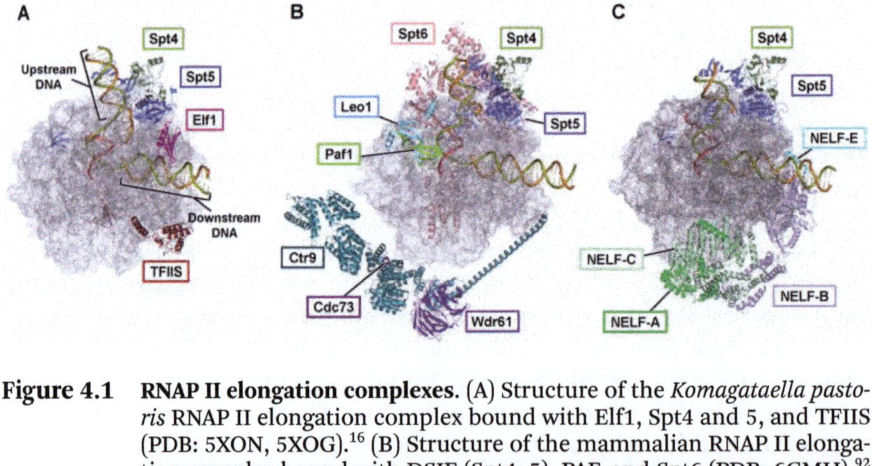

Figure 4.1 **RNAP II elongation complexes.** (A) Structure of the *Komagataella pastoris* RNAP II elongation complex bound with Elf1, Spt4 and 5, and TFIIS (PDB: 5XON, 5XOG).[16] (B) Structure of the mammalian RNAP II elongation complex bound with DSIF (Spt4–5), PAF, and Spt6 (PDB: 6GMH).[92] (C) Structure of the mammalian RNAP II elongation complex bound with DSIF (Spt4–5) and NELF (PDB: 6GML).[19]

for processive transcription by counteracting pausing or premature termination (Box 4.1).

Spt6 and the PAF complex apparently associate with the RNAP II elongation complex after the recruitment of Spt4–5. The cryo-electron microscopy (cryo-EM) structure of the mammalian RNAP II elongation complex bound

Box 4.1 Transition From the Initiation to the Elongation Complex

The architectures of the RNAP II transcription complexes in the transcription initiation stage and in the elongation stage are very different (Figure 4.2).[16,17,20–22] A structural comparison revealed that Spt4–5 and Elf1 in the elongation complex share overlapping locations with general transcription factors (GTFs) and DNA in the pre-initiation complex (PIC) or the initially transcribing complex (ITC). This indicated that the binding of the elongation factors is incompatible with the binding of GTFs, and that the GTFs must be exchanged with the elongation factors during the transition from the initiation to the early elongation stages of transcription.

The exchange of protein factors rearranges the complex structure, switching its functions from those required for transcription initiation to those for transcription elongation. GTFs are required for binding to the gene promoter, melting the double-stranded DNA to incorporate the template strand into the active site, and initiating transcription by using nucleoside triphosphates. Elongation of the nascent RNA chain to a certain length triggers the dissociation of the GTFs, which are replaced by elongation factors. The elongation factors should stabilize the established elongation complex, preventing possible bubble collapse and dissociation of RNA or DNA during transcription. The precise order and mechanism of the dissociation of the GTFs and the replacement with the elongation factors are currently unclear, and remain as an open question.

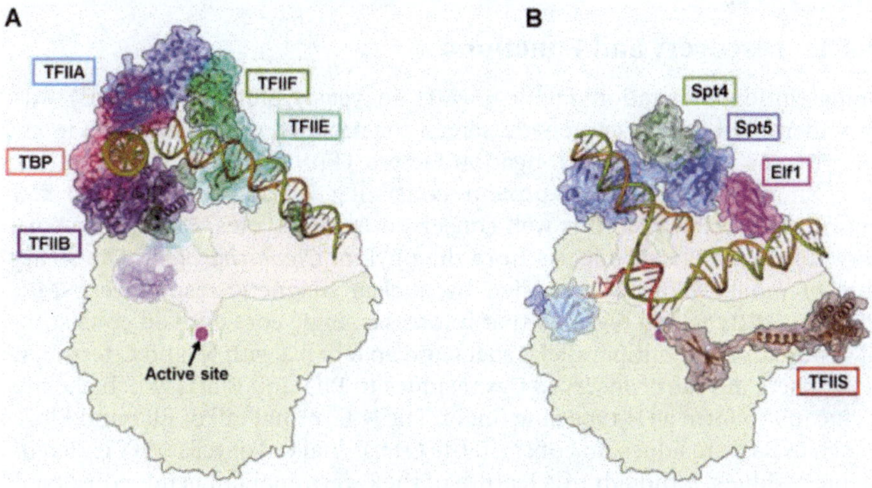

Figure 4.2 **Structural transition from the initiation to elongation complex.** (A) Structure of the *S. cerevisiae* RNAP II pre-initiation complex bound with general transcription factors (PDB: 5FZ5).[21] (B) Structure of the *K. pastoris* RNAP II elongation complex bound with Elf1, Spt4 and 5, and TFIIS (PDB: 5XON, 5XOG).[16]

with DSIF (Spt4–5), Spt6, and PAF revealed a more complex form of the elongation complex (Figure 4.1).[18] Spt6 binds to the RNAP II Rpb4/7 stalk and the Rpb1 CTD-linker, which is a loop connecting the heptad repeats of the Rpb1 C-terminal domain (CTD) to the Rpb1 body. It expands the RNA-exit tunnel, directing the nascent RNA outward. PAF covers a large area of the RNAP II face opposite from the Spt4–5 binding face, encompassing the DNA-entry tunnel to the DNA-exit tunnel. At the DNA-exit site, the left open reading frame 1 homologue (Leo1) subunits of PAF interact with the upstream DNA duplex and Spt4/5. At the DNA-entry site, the C-terminal extension of the cyclin three - requiring 9 (Ctr9) subunit could interact with the downstream DNA duplex. This form of the complex is involved in various chromatin functions during transcription elongation.

The cryo-EM structure of the mammalian RNAP II elongation complex bound with human DSIF (Spt4–5) and NELF represents the maintenance mechanism of a promoter-proximally paused RNAP II.[19] NELF is a large protein complex consisting of four subunits. It covers a wide area of the Rpb1 subunit of RNAP II (Figure 4.1), and seemingly stabilizes an off-pathway structure of the elongation complex. NELF maintains a "tilted" DNA–RNA hybrid and an "open" trigger loop, which represent an inactive state of the RNAP II active site. NELF also covers the entrance of the RNAP II secondary channel, hindering the TFIIS binding. The presence of bound NELF is also incompatible with PAF binding, and thus it represents the mechanism to prevent the elongation complex from converting into a processive form.[18] We will discuss these elongation factors in more detail in the following sections.

4.4 Elf1

4.4.1 Discovery and Functions

Transcription elongation factor 1 (Elf1 in yeast; Elof1 in humans) was first identified in a yeast genetic screen for proteins that genetically interact with the transcription elongation factors TFIIS and Spt6.[8] Elf1 or Elof1 is a small, single-domain protein consisting of approximately 80–150 amino-acid residues, and is well conserved in eukaryotes. Elf1 is also conserved in the archaeal species from the phylum *Crenarchaeota*.[23] The structure of mouse Elof1 was studied by nuclear magnetic resonance (NMR; PDB ID: 1WII), which revealed that it forms a single core domain consisting of a four-stranded antiparallel β sheet and an α helix, with N- and C-terminal tails. There are four conserved Cys residues in Elf1 and Elof1, which chelate a zinc ion to form a C4-type zinc finger. The N-terminal tail of Elf1 and Elof1 is rich in basic residues. In contrast, the C-terminal tail of yeast Elf1 is rich in acidic residues, although this feature of the C-terminal tail is not conserved in the Elof1 proteins of higher eukaryotes.

Despite its conservation in Eukarya and Archaea, little is known about the biological functions of Elf1 and Elof1. Yeast genetic studies revealed that Elf1 genetically interacts with genes involved in transcription elongation.[8]

The deletion of *elf1* causes synthetic lethality when combined with mutations in *spt4*, *spt5*, and *spt6* or the deletion of *spt4*. Elf also genetically interacts with PAF, as the *elf1* deletion caused synthetic lethality when combined with deletion of the Paf1, Leo1, Ctr9, and cell division cycle 73 (Cdc73) subunits of PAF. The results of ChIP experiments indicated that Elf1 is associated with the coding regions of genes.[8] This association is transcription-dependent, as it was observed only when gene expression was induced. The Elf1 association with actively transcribed genes is partially dependent on Spt4 and Spt6. Genome wide studies revealed that the Elf1 occupancy peaked just upstream of the poly-adenylation site, and Elf1 was suggested to be involved in RNA 3′-end formation and transcription termination.[10] *In vitro*, Elf1 affects RNA elongation by *K. pastoris* RNAP II.[16] These observations indicated that Elf1 is a component of the RNAP II elongation complex, and collaborates with the known elongation factors.

Elf1 is also implicated in chromatin structure maintenance in actively transcribed genes.[8] Deletion of the *elf1* gene caused a defect in the maintenance of the proper chromatin structure at the synthetic GAL1–FLO8–HIS3 gene, and allowed transcription from a cryptic promoter within the coding region of this particular gene. A tandem affinity purification revealed that Elf1 physically associates with casein kinase II (CKII). Seven Ser or Thr residues in the C-terminal tail of Elf1 are the potential targets of CKII, and among these, the two serine residues Ser95 and Ser117 were proposed to be the CKII phosphorylation sites *in vitro*, on the basis of the results of matrix-assisted laser desorption/ionization-mass spectrometry (MALDI-MS) analyses.[24] This finding indicated that the Elf1 functions are regulated by phosphorylation.

4.4.2 Structures

The direct interaction of Elf1 with RNAP II was revealed by a structural study, which yielded insights into the role of Elf1 in transcription elongation (Figure 4.1A).[16] The crystal structure of the *K. pastoris* RNAP II elongation complex bound with Elf1 and the KOW5 domain of Spt5 indicated that Elf1 binds to the RNAP II central cleft, near the downstream DNA duplex. Elf1 docks to the Rpb2 lobe domain. The α-helix of Elf1 (the RNAP II-binding helix) is rich in acidic residues, and forms an interface with the Arg-rich face at the tip of the Rpb2 lobe. The other side of Elf1 reaches the Rpb1 clamp-head domain. Consequently, Elf1 covers the RNAP II central cleft to hold the downstream DNA duplex within it. Thus, Elf1 is an integral part of the RNAP II elongation complex that forms the downstream "DNA-entry tunnel" (Figure 4.2B). Although it was disordered, the N-terminal basic tail of Elf1 could interact with the downstream DNA.

It is interesting that the basic patch of the RNAP II lobe, where Elf1 binds, also serves as the DNA-binding site in the closed pre-initiation complex of RNAP II (the "outer" DNA-binding site).[21] Therefore, Elf1 physically blocks the outer DNA-binding site used during the transcription initiation stage, and instead holds the DNA within the "inner" DNA-binding site, which forms the essential DNA path during the transcription elongation stage. Thus, Elf1 seems to maintain the established elongation complex by preventing possible bubble collapse and dissociation of the nucleic acids. A recent structural study revealed that Elf1 plays a pivotal role in the nucleosomal DNA transcription.[25] In the nucleosome-transcribing elongation complex, Elf1 forms the downstream edge that faces the nucleosome, and helps RNAP II overcome the nucleosomal barriers (Box 4.2).

Box 4.2 Mechanism of Nucleosome Transcription by RNAP II

In the eukaryotic cell nucleus, genomic DNA is packaged into chromatin, composed of multiple basic units called nucleosomes. The nucleosome core particle is a disk-like structure with about 150 base-pairs of DNA wrapped around a histone octamer, comprising two H2A–H2B dimers and two H3–H4 dimers. Although the nucleosome is necessary for packaging the DNA into the nucleus and protecting the DNA from damage, it paradoxically forms a barrier to DNA replication, repair, and transcription. The mechanism by which these important tasks are achieved in the chromatin context had been a long-standing question.

A recent cryo-EM study shed light on the mechanism by which RNAP II transcribes through a nucleosome.[84] In this study, nucleosomal DNA was transcribed by *K. pastoris* RNAP II, and the structures of multiple RNAP II-nucleosome complexes were analyzed. These structures included RNAP II stalled at four different super helical locations (SHLs) of the nucleosome: SHL(−6), SHL(−5), SHL(−2), and SHL(−1). In these complexes, the RNAP II was stalled due to tight histone–DNA contacts: the DNA–H2A–H2B contact at SHL(−5) and the DNA–H3–H4 contacts at SHL(−6), SHL(−2), and SHL(−1). Although the RNAP II stalled at four different locations of the nucleosome, the nucleosome orientations relative to RNAP II were similar. This is probably because the paused states are maintained through interactions between the nucleosome and the Rpb1 and Rpb2 subunits of RNAP II. Similar RNAP II–nucleosome interactions were found in the recent cryo-EM structure of *S. cerevisiae* RNAP II stalled at SHL(−6) of the nucleosome.[85] The structural snapshots of the RNAP II–nucleosome complexes reveal how RNAP II progresses through the nucleosome, with the sequential transitions of the nucleosome during the RNAP II passage (Figure 4.3).[84]

In cellular environments, transcription elongation factors, histone chaperones, and nucleosome remodelers help RNAP II overcome the nucleosomal barriers to efficient chromatin transcription.[86,87] Another cryo-EM study of the nucleosome-transcribing RNAP II elongation complex bound with Elf1 and Spt4–5 revealed how these conserved elongation factors function in nucleosome transcription.[25] The domain array of Elf1, Spt5 NGN, and Spt4 intervenes between RNAP II and the nucleosome to minimize the RNAP II–nucleosome interactions that cause RNAP II stalling. The array also serves as a separator to prevent the downstream DNA from reassociating with the exposed histones. At SHL(−1), the N-terminal basic tail of Elf1 competes with histones for DNA to facilitate the dissociation of the histone–DNA contact. Elf1 and Spt4–5 adjust the nucleosome position relative to RNAP II, to favor the forward progression of the elongation complex upon dissociation of the contact. In the cellular context, other elongation factors, histone chaperones, and nucleosome remodelers accompany the elongation complex and cooperatively facilitate chromatin transcription. These new structures provide solid foundations for future studies on chromatin functions and regulation.

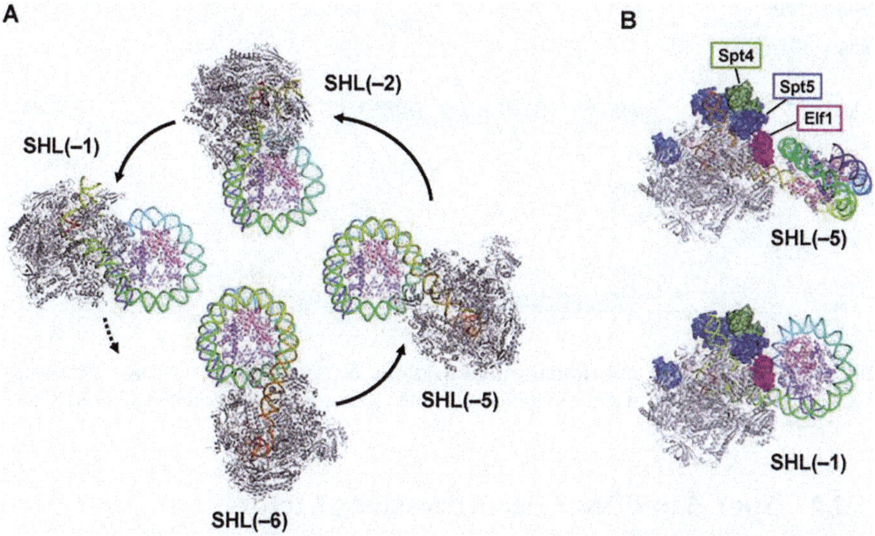

Figure 4.3 **Nucleosomal DNA transcription by RNAP II.** (A) Structural snapshots of RNAP II-nucleosome complexes stalled at four locations within a nucleosome (PDB: 6A5O, 6A5P, 6A5R, 6A5T).[84] (B) Structures of the nucleosome-transcribing RNAP II elongation complex bound with Elf1 and Spt4–5 at SHL(−5) and SHL(−1) (PDB: 6J4W, 6IR9).[25]

4.5 Spt4–5 and NusG

4.5.1 Overview

Spt4–5 or NusG is a universally conserved transcription elongation factor. Spt4–5 is a heterodimeric protein consisting of Spt4 and Spt5, and is found in Archaea and Eukarya. NusG is a bacterial monomeric protein corresponding to Spt5, while there is no Spt4 counterpart in bacteria. Spt5 and NusG are essential in organisms from bacteria to higher eukaryotes.[26,27]

Spt4 and Spt5, together with Spt6, were first identified as negative regulators of the Ty element-mediated expression of the HIS4 gene in the yeast *S. cerevisiae*.[4] Spt5 is a large protein required for the normal transcription of many genes,[26] and shares a NGN domain with the bacterial NusG (Figure 4.4).[28] In contrast, Spt4 is a small approximately 100 amino-acid residue protein containing four Cys residues, which form a zinc finger required for its function.[29] While Spt5 is essential for yeast cell viability, Spt4 is not. Spt4 and Spt5 are tightly coupled with each other, and Spt5 is also associated with RNAP II *in vivo*.[28] The results of several ChIP studies indicated that Spt4–5 interacts with transcribing RNAP II, and that its association begins downstream of the transcription start site and lasts until the termination site.[10,30–32]

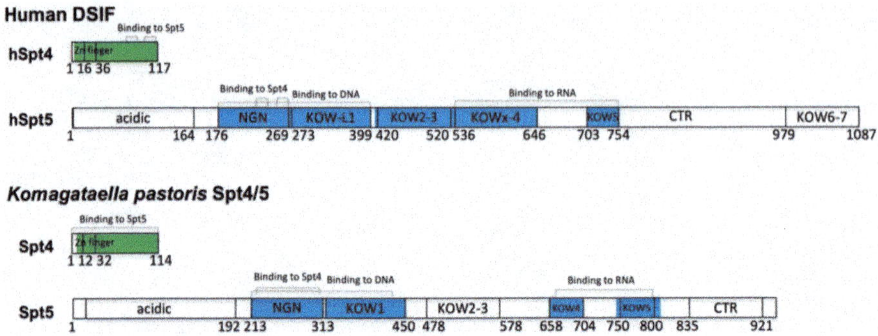

Figure 4.4 **Subunit and domain organizations of Spt4 and 5.** Domains that were revealed in the elongation complex structures are colored as in Figure 4.1.

4.5.2 Spt4–5 and NusG as Processivity Factors

Spt4–5 or NusG is generally recognized as a factor that stimulates the transcription elongation rate and the RNAP processivity.[33–35] NusG inhibits RNAP backtracking and reduces the pause lifetime, by shifting the equilibrium of the RNAP translocation state to the post-translocated state.[36] Results from a single-molecule study indicated that NusG increases the pause-free velocity of transcription and suppresses the entry into the long-lifetime pause.[37] Eukaryotic Spt4–5 also stimulates transcription elongation by reducing the frequency of pausing or arrest.[34,38] Mutations of Spt4, Spt5, and Spt6 cause phenotypes related to elongation defects *in vivo*, and these proteins seem to coordinate the elongation rate by RNAP II for normal yeast growth.[28] Mutations in Spt4 and 5 affect the RNAP II processivity *in vivo*.[39]

4.5.3 Other Functions of Spt4–5 and NusG

Spt4/5 is also significant for chromatin functions. Early studies revealed that mutations of the *spt4* and *spt5* genes are phenotypically relevant to mutations of genes encoding histones H2A and H2B.[40] Further studies revealed that Spt5 interacts with the ATP-dependent chromatin remodeler chromodomain-helicase DNA-binding 1 (Chd1), PAF, and the Rpd3S histone deacetylase complex.[41–44] A single-molecule fluorescence resonance energy transfer study demonstrated that Spt4–5 affects RNAP II processivity through a nucleosome.[45] Spt4–5 suppresses RNAP II pausing at a major pausing point, while it changes the pausing position with a different pausing frequency. A recent structural study revealed how Spt4–5 cooperates with Elf1 to facilitate the nucleosomal DNA transcription by RNAP II (Box 4.2).[25]

In higher eukaryotes, including humans, Spt4–5 is also known as a DSIF, which cooperates with NELF to maintain the promoter-proximal pausing of RNAP II.[34] DRB is an inhibitor of protein kinases, including the positive transcription elongation factor b (P-TEFb),[46] and inhibits the RNAP II pause release, which requires phosphorylation of DSIF, NELF, and RNAP II.

Spt4–5 is also an adaptor molecule that binds and recruits other factors. Spt5 has an intrinsically disordered region containing hexapeptide repeat sequences following the KOW5 domain (the C-terminal repeat region, CTR).[34] Similar to the C-terminal domain (CTD) of the RNAP II Rpb1, the CTR contains conserved Ser and Thr residues, which are the phosphorylation targets of the kinases P-TEFb or bypass upstream activating sequence requirement protein 1 (Bur1).[47–49] Depending on their phosphorylation states, the CTR interacts with various factors, including the mRNA-capping enzymes, PAF, the facilitates chromatin transcription that enhances the engagement of Pol II into transcriptional elongation (FACT) protein, and RNA 3′-end processing factors, for the coordination of various events with transcription,[48,50–54] although the deletion of the Spt5 CTR is not lethal in yeasts. In contrast, the CTR in *Drosophila* DSIF does not seem to be essential for promoter-proximal pausing,[55] while its phosphorylation is critical for human DSIF to stimulate transcription elongation.[52]

NusG also serves as an interface with other factors. NusG is implicated in transcription termination by recruiting the termination factor Rho.[56,57] It is also known as one of the components of the λ bacteriophage protein N-dependent antitermination complex.[58] NusG and its homolog replication factor H (RfaH) reportedly serve as adaptor molecules connecting RNAP and the ribosome, although the NusG binding site appears to be too distant from the ribosomal S10 subunit in the reported RNAP–ribosome complex (expressome) structure.[59]

4.5.4 Structures

Eukaryotic Spt5 is a large protein consisting of a tandem array of multiple domains connected by flexible linkers (Figure. 4.4). It comprises the NGN domain, several KOW domains,[60] and an intrinsically disordered CTR. The NGN domain is conserved across all three domains of life, and forms a tight complex with the partner protein Spt4 in Archaea and Eukarya.[61] Archaeal Spt5 and bacterial NusG typically contain only a single KOW domain, while the eukaryotic Spt5 has 5–7 repeats of the KOW domains. The CTR repeat sequences contain Ser, Thr, and Tyr residues, which can be phosphorylated.[62]

4.5.4.1 The NGN Domain

The NGN domain of Spt5/NusG is composed of a four-stranded anti-parallel β sheet sandwiched by α helices. Spt4 also contains a four-stranded anti-parallel β sheet, which is flanked by a C4-zinc finger. The NGN domain of Spt5 interacts with Spt4 to form a tight heterodimer of Spt4–5. The β sheets of the NGN domain and Spt4 join together to form an eight-stranded anti-parallel β sheet, which becomes the core of the heterodimer. Although Spt4 is not essential in *S. cerevisiae*,[29] it seems to be essential in *Drosophila.*[63]

Early studies revealed that *Escherichia coli* NusG and its paralogue RfaH, and the archaeal Spt4–5 bind to the mobile clamp module of RNAP.[35,64,65] The tip of the coiled-coil (CC) domain of the clamp module is the common binding site for a hydrophobic concave surface on the NGN domain. The 3.3 Å crystal structure of the isolated clamp module of *Pyrococcus furiosus* RNAP complexed with Spt4 and 5 confirmed the direct interaction between the NGN domain and the clamp coiled coil.[66] This was also confirmed by the 13 Å cryo-EM single-particle reconstruction of the complex of RNAP and Spt4 and 5 from the same archaeon.[67] The Spt5 NGN domain and Spt4 form a basic patch, through which Spt4–5 binds double-stranded DNA (dsDNA).[68] Transcription stimulation by archaeal Spt4 and 5 requires the NGN domain and either Spt4 or the KOW domain. Archaeal Spt4–5 does not depend on interactions with the DNA non-template strand for its stimulation function.[35]

The elongation complex structures of the *K. pastoris* and bovine RNAP II and the *E. coli* RNAP confirmed that the NGN domain bridges the clamp and protrusion domains, contacting the DNA (Figure 4.1A).[58,66,67,69,70] Many basic residues in the Spt5 NGN domain interact with the upstream DNA duplex and the non-template DNA strand within the transcription bubble. In the *K. pastoris* elongation complex structure, Spt4, the Spt5 NGN, and Elf1 form a protein domain array, which completely seals the main cleft of RNAP II.[16] The domain array overlaps with the DNA duplex in the closed pre-initiation complex (PIC),[21] and should maintain the transcription bubble by preventing the DNA from rewinding into a duplex. Spt4 and the Spt5 NGN domain are adjacent to the Spt5 KOW1 domain, and they encircle the upstream DNA duplex, establishing the "DNA-exit tunnel", which defines the upstream DNA orientation and the size of the transcription bubble (Figure 4.2B).[16] This should be significant because the extension of the DNA–RNA hybrid or the formation of an R-loop impairs transcription elongation.[71,72] The DNA-exit site is another important place where transcription is connected to other critical events, such as DNA repair.[73]

4.5.4.2 The KOW Domains

The KOW domain is a β barrel-like nucleic-acid binding domain, and is a close structural homolog of the tudor protein–protein interaction domain.[74] The KOW domain is present in a ribosome-binding protein, and, besides the ribosome,[75] it also reportedly interacts with other bio-molecules, such as the termination factor Rho and the factors involved in N-dependent anti-termination (NusA, B, and E).[74,76]

The KOW1 domain (including the eukaryote-specific insertion) from *S. cerevisiae* Spt5 forms a positively charged patch on its surface, and binds DNA and RNA *in vitro*.[77] As mutations of the basic residues in the positively charged patch cause yeast growth defects when combined with the deletion of Spt4, the KOW1 domain had been proposed to play an overlapping role with Spt4, probably for the nucleic acid interaction. As described in the previous section, the KOW1 domain collaborates with Spt4 and the Spt5 NGN domain to form the DNA-exit tunnel.[16] KOW2 and KOW3 are tandem domains that are tightly packed together. A crosslinking study revealed that the KOW3 domain of yeast Spt5 interacts with the clamp of RNAP II.[78] The structure of the bovine RNAP II elongation complex bound with DSIF (Spt4–5) revealed that a tandem array of KOW domains (KOW2–3–X–4) is located between the clamp and the stalk of RNAP II.[17]

The KOW4–KOW5 region of Spt4–5 is required for the elongation complex assembly in yeast,[79] as well as the promoter-proximal pausing in *Drosophila*.[55] Although archaeal Spt4–5 binds RNAP independently of the bound nucleic acids, the RNAP II binding by human and *Drosophila* Spt4–5 requires a certain length of the nascent RNA transcript.[80,81] A crosslinking study revealed that the KOW5 domain of *Drosophila* Spt5 is crosslinked to the nascent RNA[55] and important for transcription processivity.[79] *In vitro* primer extension assays, using a series of *K. pastoris* Spt4–5 deletion variants, revealed that only KOW5 is important for the stimulation of RNA elongation by RNAP II.[16]

The structure of the RNAP II elongation complex bound with Spt4–5 revealed that the Spt5 KOW5 domain binds near the RNA exit site.[16] The Spt5 KOW4 domain is located at the stalk of RNAP II, and part of the KOW4–KOW5 linker contacts the RNA. Thus, the Spt5 KOW4–KOW5 domains seem to support the efficient exit of the nascent RNA. The Spt5 KOW5 domain might also adjust the complex conformation to be more favorable for RNA elongation, by restricting the inter-subunit movements within RNAP II.[82,83] The structure of the bovine RNAP II elongation complex bound with Spt4–5 revealed the tandem array structure of four KOW domains (KOW2–3–X–4), which intervenes between the RNAP II body and the stalk.[17] The tandem array of KOW domains was not visible in the *K. pastoris* elongation complex structure.[16] The observed structural variations may reflect different complex states or species-specific characteristics. Finally, the Spt5 CTR should extrude from KOW5 near the RNA exit, where it would participate in various co-transcriptional functions, such as mRNA processing.

4.6 NELF

4.6.1 Overview

Promoter-proximal pausing of RNAP II is a common event during the transition between promoter escape and processive transcription elongation. RNAP II pausing and release are a major step in gene control in metazoans. In many genes, RNAP II stalls approximately 20–60 nucleotides downstream of the transcription start site (TSS) after the establishment of the transcription elongation complex. In this mechanism, negative elongation factor (NELF) and DSIF (another name for Spt4–5) are considered to play key roles in the maintenance of the paused RNAP II.[9] NELF is a large complex consisting of four protein subunits: NELF-A, B, C/D, and E (Figure 4.5). The NELF-A subunit is identical to the *WHSC2* gene product in humans, related to Wolf–Hirschhorn syndrome.[88] The NELF-B subunit is also known as COBRA1, which is a cofactor of the tumor suppressor breast cancer type 1 susceptibility protein (BRCA1).[89] The existence of NELF is limited to several eukaryotic classes; it exists in metazoans, including humans, mice, and *Drosophila*, while it has not been identified in *Caenorhabditis elegans*, *Arabidopsis thaliana*, and *Saccharomyces cerevisiae*.[89]

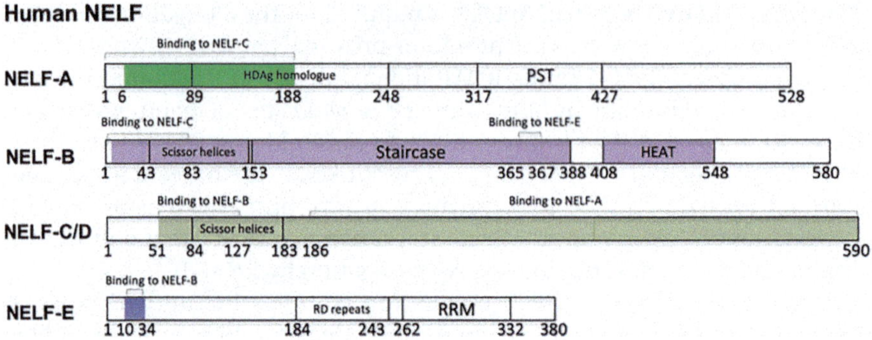

Figure 4.5 **Subunit and domain organizations of NELF.** HEAT: Huntingtin, elongation factor 3 (EF3), protein phosphatase 2A (PP2A), and yeast kinase TOR1 motif.

Human NELF strongly inhibits transcription elongation depending on DSIF *in vitro*.[9] This inhibition is due to pausing, but not termination, as a high concentration of NTPs allows RNA elongation to resume.[34] Consistently, a global run-on sequencing study revealed that the majority of RNAP II at *Drosophila* promoters is transcriptionally engaged.[90] A genome-wide location analysis revealed that NELF colocalizes with the paused RNAP II in *Drosophila* cells,[91] and the knockdown of NELF subunits reduces both the levels of promoter-proximal RNAP II and transcription at many genes.[90] Several studies have demonstrated that NELF has an ability to bind RNA.[81,92,93] The NELF-E subunit contains an RNA recognition motif (RRM), which potentially binds nascent RNA.[89,94] However, the significance of the RNA-binding ability of NELF in the RNAP II pausing remains unclear.[81]

Release from the pause into productive elongation is triggered by the positive transcription elongation factor b (P-TEFb), a heterodimer of cyclin-dependent kinase 9 (CDK9) and either cyclin T1 or T2. It phosphorylates NELF (the NELF-A and NELF-E subunits), DSIF, and RNAP II CTD. The phosphorylations induce NELF dissociation, DSIF activation, and RNAP II pause release,[94,95] although it is unclear whether the CTD phosphorylation is required for the pause release.[96] P-TEFb is recruited to the paused RNAP II in various ways, such as *via* histone-associated BRD4, Mediator, and activator proteins including HIV transactivator of transcription Tat, the cellular homolog of the myelocytomatosis virus oncogene (c-Myc), and nuclear factor (NF)-κB.[96,97] The enhancer RNA (eRNA) facilitates the release of the paused RNAP II in a P-TEFb-independent manner, by acting as a decoy for NELF.[98]

NELF also participates in co-transcriptional RNA processing. The m^7GpppN cap structure is added to the RNA 5′-end, during the RNAP II pausing proximal to the transcription start site.[99–101] NELF interacts with the nuclear cap-binding complex (CBC) through the NELF-E subunit, and mediates the 3′-end processing of the polyA(−) histone-coding mRNA.[102,103] It is also implicated in the 3′-end processing of small nuclear RNAs (snRNAs), through its interaction with the Integrator complex, together with DSIF.[104]

4.6.2 Structures

The crystal structure of the human NELF core subcomplex, formed between NELF-A and NELF-C, has been reported.[92] A recent cryo-EM structure of an RNAP II paused elongation complex containing NELF and DSIF revealed almost the entire structure of NELF, except for several flexible regions (Figure 4.1C).[19] Overall, NELF is composed predominantly of α helices, and covers a large area of an RNAP II face. The NELF-A and NELF-C subunits form a stable subcomplex, and NELF-C mediates the interaction with NELF-B, which binds NELF-E. The subcomplex of NELF-A and NELF-C binds Rpb8 and the funnel and foot domains of Rpb1, masking the secondary channel of RNAP II. NELF-B interacts with NELF-C near the foot domain, and stretches toward the stalk module of RNAP II. It directly contacts Rpb5 and Rpb6. NELF-A and NELF-E have flexible C-terminal regions named "tentacles". Crosslinking-mass spectrometry analyses revealed that the NELF-A tentacle extends along the protrusion domain of Rpb2 to reach Spt4 and the NGN domain of Spt5, near the DNA exit site. The NELF-E tentacle has an RNA-recognition motif (RRM) and extends along the KOW domains of Spt5 toward the RNA exit site. Although the NELF-E RRM binds the hairpin structure of the HIV1 TAR RNA,[92,93,105] it is not required for the association with RNAP II,[93] and the RNA binding by NELF may not be necessary for the RNAP II pausing.[81]

The structure of the paused elongation complex elegantly explained the mechanism of RNAP II pausing invoked by NELF binding.[19] It revealed the presence of a tilted DNA–RNA hybrid in the RNAP II active site, where DNA and RNA adopt pre-translocated and post-translocated positions, respectively, as observed in bacterial paused RNAP structures.[83,106] In addition, part of NELF-C interacts with the trigger loop in an open conformation. Furthermore, NELF bridges the shelf and core modules of RNAP II, and appears to restrict the mobility of RNAP II.[82,83,107] These features are consistent with the incompatibility of this complex with the nucleotide addition reaction. NELF binding to RNAP II is also incompatible with TFIIS binding to the secondary channel, consistent with the previous report that NELF and DSIF inhibit the RNA cleavage-stimulation activity of TFIIS.[108] The presence of NELF is also incompatible with PAF binding to RNAP II, preventing the elongation complex from converting into a processive form,[18] as described in the following sections.

4.7 Spt6

4.7.1 Overview

Spt6 was identified by a genetic screen that also discovered Spt4, Spt5, and Spt16.[4,28,109,110] It is conserved in eukaryotes, and composed of roughly 1400–1800 amino-acid residues. Spt6 genetically interacts with Spt4 and 5 and PAF,[6,109] and is essential for yeast cell viability.[110] Its physical interactions with Spt5 and RNAP II have also been reported.[28,109,111] Spt6 colocalizes with

Spt5 and transcribing RNAP II to actively transcribed regions in *Drosophila* chromosomes.[112,113] A yeast genome-wide profiling study revealed that Spt6 is associated with RNA-elongating RNAP II on gene bodies.[10] Spt6 functions as an elongation factor, as human Spt6 stimulates RNA elongation by RNAP II *in vitro*, either alone or in the presence of DSIF.[111] Consistently, the yeast *spt6* mutant (*spt4* and *spt5* as well) showed phenotypes of impaired transcription elongation *in vivo*.[28]

Besides its function as a stimulator of transcription elongation, Spt6 plays an essential role in chromatin structure maintenance as a histone chaperone. Mutations in the *spt6* gene cause spurious transcription initiation from cryptic promoters within genes.[114] Therefore, Spt6 is implicated in the reassembly of nucleosomes in the wake of RNAP II passage, thus preventing the spurious intragenic transcription initiation. Spt6 may cooperate with another important histone chaperone, FACT, for the chromatin structure maintenance during transcription elongation.[1] Spt6 primarily interacts with histone H3, and promotes nucleosome assembly *in vitro*.[115] It can also bind dsDNA *in vitro*.[116] Spt6 tightly associates with its binding partner suppresses postrecruitment functions gene number 1 (Spn1) [interacts with spt6, CTD assembly factor 1 (Iws1 in higher eukaryotes)], which is a nucleosome-binding protein that modulates the nucleosome-binding ability of Spt6.[114]

4.7.2 Structures

Spt6 comprises three distinct regions: the N-terminal region, the central core, and the C-terminal tandem sarcoma oncoprotein homology 2 (SH2) domain (Figure 4.6). The N-terminal region is composed of a highly acidic stretch of 200–300 amino acid residues, and predicted to have a disordered structure. Although disordered, this region is well conserved, and genetically interacts with PAF.[117] This N-terminal region contains sequence features commonly found in various histone chaperones,[118] and is necessary for the nucleosome binding by Spt6.[119] This region also serves as a platform to recruit other proteins to the RNAP II elongation complex. For example, the N-terminal region contains a binding site for Spn1 (Iws1),[119,120] which also has the ability to bind and assemble nucleosomes.[121]

The Spt6 core is composed of approximately 1000 amino acid residues, and has a flat and elongated structure rich in α helices.[116] It comprises multiple domains: the helix-turn-helix (HtH) domain, the YqgF homologous domain, the helix-hairpin-helix (HhH) domain, the death-like domain (DLD), and the S1 domain. While most of the domains seem to be rigidly packed

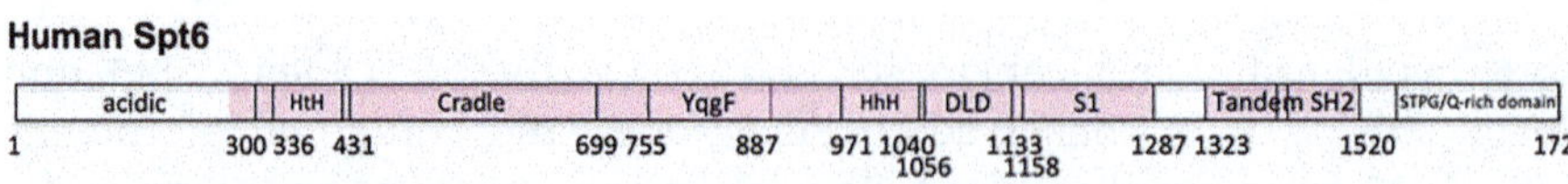

Figure 4.6 Domain organizations of Spt6. STPG/Q-rich: serine, threonine, proline and glycine/glutamine rich.

with each other, the S1 domain displayed more flexibility, indicating that it is only weakly tethered to the rest of the core.[116,122] The Spt6 core has the ability to bind dsDNA.[116] Interestingly, some bacteria and archaea possess the toxin expression (Tex) protein, a putative transcription factor, which is highly homologous to the eukaryotic Spt6 core.[122]

The C-terminal tandem SH2 domain (tSH2) is connected to the preceding core with a flexible linker.[116] The tSH2 domain contains two SH2 motifs that are tightly packed with each other.[116,123,124] The tSH2 domain is functionally important, as the deletion of this domain caused widespread defects attributed to Spt6.[28,123–125] The tSH2 domain has affinity for phosphorylated peptides, and thus was speculated to bind the phosphorylated heptad repeats within the RNAP II CTD.[126,127] However, it was later found that the tSH2 domain binds more tightly to the linker region between the CTD repeats and the main body of Rpb1 (CTD-linker).[18,128] The CTD-linker contains several residues that can be phosphorylated *in vivo*, and recruits the Spt6 tSH2 domain in a phosphorylation-dependent manner. While the deletion of the tSH2 domain or mutations in the CTD-linker interacting residues reduced the Spt6 occupancy accompanying RNAP II in ChIP experiments, these modifications did not completely abolish the Spt6 functions or its recruitment to RNAP II, indicating the presence of additional RNAP II-interaction sites other than the tSH2 domain or another Spt6 function that is not co-transcriptional.[10,128,129]

4.7.3 Spt6 in the RNAP II Elongation Complex

The stable *in vitro* reconstitution of the mammalian RNAP II elongation complex containing Spt6 was accomplished by the simultaneous addition of DSIF, Spt6, and PAF to the RNAP II-EC, followed by *in vitro* phosphorylation by P-TEFb kinase.[18] The cryo-EM structure of the reconstituted complex revealed that Spt6 was located far from the RNAP II active site, and bound to the RNA-exit face of RNAP II (Figure 4.1B).[18] Spt6 interacted with Spt5, PAF, the nascent RNA, and RNAP II, which explained the cooperative binding of these factors. Spt6 binding caused movements of the RNAP II stalk and the Spt5 KOW domains, which might allosterically activate the elongation complex.

The Spt6 core region interacts with the distal part of the Rpb4/7 stalk, and also contacts the KOWx-4 domains of Spt5, which is located between the stalk, Rpb1 dock, and Rpb1 clamp. The YqgF domain and the S1 domain of the Spt6 core form a positively-charged cleft, which binds the exiting RNA. While it was proposed that the nascent RNA is associated with the RNAP II stalk in the absence of elongation factors,[130–132] the Spt6 binding and the creation of the new RNA binding surface redefine the exit path of the nascent RNA. This Spt6-mediated RNA path could be implicated in guiding the nascent RNA to other RNA processing machineries, and may partially explain the roles of Spt6 in RNA processing.[126,127,129,133] The tSH2 domain is tethered to the core region, and located at the base of the Rpb4/7 stalk, where it is bound to the

CTD-linker. While the disordered N-terminal region of Spt6 was not visible in the structure, it should be located around the downstream DNA and the upstream DNA, consistent with its role in the histone-chaperone activity.

4.8 PAF

4.8.1 Overview

The Paf1 complex (PAF) was originally identified in yeast as an RNA polymerase II-associated factor.[134] Yeast PAF comprises five subunits, Paf1, Leo1, Ctr9, Cdc73, and Rtf1,[135] while higher eukaryotes, including humans, contain the additional subunit superkiller 8 homolog (Ski8) [WD repeat domain 61 (Wdr61)][136] (Figure 4.7). This highly conserved protein complex has genetic and physical connections to other transcription elongation factors or histone chaperones, including TFIIS, Spt4/5, and FACT.[6,7,42] PAF is recruited to the transcription elongation complex *via* RNAP II interactions with the Ctr9, Cdc73, Leo1, and Rtf1 subunits.[137–139] Kinases that can phosphorylate the RNAP II CTD and the Spt5 CTR are required for PAF recruitment in yeast and human cells.[53,140,141] Genomic studies revealed that PAF is associated with RNAP II with the unphosphorylated, Ser5-phosphorylated, and Ser2-phosphorylated forms of its CTD near promoters and throughout the coding regions of genes.[7,142–145] At most genes in yeast, the PAF occupancy peaks downstream of the transcription start site near the +2 and +3 nucleosomes.[10,146] In higher eukaryotes, PAF is recruited to enhancers and promoters by transcriptional activators, and its occupancy is the highest near the transcription start site (TSS) and transcription end site (TES).[140,142,147]

PAF is recruited to the open reading frames (ORFs) of actively transcribed genes,[145] and has a direct, positive effect on transcription elongation. A study using yeast cell extracts from mutants with the deletion of each PAF subunit revealed a positive effect of PAF on *in vitro* transcription of naked DNA.[148]

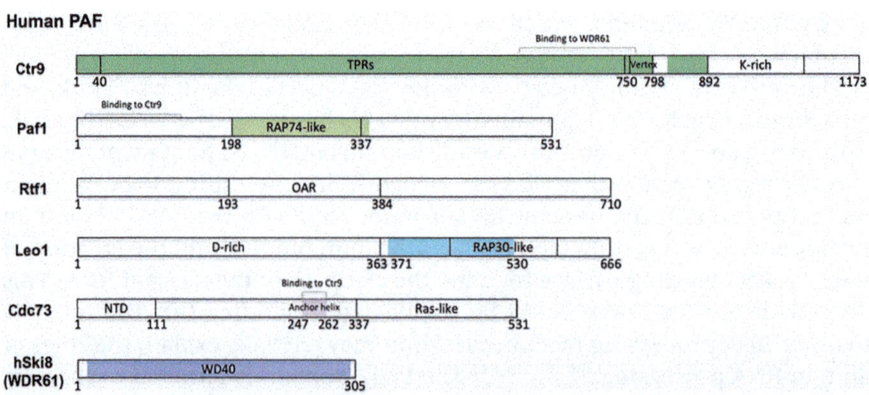

Figure 4.7 Subunit and domain organizations of PAF. OAR: ORF association region.

The reconstituted human PAF stimulated transcription elongation on chromatin templates, independent of its effects on histone modifications.[138] This study also revealed that human PAF directly interacts with TFIIS, with strong synergy on transcription promotion, consistent with the observation that the simultaneous mutations of PAF1 and TFIIS are lethal in yeast.[42] Human PAF also collaborates with DSIF (Spt4–5) to stimulate RNA elongation *in vitro*.[149]

4.8.2 Other Functions of PAF

Besides its function as an elongation stimulator, PAF is implicated in many important functions relevant to chromatin, gene regulation, and mRNA processing. PAF is essential for the ubiquitylation of a lysine residue (K123 in yeast; K120 in humans) in the C-terminal tail of histone H2B (H2Bub), which is required for the methylation of H3 K4 and H3 K79.[150–152] H2Bub disrupts chromatin compaction,[153] and regulates transcription positively and negatively, depending on the gene.[154] H2B is also ubiquitylated on K34 in its N-terminal tail in a PAF-dependent manner, and this ubiquitylation also stimulates the H3 K4 and H3 K79 methylations.[155,156] In yeast, the P-TEFb homolog Bur1,[157,158] Spt4,[159] and the Spt5 CTR[50,53] are required for the PAF recruitment to transcribing RNAP II and for the H3 K4 methylation.[43]

PAF is also implicated in the acetylation of H4 K16.[156] The deletion or mutations of the Leo1 subunit of PAF reduced the H4 K16 acetylation, indicating that PAF antagonizes heterochromatin spreading.[160] The Paf1, Ctr9, and Cdc73 subunits are required for the trimethylation of H3 K36, which represses the acetylation of H3 and H4 within coding regions.[161] This effect depends on both the recruitment of a histone deacetylase complex and the inhibition of the co-transcriptional exchange of H3, thus preventing the incorporation of pre-acetylated histones.[162]

PAF promotes the RNAP II release from promoter-proximal pausing at more than 5800 genes in human monocytic leukemia (THP1) cells.[140] PAF facilitates pause release by recruiting P-TEFb in various forms of the super elongation complex (SEC), in a manner mediated by the direct interaction between the Paf1 subunit of PAF and the AF9 subunit of the SEC.[163] Paradoxically, PAF inhibits pause release at a large number of genes, as revealed by a Paf1 depletion experiment in human colon cancer cells.[142] PAF promotes pausing by inhibiting the recruitment of P-TEFb or SEC.[142] Thus, the role of PAF in prompter-proximal pausing is complex. PAF is also implicated in more complex biological phenomena, such as development, cell differentiation, maintenance of stem cell pluripotency, and cancer.[43,164,165]

PAF is involved in co-transcriptional RNA processing and nuclear RNA export. PAF stimulates transcript cleavage and polyadenylation *in vitro*.[166] In yeast, the loss of PAF subunits leads to shortened poly(A) tails in mRNA and alternative poly-adenylation site selection.[164] Interactions between the Cdc73 and Rtf1 subunits of yeast PAF and a subunit of the cleavage and polyadenylation apparatus have been observed.[167,168] PAF is also implicated in the nuclear export of mRNAs, as PAF depletion leads to their nuclear retention and degradation.[169]

4.8.3 Structures

Although only partial information was previously available for the PAF structure, recent cryo-EM studies have uncovered nearly the entire structures of the yeast and human PAFs in complex with the RNAP II elongation complex (Figure 4.1B).[18,170] Ctr9 is the largest subunit of PAF, and serves as a scaffold of the PAF complex. Its mutations are known to be relevant in Wilms tumor development.[144] The cryo-EM structure of the RNAP II elongation complex bound with human PAF and Spt6 revealed that Ctr9 comprises a huge helical domain composed of multiple tetratricopeptide repeat (TPR) motifs, followed by a long trestle helix and a flexible C-terminal region.[18] Consistently, the subcomplex structures of human Ctr9–Paf1–Cdc73 [171] and human and yeast Ctr9–Paf1 [172] also revealed the multiple TPR motifs of Ctr9. Ctr9 wraps one side of the RNAP II surface, contacting the Rpb5, Rpb1, Rpb8, and Rpb11 subunits of RNAP II.[18] The trestle is a long helix spanning approximately 100 Å from the Rpb1 foot domain to Rpb5, near the downstream DNA binding site. The vertex formed between the TPRs and the trestle helix interacts with the C-terminal extension from the tSH2 domain of Spt6. The higher-eukaryote specific subunit Ski8 (WDR61) forms a seven-bladed β-propeller structure, and binds at the junction of the Ctr9 TPRs and the trestle. The C-terminal flexible region of Ctr9 is able to reach the downstream DNA or nucleosome, and could be implicated in chromatin functions by recruiting histone modification enzymes or chromatin remodelers.

Cdc73 directly interacts with RNAP II, and is important for PAF recruitment.[134,173] Mutations in Cdc73 are associated with hyperparathyroidism–jaw tumor syndrome.[174] Cdc73 consists of the C-terminal Ras-like domain,[175,176] the rigid N-terminal domain,[177] and an "anchor helix" intervening between them.[18] The conserved C-terminal Ras-like domain interacts with the phosphorylated CTD of RNAP II *in vitro*.[141,176] In the cryo-EM structure of the elongation complex, only the anchor helix of Cdc73 is visible, and it is bound to the surface groove of one of the TPRs of Ctr9.[18]

The Paf1 and Leo1 subunits form a tight double β-barrel complex,[178] which is bound to the Rpb2 external domain of RNAP II.[18] The Paf1 subunit is far apart from the Ctr9 subunit on the RNAP II surface. However, Paf1 interacts with Ctr9 *via* its N-terminal flexible region, which folds into an extended conformation and is embedded within the spiral of the Ctr9 TPR domain.[171] This interaction between the Paf1 and Ctr9 subunits is the key for the assembly of all of the PAF subunits into the PAF complex, and is essential for yeast cell viability. The C-terminal extension of Leo1 reaches the upstream DNA, and could participate in the DNA-exit tunnel.[18]

Rtf1 consists of the central Plus3 domain (a domain containing 3 positively charged residues), flanked by highly Ser-, Glu-, and Lys-rich regions. The central Plus3 domain binds single-stranded DNA and the phosphorylated CTR of Spt5,[47,179] which should support the PAF association with the elongation complex. Rtf1 has the histone modification domain (HMD), which recruits radiation sensitivity gene 6 homolog (Rad6) to the elongation complex for the H2B ubiquitylation and H3K4 methylation.[146] Rtf1 is loosely associated with the PAF complex in higher eukaryotes, and consistently, the recent cryo-EM structure of the elongation complex lacked the bound Rtf1 subunit.[18] Therefore, the means by which the functions ascribed to Rtf1 are achieved, in the context of chromatin transcription, remain to be addressed.

4.9 Conclusion

In this chapter, we have focused on the "basal" transcription elongation factors that are tightly associated with RNA-elongating RNAP II. Besides these factors, eukaryotes have developed numerous proteins associated with transcribing RNAP II, including those recruited *via* the CTD of RNAP II.[1] These include factors involved in chromatin remodeling, histone modifications, DNA-damage repair, mRNA maturation, including 5′-capping, splicing, 3′-end cleavage, and polyadenylation, and mRNA export.[86,180–183] The described elongation factors not only ensure the RNAP processivity, but also serve as scaffolds for these essential transcription-coupled mechanisms. Further studies are awaited to clarify our understanding of these complex mechanisms during gene expression.

Acknowledgements

This work was supported by JSPS KAKENHI Grant Number JP20H05690 to S.S.

References

1. R. J. Sims 3rd, R. Belotserkovskaya and D. Reinberg, *Genes Dev.*, 2004, **18**, 2437–2468.
2. C. N. Tennyson, H. J. Klamut and R. G. Worton, *Nat. Genet.*, 1995, **9**, 184–190.
3. K. Sekimizu, N. Kobayashi, D. Mizuno and S. Natori, *Biochemistry*, 1976, **15**, 5064–5070.
4. F. Winston, D. T. Chaleff, B. Valent and G. R. Fink, *Genetics*, 1984, **107**, 179–197.
5. L. A. Stolinski, D. M. Eisenmann and K. M. Arndt, *Mol. Cell. Biol.*, 1997, **17**, 4490–4500.
6. P. J. Costa and K. M. Arndt, *Genetics*, 2000, **156**, 535–547.
7. N. J. Krogan, M. Kim, S. H. Ahn, G. Zhong, M. S. Kobor, G. Cagney, A. Emili, A. Shilatifard, S. Buratowski and J. F. Greenblatt, *Mol. Cell. Biol.*, 2002, **22**, 6979–6992.
8. D. Prather, N. J. Krogan, A. Emili, J. F. Greenblatt and F. Winston, *Mol. Cell. Biol.*, 2005, **25**, 10122–10135.
9. Y. Yamaguchi, T. Takagi, T. Wada, K. Yano, A. Furuya, S. Sugimoto, J. Hasegawa and H. Handa, *Cell*, 1999, **97**, 41–51.
10. A. Mayer, M. Lidschreiber, M. Siebert, K. Leike, J. Soding and P. Cramer, *Nat. Struct. Mol. Biol.*, 2010, **17**, 1272–1278.
11. H. Kettenberger, K. J. Armache and P. Cramer, *Cell*, 2003, **114**, 347–357.

12. D. Wang, D. A. Bushnell, X. Huang, K. D. Westover, M. Levitt and R. D. Kornberg, *Science*, 2009, **324**, 1203–1206.
13. A. C. Cheung and P. Cramer, *Nature*, 2011, **471**, 249–253.
14. H. Kettenberger, K. J. Armache and P. Cramer, *Mol. Cell*, 2004, **16**, 955–965.
15. B. P. Hogan, T. Hartsch and D. A. Erie, *J. Biol. Chem.*, 2002, **277**, 967–975.
16. H. Ehara, T. Yokoyama, H. Shigematsu, S. Yokoyama, M. Shirouzu and S. I. Sekine, *Science*, 2017, **357**, 921–924.
17. C. Bernecky, J. M. Plitzko and P. Cramer, *Nat. Struct. Mol. Biol.*, 2017, **24**, 809–815.
18. S. M. Vos, L. Farnung, M. Boehning, C. Wigge, A. Linden, H. Urlaub and P. Cramer, *Nature*, 2018, **560**, 607–612.
19. S. M. Vos, L. Farnung, H. Urlaub and P. Cramer, *Nature*, 2018, **560**, 601–606.
20. C. Plaschka, L. Lariviere, L. Wenzeck, M. Seizl, M. Hemann, D. Tegunov, E. V. Petrotchenko, C. H. Borchers, W. Baumeister, F. Herzog, E. Villa and P. Cramer, *Nature*, 2015, **518**, 376–380.
21. C. Plaschka, M. Hantsche, C. Dienemann, C. Burzinski, J. Plitzko and P. Cramer, *Nature*, 2016, **533**, 353–358.
22. H. Ehara and S. I. Sekine, *Transcription*, 2018, 286–291.
23. J. P. Daniels, S. Kelly, B. Wickstead and K. Gull, *Biol. Direct*, 2009, **4**, 24.
24. K. Kubinski, R. Zielinski, U. Hellman, E. Mazur and R. Szyszka, *J. Biochem. Mol. Biol.*, 2006, **39**, 311–318.
25. H. Ehara, T. Kujirai, Y. Fujino, M. Shirouzu, H. Kurumizaka and S. I. Sekine, *Science*, 2019, **363**, 744–747.
26. M. S. Swanson, E. A. Malone and F. Winston, *Mol. Cell. Biol.*, 1991, **11**, 3009–3019.
27. B. R. Keegan, J. L. Feldman, D. H. Lee, D. S. Koos, R. K. Ho, D. Y. Stainier and D. Yelon, *Development*, 2002, **129**, 1623–1632.
28. G. A. Hartzog, T. Wada, H. Handa and F. Winston, *Genes Dev.*, 1998, **12**, 357–369.
29. E. A. Malone, J. S. Fassler and F. Winston, *Mol. Gen. Genet.*, 1993, **237**, 449–459.
30. K. Glover-Cutter, S. Kim, J. Espinosa and D. L. Bentley, *Nat. Struct. Mol. Biol.*, 2008, **15**, 71–78.
31. R. Pavri, A. Gazumyan, M. Jankovic, M. Di Virgilio, I. Klein, C. Ansarah-Sobrinho, W. Resch, A. Yamane, B. Reina San-Martin, V. Barreto, T. J. Nieland, D. E. Root, R. Casellas and M. C. Nussenzweig, *Cell*, 2010, **143**, 122–133.
32. P. B. Rahl, C. Y. Lin, A. C. Seila, R. A. Flynn, S. McCuine, C. B. Burge, P. A. Sharp and R. A. Young, *Cell*, 2010, **141**, 432–445.
33. E. Burova, S. C. Hung, V. Sagitov, B. L. Stitt and M. E. Gottesman, *J. Bacteriol.*, 1995, **177**, 1388–1392.
34. T. Wada, T. Takagi, Y. Yamaguchi, A. Ferdous, T. Imai, S. Hirose, S. Sugimoto, K. Yano, G. A. Hartzog, F. Winston, S. Buratowski and H. Handa, *Genes Dev.*, 1998, **12**, 343–356.

35. A. Hirtreiter, G. E. Damsma, A. C. Cheung, D. Klose, D. Grohmann, E. Vojnic, A. C. Martin, P. Cramer and F. Werner, *Nucleic Acids Res.*, 2010, **38**, 4040–4051.
36. G. Bar-Nahum, V. Epshtein, A. E. Ruckenstein, R. Rafikov, A. Mustaev and E. Nudler, *Cell*, 2005, **120**, 183–193.
37. K. M. Herbert, J. Zhou, R. A. Mooney, A. L. Porta, R. Landick and S. M. Block, *J. Mol. Biol.*, 2010, **399**, 17–30.
38. W. Zhu, T. Wada, S. Okabe, T. Taneda, Y. Yamaguchi and H. Handa, *Nucleic Acids Res.*, 2007, **35**, 4064–4075.
39. P. B. Mason and K. Struhl, *Mol. Cell*, 2005, **17**, 831–840.
40. J. S. Fassler and F. Winston, *Genetics*, 1988, **118**, 203–212.
41. R. Simic, D. L. Lindstrom, H. G. Tran, K. L. Roinick, P. J. Costa, A. D. Johnson, G. A. Hartzog and K. M. Arndt, *EMBO J.*, 2003, **22**, 1846–1856.
42. S. L. Squazzo, P. J. Costa, D. L. Lindstrom, K. E. Kumer, R. Simic, J. L. Jennings, A. J. Link, K. M. Arndt and G. A. Hartzog, *EMBO J.*, 2002, **21**, 1764–1774.
43. J. A. Jaehning, *Biochim. Biophys. Acta*, 2010, **1799**, 379–388.
44. T. K. Quan and G. A. Hartzog, *Genetics*, 2010, **184**, 321–334.
45. J. B. Crickard, J. Lee, T. H. Lee and J. C. Reese, *Nucleic Acids Res.*, 2017, **45**, 6362–6374.
46. L. A. Chodosh, A. Fire, M. Samuels and P. A. Sharp, *J. Biol. Chem.*, 1989, **264**, 2250–2257.
47. A. D. Wier, M. K. Mayekar, A. Heroux, K. M. Arndt and A. P. VanDemark, *Proc. Natl. Acad. Sci. U. S. A.*, 2013, **110**, 17290–17295.
48. A. Mayer, A. Schreieck, M. Lidschreiber, K. Leike, D. E. Martin and P. Cramer, *Mol. Cell. Biol.*, 2012, **32**, 1321–1331.
49. B. Ding, D. LeJeune and S. Li, *J. Biol. Chem.*, 2010, **285**, 5317–5326.
50. K. Zhou, W. H. Kuo, J. Fillingham and J. F. Greenblatt, *Proc. Natl. Acad. Sci. U. S. A.*, 2009, **106**, 6956–6961.
51. S. Schneider, Y. Pei, S. Shuman and B. Schwer, *Mol. Cell. Biol.*, 2010, **30**, 2353–2364.
52. T. Yamada, Y. Yamaguchi, N. Inukai, S. Okamoto, T. Mura and H. Handa, *Mol. Cell*, 2006, **21**, 227–237.
53. Y. Liu, L. Warfield, C. Zhang, J. Luo, J. Allen, W. H. Lang, J. Ranish, K. M. Shokat and S. Hahn, *Mol. Cell. Biol.*, 2009, **29**, 4852–4863.
54. Y. Pei and S. Shuman, *J. Biol. Chem.*, 2002, **277**, 19639–19648.
55. Y. Qiu and D. S. Gilmour, *J. Biol. Chem.*, 2017, **292**, 5555–5570.
56. Z. Pasman and P. H. von Hippel, *Biochemistry*, 2000, **39**, 5573–5585.
57. S. L. Sullivan and M. E. Gottesman, *Cell*, 1992, **68**, 989–994.
58. N. Said, F. Krupp, E. Anedchenko, K. F. Santos, O. Dybkov, Y. H. Huang, C. T. Lee, B. Loll, E. Behrmann, J. Burger, T. Mielke, J. Loerke, H. Urlaub, C. M. T. Spahn, G. Weber and M. C. Wahl, *Nat. Microbiol.*, 2017, **2**, 17062.
59. R. Kohler, R. A. Mooney, D. J. Mills, R. Landick and P. Cramer, *Science*, 2017, **356**, 194–197.
60. N. C. Kyrpides, C. R. Woese and C. A. Ouzounis, *Trends Biochem. Sci.*, 1996, **21**, 425–426.

61. K. Smollett, F. Blombach, R. Reichelt, M. Thomm and F. Werner, *Nat. Microbiol.*, 2017, **2**, 17021.
62. C. P. Ponting, *Nucleic Acids Res.*, 2002, **30**, 3643–3652.
63. R. Deuring, L. Fanti, J. A. Armstrong, M. Sarte, O. Papoulas, M. Prestel, G. Daubresse, M. Verardo, S. L. Moseley, M. Berloco, T. Tsukiyama, C. Wu, S. Pimpinelli and J. W. Tamkun, *Mol. Cell*, 2000, **5**, 355–365.
64. B. E. Nickels, *Methods*, 2009, **47**, 53–62.
65. G. A. Belogurov, M. N. Vassylyeva, V. Svetlov, S. Klyuyev, N. V. Grishin, D. G. Vassylyev and I. Artsimovitch, *Mol. Cell*, 2007, **26**, 117–129.
66. F. W. Martinez-Rucobo, S. Sainsbury, A. C. Cheung and P. Cramer, *EMBO J.*, 2011, **30**, 1302–1310.
67. B. J. Klein, D. Bose, K. J. Baker, Z. M. Yusoff, X. Zhang and K. S. Murakami, *Proc. Natl. Acad. Sci. U. S. A.*, 2011, **108**, 546–550.
68. G. Guo, Y. Gao, Z. Zhu, D. Zhao, Z. Liu, H. Zhou, L. Niu and M. Teng, *J. Struct. Biol.*, 2015, **192**, 418–425.
69. B. Liu and T. A. Steitz, *Nucleic Acids Res.*, 2017, **45**, 968–974.
70. J. Y. Kang, R. A. Mooney, Y. Nedialkov, J. Saba, T. V. Mishanina, I. Artsimovitch, R. Landick and S. A. Darst, *Cell*, 2018, **173**, 1650–1662. e1614.
71. M. L. Kireeva, N. Komissarova and M. Kashlev, *J. Mol. Biol.*, 2000, **299**, 325–335.
72. P. Huertas and A. Aguilera, *Mol. Cell*, 2003, **12**, 711–721.
73. J. Xu, I. Lahiri, W. Wang, A. Wier, M. A. Cianfrocco, J. Chong, A. A. Hare, P. B. Dervan, F. DiMaio, A. E. Leschziner and D. Wang, *Nature*, 2017, **551**, 653–657.
74. T. Steiner, J. T. Kaiser, S. Marinkovic, R. Huber and M. C. Wahl, *EMBO J.*, 2002, **21**, 4641–4653.
75. G. O. Bylund, B. C. Persson, L. A. Lundberg and P. M. Wikstrom, *J. Bacteriol.*, 1997, **179**, 4567–4574.
76. R. A. Mooney, K. Schweimer, P. Rosch, M. Gottesman and R. Landick, *J. Mol. Biol.*, 2009, **391**, 341–358.
77. P. A. Meyer, S. Li, M. Zhang, K. Yamada, Y. Takagi, G. A. Hartzog and J. Fu, *Mol. Cell. Biol.*, 2015, **35**, 3354–3369.
78. W. Li, C. Giles and S. Li, *Nucleic Acids Res.*, 2014, **42**, 7069–7083.
79. J. B. Crickard, J. Fu and J. C. Reese, *J. Biol. Chem.*, 2016, **291**, 9853–9870.
80. B. Cheng and D. H. Price, *Nucleic Acids Res.*, 2008, **36**, e135.
81. A. Missra and D. S. Gilmour, *Proc. Natl. Acad. Sci. U. S. A.*, 2010, **107**, 11301–11306.
82. S. Sekine, Y. Murayama, V. Svetlov, E. Nudler and S. Yokoyama, *Mol. Cell*, 2015, **57**, 408–421.
83. J. Y. Kang, T. V. Mishanina, M. J. Bellecourt, R. A. Mooney, S. A. Darst and R. Landick, *Mol. Cell*, 2018, **69**, 802–815.e801.
84. T. Kujirai, H. Ehara, Y. Fujino, M. Shirouzu, S. I. Sekine and H. Kurumizaka, *Science*, 2018, **362**, 595–598.
85. L. Farnung, S. M. Vos and P. Cramer, *Nat. Commun.*, 2018, **9**, 5432.
86. S. J. Petesch and J. T. Lis, *Trends Genet.*, 2012, **28**, 285–294.
87. B. Li, M. Carey and J. L. Workman, *Cell*, 2007, **128**, 707–719.

88. Y. Yamaguchi, J. Filipovska, K. Yano, A. Furuya, N. Inukai, T. Narita, T. Wada, S. Sugimoto, M. M. Konarska and H. Handa, *Science*, 2001, **293**, 124–127.
89. T. Narita, Y. Yamaguchi, K. Yano, S. Sugimoto, S. Chanarat, T. Wada, D. K. Kim, J. Hasegawa, M. Omori, N. Inukai, M. Endoh, T. Yamada and H. Handa, *Mol. Cell. Biol.*, 2003, **23**, 1863–1873.
90. L. J. Core, J. J. Waterfall, D. A. Gilchrist, D. C. Fargo, H. Kwak, K. Adelman and J. T. Lis, *Cell Rep.*, 2012, **2**, 1025–1035.
91. C. Lee, X. Li, A. Hechmer, M. Eisen, M. D. Biggin, B. J. Venters, C. Jiang, J. Li, B. F. Pugh and D. S. Gilmour, *Mol. Cell. Biol.*, 2008, **28**, 3290–3300.
92. S. M. Vos, D. Pollmann, L. Caizzi, K. B. Hofmann, P. Rombaut, T. Zimniak, F. Herzog and P. Cramer, *Elife*, 2016, **5**, e14981.
93. Y. Yamaguchi, N. Inukai, T. Narita, T. Wada and H. Handa, *Mol. Cell. Biol.*, 2002, **22**, 2918–2927.
94. K. Fujinaga, D. Irwin, Y. Huang, R. Taube, T. Kurosu and B. M. Peterlin, *Mol. Cell. Biol.*, 2004, **24**, 787–795.
95. X. Lu, X. Zhu, Y. Li, M. Liu, B. Yu, Y. Wang, M. Rao, H. Yang, K. Zhou, Y. Chen, M. Chen, S. Zhuang, L. F. Chen, R. Liu and R. Chen, *Nucleic Acids Res.*, 2016, **44**, 6853–6867.
96. H. Kwak and J. T. Lis, *Annu. Rev. Genet.*, 2013, **47**, 483–508.
97. F. X. Chen, E. R. Smith and A. Shilatifard, *Nat. Rev. Mol. Cell Biol.*, 2018, **19**, 464–478.
98. K. Schaukowitch, J. Y. Joo, X. Liu, J. K. Watts, C. Martinez and T. K. Kim, *Mol. Cell*, 2014, **56**, 29–42.
99. Y. L. Chiu, C. K. Ho, N. Saha, B. Schwer, S. Shuman and T. M. Rana, *Mol. Cell*, 2002, **10**, 585–597.
100. E. B. Rasmussen and J. T. Lis, *Proc. Natl. Acad. Sci. U. S. A.*, 1993, **90**, 7923–7927.
101. K. Adelman and J. T. Lis, *Nat. Rev. Genet.*, 2012, **13**, 720–731.
102. T. Narita, T. M. Yung, J. Yamamoto, Y. Tsuboi, H. Tanabe, K. Tanaka, Y. Yamaguchi and H. Handa, *Mol. Cell*, 2007, **26**, 349–365.
103. W. M. Schulze and S. Cusack, *Nat. Commun.*, 2017, **8**, 1302.
104. J. Yamamoto, Y. Hagiwara, K. Chiba, T. Isobe, T. Narita, H. Handa and Y. Yamaguchi, *Nat. Commun.*, 2014, **5**, 4263.
105. J. M. Pagano, H. Kwak, C. T. Waters, R. O. Sprouse, B. S. White, A. Ozer, K. Szeto, D. Shalloway, H. G. Craighead and J. T. Lis, *PLoS Genet.*, 2014, **10**, e1004090.
106. X. Guo, A. G. Myasnikov, J. Chen, C. Crucifix, G. Papai, M. Takacs, P. Schultz and A. Weixlbaumer, *Mol. Cell*, 2018, **69**, 816–827.e814.
107. S. Tagami, S. Sekine, T. Kumarevel, N. Hino, Y. Murayama, S. Kamegamori, M. Yamamoto, K. Sakamoto and S. Yokoyama, *Nature*, 2010, **468**, 978–982.
108. M. Palangat, D. B. Renner, D. H. Price and R. Landick, *Proc. Natl. Acad. Sci. U. S. A.*, 2005, **102**, 15036–15041.
109. M. S. Swanson and F. Winston, *Genetics*, 1992, **132**, 325–336.
110. C. D. Clark-Adams and F. Winston, *Mol. Cell. Biol.*, 1987, **7**, 679–686.

111. M. Endoh, W. Zhu, J. Hasegawa, H. Watanabe, D. K. Kim, M. Aida, N. Inukai, T. Narita, T. Yamada, A. Furuya, H. Sato, Y. Yamaguchi, S. S. Mandal, D. Reinberg, T. Wada and H. Handa, *Mol. Cell. Biol.*, 2004, **24**, 3324–3336.
112. E. D. Andrulis, E. Guzman, P. Doring, J. Werner and J. T. Lis, *Genes Dev.*, 2000, **14**, 2635–2649.
113. C. D. Kaplan, J. R. Morris, C. Wu and F. Winston, *Genes Dev.*, 2000, **14**, 2623–2634.
114. C. D. Kaplan, L. Laprade and F. Winston, *Science*, 2003, **301**, 1096–1099.
115. A. Bortvin and F. Winston, *Science*, 1996, **272**, 1473–1476.
116. D. Close, S. J. Johnson, M. A. Sdano, S. M. McDonald, H. Robinson, T. Formosa and C. P. Hill, *J. Mol. Biol.*, 2011, **408**, 697–713.
117. C. D. Kaplan, M. J. Holland and F. Winston, *J. Biol. Chem.*, 2005, **280**, 913–922.
118. C. Warren and D. Shechter, *J. Mol. Biol.*, 2017, **429**, 2401–2426.
119. S. M. McDonald, D. Close, H. Xin, T. Formosa and C. P. Hill, *Mol. Cell*, 2010, **40**, 725–735.
120. M. L. Diebold, M. Koch, E. Loeliger, V. Cura, F. Winston, J. Cavarelli and C. Romier, *EMBO J.*, 2010, **29**, 3979–3991.
121. S. Li, A. R. Almeida, C. A. Radebaugh, L. Zhang, X. Chen, L. Huang, A. K. Thurston, A. A. Kalashnikova, J. C. Hansen, K. Luger and L. A. Stargell, *Nucleic Acids Res.*, 2018, **46**, 2321–2334.
122. S. J. Johnson, D. Close, H. Robinson, I. Vallet-Gely, S. L. Dove and C. P. Hill, *J. Mol. Biol.*, 2008, **377**, 1460–1473.
123. M. L. Diebold, E. Loeliger, M. Koch, F. Winston, J. Cavarelli and C. Romier, *J. Biol. Chem.*, 2010, **285**, 38389–38398.
124. M. Sun, L. Lariviere, S. Dengl, A. Mayer and P. Cramer, *J. Biol. Chem.*, 2010, **285**, 41597–41603.
125. L. McCullough, Z. Connell, C. Petersen and T. Formosa, *Genetics*, 2015, **201**, 1031–1045.
126. S. M. Yoh, H. Cho, L. Pickle, R. M. Evans and K. A. Jones, *Genes Dev.*, 2007, **21**, 160–174.
127. S. M. Yoh, J. S. Lucas and K. A. Jones, *Genes Dev.*, 2008, **22**, 3422–3434.
128. M. A. Sdano, J. M. Fulcher, S. Palani, M. B. Chandrasekharan, T. J. Parnell, F. G. Whitby, T. Formosa and C. P. Hill, *Elife*, 2017, **6**, e28723.
129. R. Dronamraju, A. J. Hepperla, Y. Shibata, A. T. Adams, T. Magnuson, I. J. Davis and B. D. Strahl, *Mol. Cell*, 2018, **70**, 1054–1066.e1054.
130. S. M. Orlicky, P. T. Tran, M. H. Sayre and A. M. Edwards, *J. Biol. Chem.*, 2001, **276**, 10097–10102.
131. H. Meka, F. Werner, S. C. Cordell, S. Onesti and P. Brick, *Nucleic Acids Res.*, 2005, **33**, 6435–6444.
132. A. Ujvari and D. S. Luse, *Nat. Struct. Mol. Biol.*, 2006, **13**, 49–54.
133. E. D. Andrulis, J. Werner, A. Nazarian, H. Erdjument-Bromage, P. Tempst and J. T. Lis, *Nature*, 2002, **420**, 837–841.
134. X. Shi, M. Chang, A. J. Wolf, C. H. Chang, A. A. Frazer-Abel, P. A. Wade, Z. F. Burton and J. A. Jaehning, *Mol. Cell. Biol.*, 1997, **17**, 1160–1169.

135. C. L. Mueller and J. A. Jaehning, *Mol. Cell. Biol.*, 2002, **22**, 1971–1980.
136. B. Zhu, S. S. Mandal, A. D. Pham, Y. Zheng, H. Erdjument-Bromage, S. K. Batra, P. Tempst and D. Reinberg, *Genes Dev.*, 2005, **19**, 1668–1673.
137. O. Rozenblatt-Rosen, C. M. Hughes, S. J. Nannepaga, K. S. Shanmugam, T. D. Copeland, T. Guszczynski, J. H. Resau and M. Meyerson, *Mol. Cell. Biol.*, 2005, **25**, 612–620.
138. J. Kim, M. Guermah and R. G. Roeder, *Cell*, 2010, **140**, 491–503.
139. J. L. Dermody and S. Buratowski, *J. Biol. Chem.*, 2010, **285**, 33671–33679.
140. M. Yu, W. Yang, T. Ni, Z. Tang, T. Nakadai, J. Zhu and R. G. Roeder, *Science*, 2015, **350**, 1383–1386.
141. H. Qiu, C. Hu, N. A. Gaur and A. G. Hinnebusch, *EMBO J.*, 2012, **31**, 3494–3505.
142. F. X. Chen, A. R. Woodfin, A. Gardini, R. A. Rickels, S. A. Marshall, E. R. Smith, R. Shiekhattar and A. Shilatifard, *Cell*, 2015, **162**, 1003–1015.
143. L. Sadeghi, P. Prasad, K. Ekwall, A. Cohen and J. P. Svensson, *EMBO Rep.*, 2015, **16**, 1673–1687.
144. S. Hanks, E. R. Perdeaux, S. Seal, E. Ruark, S. S. Mahamdallie, A. Murray, E. Ramsay, S. Del Vecchio Duarte, A. Zachariou, B. de Souza, M. Warren-Perry, A. Elliott, A. Davidson, H. Price, C. Stiller, K. Pritchard-Jones and N. Rahman, *Nat. Commun.*, 2014, **5**, 4398.
145. D. K. Pokholok, N. M. Hannett and R. A. Young, *Mol. Cell*, 2002, **9**, 799–809.
146. S. B. Van Oss, M. K. Shirra, A. R. Bataille, A. D. Wier, K. Yen, V. Vinayachandran, I. L. Byeon, C. E. Cucinotta, A. Heroux, J. Jeon, J. Kim, A. P. VanDemark, B. F. Pugh and K. M. Arndt, *Mol. Cell*, 2016, **64**, 815–825.
147. Y. Yang, W. Li, M. Hoque, L. Hou, S. Shen, B. Tian and B. D. Dynlacht, *PLoS Genet.*, 2016, **12**, e1005794.
148. A. G. Rondon, M. Gallardo, M. Garcia-Rubio and A. Aguilera, *EMBO Rep.*, 2004, **5**, 47–53.
149. Y. Chen, Y. Yamaguchi, Y. Tsugeno, J. Yamamoto, T. Yamada, M. Nakamura, K. Hisatake and H. Handa, *Genes Dev.*, 2009, **23**, 2765–2777.
150. H. H. Ng, S. Dole and K. Struhl, *J. Biol. Chem.*, 2003, **278**, 33625–33628.
151. A. Wood, J. Schneider, J. Dover, M. Johnston and A. Shilatifard, *J. Biol. Chem.*, 2003, **278**, 34739–34742.
152. N. J. Krogan, J. Dover, A. Wood, J. Schneider, J. Heidt, M. A. Boateng, K. Dean, O. W. Ryan, A. Golshani, M. Johnston, J. F. Greenblatt and A. Shilatifard, *Mol. Cell*, 2003, **11**, 721–729.
153. B. Fierz, C. Chatterjee, R. K. McGinty, M. Bar-Dagan, D. P. Raleigh and T. W. Muir, *Nat. Chem. Biol.*, 2011, **7**, 113–119.
154. K. Batta, Z. Zhang, K. Yen, D. B. Goffman and B. F. Pugh, *Genes Dev.*, 2011, **25**, 2254–2265.
155. L. Wu, B. M. Zee, Y. Wang, B. A. Garcia and Y. Dou, *Mol. Cell*, 2011, **43**, 132–144.
156. L. Wu, L. Li, B. Zhou, Z. Qin and Y. Dou, *Mol. Cell*, 2014, **54**, 920–931.
157. A. Wood, J. Schneider, J. Dover, M. Johnston and A. Shilatifard, *Mol. Cell*, 2005, **20**, 589–599.
158. R. N. Laribee, N. J. Krogan, T. Xiao, Y. Shibata, T. R. Hughes, J. F. Greenblatt and B. D. Strahl, *Curr. Biol.*, 2005, **15**, 1487–1493.

159. H. Qiu, C. Hu, C. M. Wong and A. G. Hinnebusch, *Mol. Cell. Biol.*, 2006, **26**, 3135–3148.
160. L. Verrier, F. Taglini, R. R. Barrales, S. Webb, T. Urano, S. Braun and E. H. Bayne, *Open Biol.*, 2015, **5**, 150045.
161. Y. Chu, R. Simic, M. H. Warner, K. M. Arndt and G. Prelich, *EMBO J.*, 2007, **26**, 4646–4656.
162. S. Venkatesh and J. L. Workman, *Wiley Interdiscip. Rev.: Dev. Biol.*, 2013, **2**, 685–700.
163. N. He, C. K. Chan, B. Sobhian, S. Chou, Y. Xue, M. Liu, T. Alber, M. Benkirane and Q. Zhou, *Proc. Natl. Acad. Sci. U. S. A.*, 2011, **108**, E636–E645.
164. S. B. Van Oss, C. E. Cucinotta and K. M. Arndt, *Trends Biochem. Sci.*, 2017, **42**, 788–798.
165. B. N. Tomson and K. M. Arndt, *Biochim. Biophys. Acta*, 2013, **1829**, 116–126.
166. T. Nagaike, C. Logan, I. Hotta, O. Rozenblatt-Rosen, M. Meyerson and J. L. Manley, *Mol. Cell*, 2011, **41**, 409–418.
167. K. Nordick, M. G. Hoffman, J. L. Betz and J. A. Jaehning, *Eukaryotic Cell*, 2008, **7**, 1158–1167.
168. O. Rozenblatt-Rosen, T. Nagaike, J. M. Francis, S. Kaneko, K. A. Glatt, C. M. Hughes, T. LaFramboise, J. L. Manley and M. Meyerson, *Proc. Natl. Acad. Sci. U. S. A.*, 2009, **106**, 755–760.
169. H. Fischl, F. S. Howe, A. Furger and J. Mellor, *Mol. Cell*, 2017, **65**, 685–698.e688.
170. Y. Xu, C. Bernecky, C. T. Lee, K. C. Maier, B. Schwalb, D. Tegunov, J. M. Plitzko, H. Urlaub and P. Cramer, *Nat. Commun.*, 2017, **8**, 15741.
171. P. Deng, Y. Zhou, J. Jiang, H. Li, W. Tian, Y. Cao, Y. Qin, J. Kim, R. G. Roeder, D. J. Patel and Z. Wang, *Proc. Natl. Acad. Sci. U. S. A.*, 2018, **115**, 9998–10003.
172. Y. Xie, M. Zheng, X. Chu, Y. Chen, H. Xu, J. Wang, H. Zhou and J. Long, *Nat. Commun.*, 2018, **9**, 3795.
173. C. L. Mueller, S. E. Porter, M. G. Hoffman and J. A. Jaehning, *Mol. Cell*, 2004, **14**, 447–456.
174. J. D. Carpten, C. M. Robbins, A. Villablanca, L. Forsberg, S. Presciuttini, J. Bailey-Wilson, W. F. Simonds, E. M. Gillanders, A. M. Kennedy, J. D. Chen, S. K. Agarwal, R. Sood, M. P. Jones, T. Y. Moses, C. Haven, D. Petillo, P. D. Leotlela, B. Harding, D. Cameron, A. A. Pannett, A. Hoog, H. Heath 3rd, L. A. James-Newton, B. Robinson, R. J. Zarbo, B. M. Cavaco, W. Wassif, N. D. Perrier, I. B. Rosen, U. Kristoffersson, P. D. Turnpenny, L. O. Farnebo, G. M. Besser, C. E. Jackson, H. Morreau, J. M. Trent, R. V. Thakker, S. J. Marx, B. T. Teh, C. Larsson and M. R. Hobbs, *Nat. Genet.*, 2002, **32**, 676–680.
175. H. Chen, N. Shi, Y. Gao, X. Li, M. Teng and L. Niu, *Acta Crystallogr., Sect. D: Biol. Crystallogr.*, 2012, **68**, 953–959.
176. C. G. Amrich, C. P. Davis, W. P. Rogal, M. K. Shirra, A. Heroux, R. G. Gardner, K. M. Arndt and A. P. VanDemark, *J. Biol. Chem.*, 2012, **287**, 10863–10875.

177. W. Sun, X. L. Kuang, Y. P. Liu, L. F. Tian, X. X. Yan and W. Xu, *Sci. Rep.*, 2017, 7, 15638.
178. X. Chu, X. Qin, H. Xu, L. Li, Z. Wang, F. Li, X. Xie, H. Zhou, Y. Shen and J. Long, *Nucleic Acids Res.*, 2013, **41**, 10619–10629.
179. R. N. de Jong, V. Truffault, T. Diercks, E. Ab, M. A. Daniels, R. Kaptein and G. E. Folkers, *Structure*, 2008, **16**, 149–159.
180. J. P. Hsin and J. L. Manley, *Genes Dev.*, 2012, **26**, 2119–2137.
181. S. Hocine, R. H. Singer and D. Grunwald, *Cold Spring Harbor Perspect. Biol.*, 2010, **2**, a000752.
182. P. C. Hanawalt and G. Spivak, *Nat. Rev. Mol. Cell Biol.*, 2008, **9**, 958–970.
183. L. H. Gregersen and J. Q. Svejstrup, *Trends Biochem. Sci.*, 2018, **43**, 327–341.

CHAPTER 5

Rho Termination Factor: One Ring to Bind Them All

NICHOLAS D. SUNDAY[a], DMITRI SVETLOV[a] AND IRINA ARTSIMOVITCH*[a]

[a]Department of Microbiology and the Center for RNA Biology, The Ohio State University, USA
*E-mail: artsimovitch.1@osu.edu

5.1 Introduction

In the classical model of transcription, RNA polymerase (RNAP) initiates at a promoter and stops at a terminator, making RNAs that are useful for the cell under current physiological conditions. However, real cellular transcription occurs differently. RNAP can initiate at many spurious sites, frequently in an antisense direction relative to the coding regions, making nonfunctional RNAs that constitute pervasive transcriptional noise and may, at times, be inhibitory. Some coding RNAs may fail to be translated, due to mutations or under amino-acid stress. RNAP itself can present a problem for genome stability if it stalls during transcription, blocking replisome progression and triggering DNA breaks. Regulators that support accurate gene expression while reducing unavoidable side effects of transcription are necessary to maintain cellular homeostasis and to enable adaptation to changing environments. The bacterial Rho protein, a ring-shaped RecA-family hexameric helicase best known for its transcription termination activity, doubles as a genome sentinel that can alleviate widespread collateral damage during transcription.

Chemical Biology No. 19
RNA Polymerases as Molecular Motors: On the Road, 2nd Edition
Edited by Robert Landick, Terence Strick and Jue Wang

Published by the Royal Society of Chemistry, www.rsc.org

Escherichia coli Rho was discovered as a factor mediating termination at the end of the bacteriophage λ *cro* repressor gene.[1] Rho also terminates transcription of the *E. coli trp* operon,[2] induces premature release of untranslated RNAs (a phenomenon known as transcriptional polarity[3]), and regulates attenuation in the leader regions of *rho*[4] and the *tna* operon.[5] Extensive biochemical, structural, and genetic studies led to a textbook model[6] in which Rho binds to an unstructured C-rich Rho utilization (*rut*) site on a nascent RNA and uses energy from adenosine triphosphate (ATP) hydrolysis to translocate towards transcribing RNAP, unwind the RNA–DNA hybrid, and triggering the dissociation of the transcription elongation complex (EC). Early studies of Rho focused on a few natural and synthetic substrates, such λ *cro* tR1 and *E. coli trp* t' terminators or polyC RNA, frequently using minimal *in vitro* systems from which RNAP was missing and Rho helicase and ATPase activities were assayed instead. These painstaking experiments revealed that Rho is able to work on surprisingly diverse RNA substrates, including those that are not enriched in C residues;[7,8] uses not only ATP but other NTPs;[9] can unwind over 100-nt-long annealed duplexes;[10] and is able to disrupt a biotin-streptavidin complex.[11]

Now, the focus is shifting to cellular roles of Rho, owing to the deployment of next-generation sequencing approaches. Genome-wide studies revealed that Rho modulates expression of numerous genes and noncoding RNAs in *E. coli*[12] and *Mycobacterium tuberculosis*,[13] in which it is essential, as well as in *Bacillus subtilis*[14] and *Staphylococcus aureus*,[15] in which it is dispensable. A picture that emerges is that Rho can work on almost any RNA *in vivo* as well, particularly when aided by N-utilizing substance G (NusG),[12,16] and that its functions extend beyond determining gene boundaries and sensing translational status. Rho terminates transcription in many leader regions, including riboswitches;[17–20] silences expression of horizontally-acquired genes, clustered regularly interspaced short palindromic repeats (CRISPR) loci, and antisense RNAs;[12,20–23] prevents unwarranted excision of prophages;[24] removes stalled RNAP from the path of the replisome to maintain chromosome integrity;[25,26] inhibits R-loop formation;[27] and has profound effects on cellular adaptation to changing conditions.[28] Numerous studies of Rho over the five decades since its discovery have been recently reviewed.[6,29–32] In this chapter, we will focus primarily on key recent findings and outstanding questions.

5.2 One Ring to Rule Them All: Structure and Dynamics of Rho Alone

5.2.1 Functional Regions of Rho

Pioneering cryo-electron microscopy (cryo-EM) imaging indicated that hexameric Rho exists in two states: an open, RNA-loading state and a closed, translocation-competent state.[33] Together with contemporary studies of *rho* mutants with altered properties and different RNA substrates, this finding led to a step-wise model for Rho-dependent termination: (1) primary RNA-binding site (PBS) residues of an open ring interact with the *rut* element; (2) once the PBS is occupied, the growing RNA is directed into the central channel and makes contacts with the secondary RNA-binding site (SBS) residues; (3) in the presence of ATP, the Rho ring closes, trapping the RNA inside the hexamer; (4) powered by ATP hydrolysis, the closed hexamer translocates RNA while maintaining contacts with the *rut* element; (5) Rho catches up to RNAP while the latter is within a well-defined termination window downstream from the *rut* element; (6) upon contacting RNAP, Rho triggers EC dissociation; and (7) the ring opens to release the RNA and reset the cycle.[34]

Since the turn of this century, a series of structures of *E. coli* Rho obtained by Berger and colleagues[35–40] provided a framework for deciphering the fine details of the termination mechanism. The 419-residue Rho monomer contains two domains separated by an approximately 30-residue flexible connector region[41] (Figure 5.1). The N-terminal domain (NTD) is responsible for initial recognition of the *rut* RNA. The RecA-type ATP- binding C-terminal domain (CTD) contains Walker A (P loop) and Walker B motifs and several catalytic Glu and Arg residues, which are responsible for ATP binding and hydrolysis, and the SBS. The C-terminus interacts with NusG.[36]

The PBS (residues 22–116) recognizes dipyrimidine motifs within the *rut* site.[42] In the structure of Rho complexed with $r(CU)_{15}$ and the ATP analogue adenylyl-imidodiphosphate (AMP-PNP),[37] the first cytidine base fits into a hydrophobic pocket formed by Phe62 and Tyr80 and makes additional van der Waals contacts to Glu108, Arg109, and Tyr110. The second base stacks on the aromatic side chain of Phe64 and forms hydrogen bonds with Arg66 and Asp78. No interactions with the 2′-OH group are observed, explaining why the PBS binds to single-stranded DNA,[43] a property used to distinguish the effects of PBS and SBS ligands. A second set of contacts with Arg88 and Phe89 was revealed by NMR analysis of the NTD bound to a hexamer DNA.[44] The phenotypes of R88E and F89S variants indicated that, in addition to mild defects in RNA binding, these substitutions confer defects at a subsequent step in Rho activation, such as ring closure.[44] In addition, residues in the helical bundle have been proposed to facilitate RNA binding to the PBS.[45] Some Rho proteins contain large insertions in the NTD[41] that may modulate interactions with RNA[46] or confer prion-like properties.[47]

The connector region linking the NTD and the CTD is present in diverse Rho proteins[41] and has been identified as a binding site for the polarity suppression (Psu) protein, a phage-encoded Rho antagonist.[48] The connector's length and predicted flexibility are more conserved than its primary sequence, indicating that this region works as a mobile tether that enables rearrangements of the NTD and CTD, for example, those induced upon RNA–PBS binding.[8] In support of this idea, substitutions in the connector expected to rigidify it (G150D and G152D) have been isolated in a screen for *rho* variants defective in termination at *rut*-less regions.[49] The results of comparison of different Rho structures indicate that the connector ends, residues 126–129 and 147–152, become ordered in the closed hexamer, and an adjacent

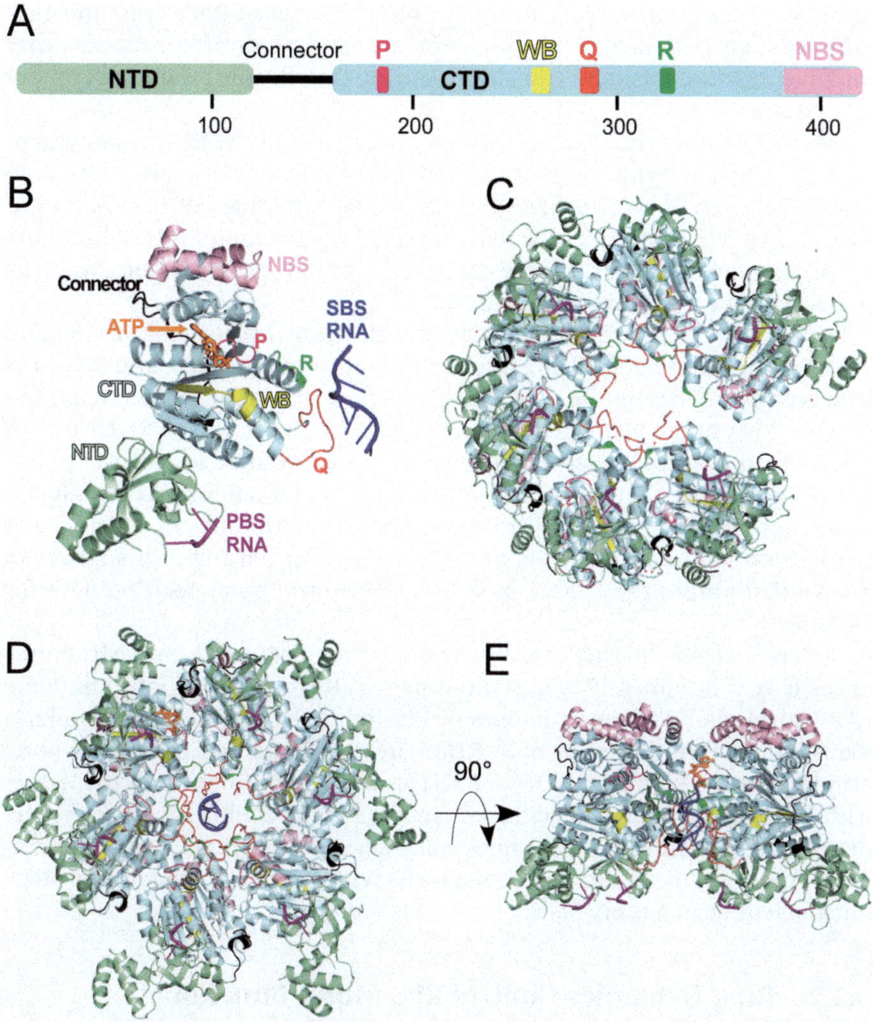

Figure 5.1 Structure of *E. coli* Rho. (A) Schematic representation of Rho with domain boundaries and key functional regions. The NTD, connector, and CTD are light green, black, and light blue respectively. Key regions are shown as colored boxes: NusG binding site (NBS, residues 391–418), light blue; P loop (P, 179–183), magenta; Q loop (Q, 278–290), Red; R loop (R, 322–326), lime green; Walker B motif (WB, 261–270), yellow. The color scheme is used in panels B–E. (B) A single protomer (subunit C, PBD ID 5JJI), with the bound ATP (represented by ADP-BeF_3 with the BeF_3 ion removed; orange) and the SBS rU7 RNA (dark blue). The PBS RNA (PBS RNA, purple) was absent from the structure and was modeled from PDB ID 1PV4. (C) An open Rho hexamer (flattened on the page) in complex with the PBS RNA (PBD ID 1PV4) represented by two cytosine residues. (D) The closed Rho hexamer (flattened on the page) bound to the SBS RNA (PBD ID 5JJI) and PBS RNA (from PDB ID 1PV4 as above). (E) A side view of the hexamer shown in D; subunits A and B were removed and the structure was rotated 90° out of the page.

α_6 helix is rotated in some[44] but not in other[35,39] closed Rho structures that differ in their RNA ligands. A possibility that the α_6-connector module may undergo ligand-dependent transitions during ring closure awaits experimental assessment.

The SBS contacts the RNA in the central channel of the Rho hexamer and is responsible for the translocase and helicase activities. Two conserved loops, Q and R, comprise the SBS.[41] In crystal structures of Rho bound to RNA and an ATP mimic,[39,40] the RNA-binding loops of protomers A through E are arranged in a spiral staircase, trapping RNA in the central pore, whereas protomer F makes no contacts with RNA.[40]

Each Q loop makes three contacts with RNA; the hydroxyl of Thr286 and the backbone amide of Gly287 interact with RNA phosphates, and Val284 is hydrogen bonded with the 2′-OH of RNA ribose, explaining a strong preference for RNA in the SBS. Each R loop makes a single contact to RNA through a conserved Lys326. Lys326 of protomers A–D are pointed toward the center of the ring, where they contact RNA phosphates and neutralize their charge in the densely packed pore.[39] However, Lys326 of protomer E, which lies at the bottom of the staircase, is pointed away from the ring. This creates a groove that guides the 3′ RNA base into the central pore, positioning it for capture.[50]

No base-specific interactions by the SBS were observed in early structures of the closed hexamer,[38,39] yet pyrimidine-rich RNAs preferentially stimulate ATP hydrolysis.[43,51] New structures of Rho bound to rA_{12} and rU_{12}[40] offer a plausible answer: although both RNAs are tightly packed inside the pore, only U_{12} forms a kink that allows a 2′ OH and two pyrimidine bases to interact with Lys326. Pyrimidine-specific contacts to Lys326 could explain sequence-dependent patterns of seven-nucleotide-spaced deoxynucleotide interference[52] that have been difficult to reconcile with a model in which Rho steps one nucleotide at a time.[39]

5.2.2 Ring Dynamics Control Rho Motor Function

Results from numerous *in vitro* studies have revealed that Rho is a self-loading helicase able to assume a translocation-competent state in the absence of accessory factors.[33,53] However, sequence preferences displayed by both the SBS and PBS indicated that the ring opening may be controlled by the RNA substrate. Structural studies captured Rho rings in both open and closed states, with C-rich RNA favoring the latter.[33,37,38,40] Biochemical analyses revealed RNA- and ATP-dependent conformational changes in Rho,[9,54] presumed to correspond to ring closure and proposed to be rate-limiting for ATPase activation.[50] What roles do RNA and ATP play in ring closure, and what states does Rho assume in solution? To answer these questions, Berger and colleagues used small-angle X-ray scattering (SAXS) and fluorescence anisotropy to monitor ring transitions. Their results indicate that Rho adopts an open state in the absence of ligands and undergoes isomerization only in the presence of both ATP and the SBS ligand.[40] Consistent with biochemical data,[43] the ring closes more readily around rU_{12} RNA than rA_{12} RNA.[35] However, saturation of the PBS with dC oligomers allowed ring closure around a suboptimal A_{12} RNA and further promoted closure around the U_{12} RNA.[35] These results demonstrate that the PBS and SBS cooperate to ensure that the ring closes around diverse sequences found in natural Rho substrates.

Recent findings implicate ring closure as a key target for regulation. Whereas preferred RNA ligands, high ATP concentration, and NusG promote ring closure,[35,36] bicyclomycin (BCM), a non-competitive inhibitor of Rho, inhibits it.[35] BCM, the only known small-molecule inhibitor of Rho, is widely used to study cellular functions of Rho.[25,55] Since the BCM-binding pocket collapses upon ring closure, the bound BCM locks the ring in an open state; consistently, BCM fails to act if added to a pre-closed ring.[35]

5.2.3 The Mechanism of Translocation

After the ring closes, Rho translocates along the RNA, hydrolyzing one to two ATP molecules per one traversed nucleotide.[56] A "tethered-tracking" model in which Rho maintains contacts with the *rut* element throughout translocation[57] is supported by the results of single-molecule trapping experiments.[58] How does Rho convert energy into directional motion? The results of functional studies indicated that Rho binds only three molecules of ATP at a time,[59] reminiscent of F_1 ATPase, the closest Rho homolog,[41] which fires its ATPase active sites sequentially around the hexamer.[60] The first model accounting for this behavior posited that the closed ring is organized as a trimer of asymmetric dimers,[61] with three strong and three weak ATP- binding sites per hexamer.[62] In this model, the functional unit is a dimer (Figure 5.2A; left), and ATP hydrolysis triggers a conformational change in which weak and strong sites interchange, causing repositioning of the RNA, with the directionality imparted by differences between the successive dimers within the hexamer.[61] This model received support from the first (3.5 Å resolution) crystal structure of closed Rho hexamer bound to the $r(CU)_{15}$ ligand.[38] In this three-fold symmetry structure, the Q loops adopted different positions in adjacent protomers in each dimer: in protomer A, the Q loop Lys283 extended into the ring pore, constricting it, whereas in protomer B the Q loop was lying flat. Reflecting this difference, weak densities for three RNA pentamers were observed at A–B interfaces. These observations led to a "hand-off" model, in which conformational changes induced by ATP hydrolysis are coupled to RNA transfer from one subunit interface to the next, with directional Q-loop movements pulling RNA through the pore.[38]

However, subsequent atomic-resolution structures of Rho bound to oligo-U ligands contradict the hand-off model.[35,39] In these complexes, the Rho hexamer was completely asymmetric (Figure 5.2A; right), with four distinct ATPase active-site architectures and different RNA contacts with the SBS residues observed in individual protomers; notably, RNA–SBS interactions were very distinct from those inferred from the earlier structure. These structures indicated an "escort" model of translocation in which, upon

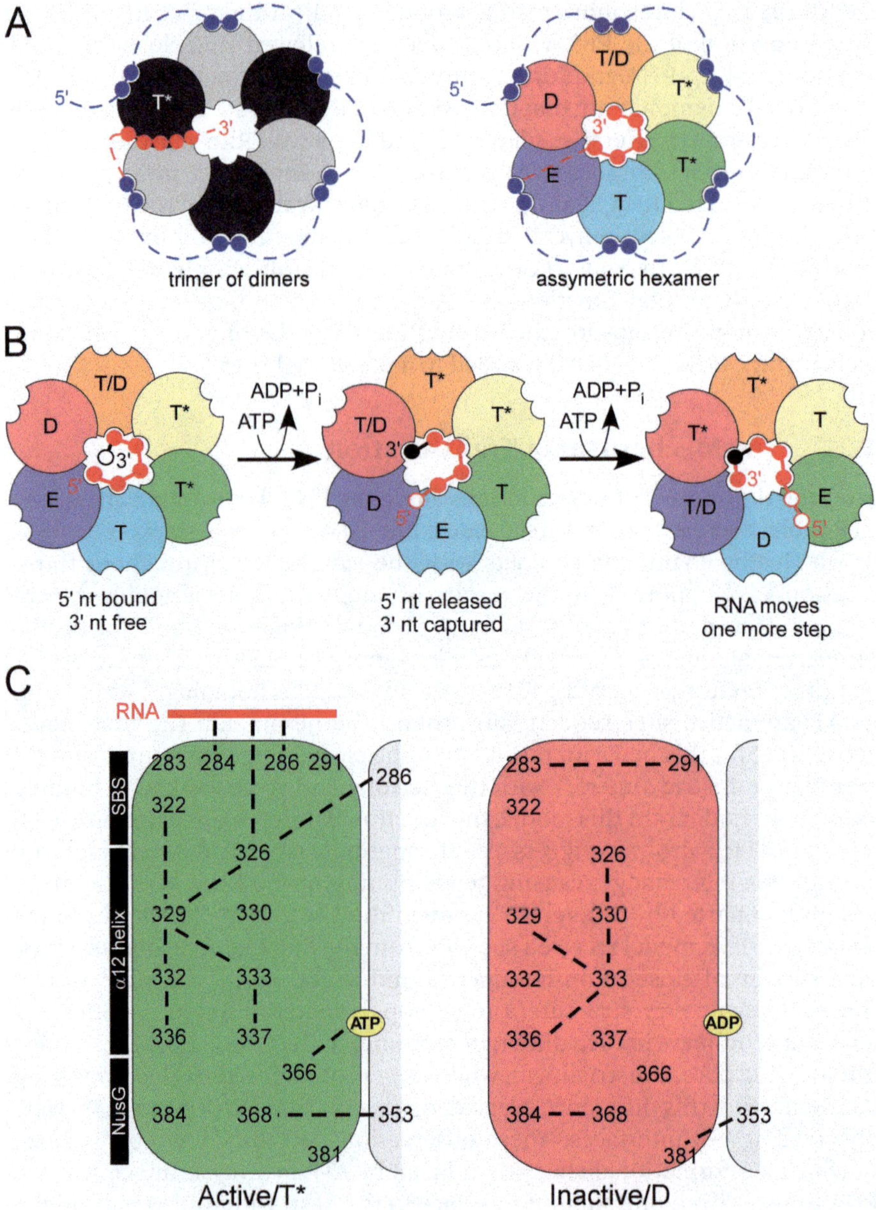

Figure 5.2 Rho–RNA interactions and the mechanism for RNA translocation. (A) Two closed ring architectures, a trimer-of-dimers (left; PDB ID 2HTI) and an asymmetric hexamer (right; PDB ID 5JJI) differ in their interactions with the SBS RNA and the proposed mechanism of translocation. The SBS RNA bound in the central pore is shown as red circles; the di-pyrimidine RNA motifs that interact with the PBS, but are absent in the closed-ring structures, are shown as blue circles. (Left) The two protomers of a repeating dimer are colored gray and black, and the

each round of ATP hydrolysis, a rotary shift of the hexamer leads to a one-nucleotide translocation along the RNA (Figure 5.2B). The results of a systematic mutational analysis of the Rho residues involved in RNA contacts in both structures indicate that the asymmetric hexamer structure, and the escort mechanism of translocation based on it, explains most experimental observations, whereas the trimer-of-dimers structure does not.[8]

5.2.4 Allosteric Networks within Rho

The majority of *rho* mutants' phenotypes can be readily explained by their position in the Rho structure. However, notable exceptions are known. Several substitutions in the NTD, G51V, G99V[63] and Y80C,[64] alter RNA binding to the SBS. Conversely, F232C[54] and T343I[65] substitutions in the CTD reduce binding to the PBS but enable Rho to utilize suboptimal sequences. A large number of relaxed specificity mutants have been characterized, including substitutions of residues in the NTD that make no contacts to RNA, L3F[65] and P103L;[66] "super Rho" variants E329C and H344R in the CTD;[67] and NusG-mimicking I168V and P235H.[16] These effects can be explained by allosteric communications among distant regions of Rho; three allosteric networks have been proposed.

First, genetic interactions between the PBS and SBS[54,63–65] and the effects of PBS ligands on ring closure and catalysis[35,43] imply allosteric links among the Rho domains.[8] The molecular details of these interactions are as yet unclear, in part because a complete path of RNA around the NTD remains unknown.

Second, sequence-dependent stimulation of ATPase and translocase activities by SBS ligands[43,52] and effects of SBS substitutions on catalysis[68]

SBS RNA is bound at the subunit interface; T* indicates the ATPase site primed to carry out hydrolysis; catalytic status of other ATP-binding sites cannot be discerned from the structure. In a hand-off model, conformational changes in the Q loops triggered by ATP hydrolysis push the RNA from one dimer interface to the next.[38] (Right) Four ATPase site states are observed in the asymmetric ring: T, bound ATP; D, product; E, nucleotide exchange; and T*, active hydrolysis. The subunits are colored from orange (A) to red (F) according to 5JJI; the SBS RNA is bound to protomers A–E in the central pore. (B) A proposed sequential RNA escort mechanism in which each round of ATP hydrolysis and product release drives a rotary change around the asymmetric hexamer accompanied by the release of the 5′-most SBS RNA nt and the capture of a new 3′ nt, a one nucleotide net translocation.[39] The filled circles indicate nucleotides bound to the SBS, the open circles indicate residues not making contacts with the protein; the 3′ unbound nucleotide in the initial complex is colored black to establish the register. (C) Examples of two interaction networks that differ in the active (T*, green, corresponding to the C chain in 5JJI) and inactive (red; chain F) protomers; this representation was inspired by Rabhi *et al.*[8] Movements of the α_{12} helix allosterically link RNA binding in the SBS to ATPase activation.[40] NusG binds to Trp381 to break autoinhibitory interactions that prevent ring closure.[36]

are indicative of communications between the SBS and the active site. This link is mediated by an α_{12}-helix that starts at Lys326 and runs towards the P loop, making contacts with the catalytic water and Arg residues.[40] The α_{12}-helix assumes different positions that correlate with the state of the ATPase center in each protomer; the ideal rU_{12} ligand makes contacts with Lys326 in B and C protomers, which adopt the active T* state (Figure 5.2B). This indicates that changes in Lys236–RNA contacts can be directly communicated to the active site, with preferred sequences promoting a conformation ideal for ATPase activity. The α_{12}-helix is present in other RecA-type motors and may play a ubiquitous role in coupling substrate recognition to catalysis.

Third, binding of NusG to the C-terminus of Rho[36] breaks *autoinhibitory interactions*, *e.g.* between Trp381 and Arg353 residues in the adjacent protomers, that correlate with an inactive Rho state (Figure 5.2C). NusG-mediated activation alleviates Rho dependence on PBS ligands, allowing the ring to close in the absence of an ideal RNA ligand, *via* a mechanism that is discussed in the following sections.

5.3 The Fellowship of the Ring: Cellular Partners of Rho

Drawing physiologically relevant conclusions from studies of Rho requires accounting for the fact that *in vivo*, Rho acts on RNAP that is bound to other factors, which, in turn, also modulate termination.

Functional interactions between Rho and Nus (N-utilization substance) factors are the best studied. The *nus* mutations were mapped to *nusA*, *nusB*, *nusC*, *nusD*, *nusE*, and *nusG* genes. The *nusC* gene (*rpoB*) encodes the β subunit of RNAP, and *nusD* encodes Rho. NusE is the ribosomal protein S10; when bound to NusB instead of the ribosome, NusE moonlights as an integral component of λ and rRNA antitermination complexes (ATCs), which also include the abundant general transcription factors NusA and NusG.[69,70] NusA and NusG are not known to be sequestered in any complexes lacking RNAP; thus most *E. coli* ECs are likely to be bound to both factors.[71] *E. coli* NusA and NusG have complex effects on Rho termination, both positive and negative, as discussed in the following sections. Furthermore, genomic DNA is often bound by nucleoid associated proteins (NAPs), and recent evidence indicates that NAPs cooperate with Rho to silence the expression of xenogenes.[12,72,73]

5.3.1 NusG Enables Rho to Act on Suboptimal RNA Sequences

NusG is a universally conserved protein that can either activate or inhibit Rho in different contexts. The results of early studies indicated that *E. coli* NusG was essential for cell viability and Rho function at a subset of terminators[74] and that the termination-promoting activity of NusG was mediated by direct interactions with Rho.[75] In contrast, when bound to S10, as part of the ATC or the ribosome,[70,76–78] NusG cannot interact with Rho and indirectly contributes to restricting Rho activity. The NusG-NTD binds to RNAP to facilitate elongation, an effect that may contribute to antitermination.[79] The CTD interacts with S10[80] to support antitermination or couple transcription to translation, or with Rho.[81] However, NusG does not affect RNA binding, ATP hydrolysis, or translocation,[82] leaving open the question of how it influences Rho function.

Genome-wide analyses of Rho terminators in *E. coli*, which were identified using BCM to inhibit Rho activity, revealed a subset of sites at which NusG cooperated with Rho to induce early termination.[12] A distinguishing feature of these signals was that, unlike ideal *rut* sequences, they were relatively C-poor. Results from a complementary study indicated that defects of substitutions in the PBS could be suppressed in the presence of NusG.[83] NusG, but not PBS–RNA contacts, was required for Rho-dependent silencing of AT-rich *Salmonella* pathogenicity islands.[84] Together with observations that NusG restored termination *in vitro* when Rho–RNA interactions were compromised,[82] these findings indicate that NusG assistance is required when RNA is a poor ligand for Rho, but are at odds with observations that NusG does not increase Rho binding to RNA.[85] How, then, does NusG work? Analysis of *rho* mutants that mimic the effect of NusG led Sen and colleagues to propose that NusG binds to the Rho-CTD and induces an allosteric signal that manifests in ring closure.[16] The phenotypes of NusG-mimicking mutants resembled those of "super Rho" variants, and one of them, P325H,[63] was also able to mimic NusG.[16] These mutations induced conformational changes in the SBS and in the ATP-binding site, indicating that, when bound to a distant site, NusG could trigger similar changes that activate Rho.

Although the NusG binding site proposed by Valabhoju *et al.*[16] does not correspond to any interface subsequently identified in the Rho–NusG-CTD structure, the structure and direct measurements of ring closure confirmed that NusG activates Rho allosterically.[36] Full-length NusG or the CTD alone strongly stimulated RNA binding to the SBS, which requires ring closure; this effect was more pronounced with A_{12} RNA than with the ideal rU_{12}. Consistent with observations that efficient PBS interactions reduce the dependence on NusG,[83] the inclusion of dC_{15} yielded fully closed Rho rings in the absence of NusG. In the binary complex, two loops in the NusG-CTD, L1 (residues 140–144) and L2 (163–167), directly interact with Rho residues Trp381, Ile382, Ile386, Phe398, Met415, Met416, Lys417, and Arg418.[36] Mutants that strongly inhibited termination *in vivo* but not *in vitro*, I382N[64] and M416K,[67] have been isolated previously, and M416K was proposed to block binding of NusG.[67] The NusG–Rho contacts were validated by mutagenesis and, most convincingly, by the demonstration that transplanting a five-residue Rho-binding loop from NusG to its paralog replication factor H (RfaH), an inhibitor of Rho,[86] yields a potent activator of ring closure.[36]

How does NusG promote changes around the SBS by binding to the C-terminus of Rho? The structure and molecular dynamics simulations indicated that NusG favors an active Rho conformation (Figure 5.2C). The NusG-CTD appears to break a salt bridge between Glu368 and Arg384 and a cation-π interaction between Arg353 and Trp381; these contacts are observed in Rho subunits poised for product exchange and substrate binding[40] and thus must be disrupted to allow ring closure. In contrast, the presence of the NusG-CTD appears to favor a salt bridge between Arg353 and Glu368, observed in active-site interfaces where the nucleotide is productively engaged. Alanine substitutions of Trp381 and Arg384 yielded Rho mutants that closed more readily than wild-type Rho in the absence of NusG, supporting the view that these residues engage in autoinhibitory interactions.

The available genetic, biochemical, and structural data[12,16,36] support a model in which NusG plays a key role in channeling Rho activity. NusG actively assists Rho in terminating futile transcription by promoting sequence-blind isomerization of the Rho ring into a translocation-competent state, yet helps to protect functional RNAs from Rho by engaging in contacts with Nus complexes or translating ribosomes. It remains to be determined if NusG action is modulated principally by the availability of S10 or whether NusG is deployed at specific sites, as indicated by observations that some NusG-dependent terminators show stronger responses than others.[83]

5.3.2 NusA has Complex Effects on Rho Termination

NusA is a multi-domain protein composed of the RNAP-binding NTD, three nascent-RNA-binding domains (S1, KH1, and KH2), and (in some bacteria, including *E. coli*) additional regulatory domains.[87] NusA is ubiquitous in bacteria and has many functions: it (i) enhances RNAP pausing and termination at hairpin-dependent sites;[88,89] (ii) modulates Rho-dependent termination;[90–92] (iii) participates in assembly of ATCs;[69,70] and (iv) facilitates co-transcriptional folding of structural RNAs.[93]

NusA-NTD, which is sufficient for NusA effects on hairpin-stabilized pausing,[87] binds near the RNA exit channel of the EC *via* interactions with the β subunit flap domain and the α-CTD.[94] Anchored though the NTD–RNAP contacts, five additional mobile domains can occupy very different positions in regulatory ECs to establish contacts with the nascent RNA, with other RNAP domains, and with accessory factors.[69,70,94] Together, these interactions remodel the RNA exit channel, affecting formation of RNA secondary structures therein, and alter the path of the upstream nascent RNA.[69,70,94] Conditions under which *E. coli nusA* becomes dispensable have not been identified, hindering NusA analysis *in vivo*, but it has been extensively studied *in vitro.*

NusA effects on transcription are the opposite of those of NusG: NusA enhances pausing and termination at hairpin-dependent sites,[89,95,96] presumably because NusA stabilizes the RNA duplex in the exit channel,[94] and slows down RNA synthesis.[95] As is the case with NusG, NusA effects on Rho termination are context-dependent and likely to reflect a combination of direct and indirect interactions. Direct contacts between NusA and Rho have been proposed to mediate Rho binding to RNAP[97] and observed by crosslinking;[98]

mutations in *nusA* have been shown to modulate Rho-dependent termination.[99–101] The modular structure of NusA, with each of its domains capable of accommodating diverse RNA and proteins, makes NusA a plausible candidate for bridging RNAP and Rho, but cannot explain why NusA inhibits Rho-dependent termination *in vitro*, an effect that is manifested by a shift of the termination window further downstream from the *rut* site.[90,91]

5.3.3 Cooperative Gene Silencing by NAPs and Rho

NAPs act as xenogeneic silencers that enable bacteria to safely maintain newly acquired foreign DNA and downregulate expression of genes needed under specific conditions.[102] In *E. coli* and *Salmonella*, heat-stable nucleoid-structuring protein (H-NS) and suppression of thymidylate synthase^{-} phenotype A (StpA) protein directly bind to DNA and assemble into long nucleoprotein filaments of various structures,[72,103] whereas *Yersinia* modulator A (YmoA) family proteins Hha and YdgT lack DNA-binding domains and are incorporated into these filaments through protein–protein contacts with H-NS and StpA.[72,104] While NAP filaments are best known for their ability to inhibit transcription initiation, they also play important roles during elongation and termination.

The available genetic, genomic, and biochemical data indicate that NAPs promote Rho termination. *First*, genes which are upregulated by BCM treatment are bound by H-NS.[12,105] *Second*, deletion of *hns* sensitizes *E. coli* to BCM,[106] augments growth defects of *rho* alleles,[12] and restores the expression of a Rho-silenced *rfa* operon.[49] On the basis of genetic evidence, Gowrishankar and colleagues proposed a model in which filaments assembled with H-NS and YmoA proteins alter the properties of the EC such that it becomes more susceptible to termination by Rho.[107] Biochemical experiments in the Landick lab demonstrate that bridged nucleoprotein filaments assembled *in vitro* with H-NS, StpA and H-NS/Hha induce RNAP pausing and backtracking and promote termination by Rho.[72,103]

In turn, efficient Rho-dependent termination is critical for H-NS silencing in *Salmonella*: depletion of NusG or *rho* K130Q substitution triggers dramatic upregulation of numerous H-NS-silenced loci, including pathogenicity islands,[84] which will be once again downregulated in an abundance of NusG. Thus, cooperative and cyclical action of Rho, NusG, and NAPs ensures that the resource-intensive *Salmonella* virulence program, which entails the expression of many long operons, is turned on only when needed.

5.4 One Ring to Find Them: Alternative Models of Rho Recruitment and RNA Release

Two classes of models have been proposed to explain how Rho terminates transcription (Figure 5.3). In one model,[53] which dominated the literature and the classroom for decades, Rho's exceptional motor properties explain

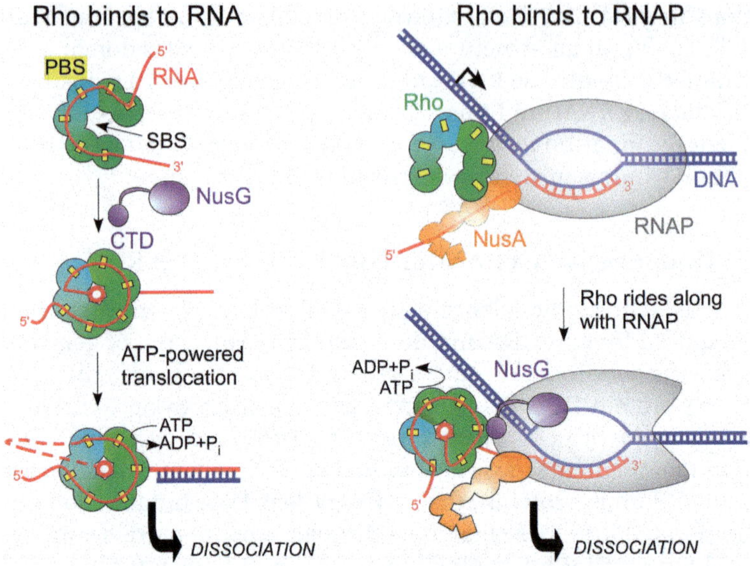

Figure 5.3 Alternative modes of Rho recruitment. Rho can be recruited to RNA, with or without the attached RNAP (left) or to the EC, possibly *via* contacts with both RNAP and NusA (right). Direct action on free RNAs may mediate Rho activities beyond transcription, whereas obligatory recruitment to RNAP would restrict Rho action to actively synthesized RNAs. In either scenario, following filling of the PBS (yellow) with RNA (red), the ring closes around the SBS RNA. The ring closure is favored by C-rich RNAs bound to the PBS and SBS and by NusG–Rho interactions.[35,36] Once the ring closes, Rho ATPase is activated and Rho translocates on the RNA while maintaining contacts to the *rut* element; upon reaching an obstacle, such as an annealed DNA oligonucleotide or the EC, Rho triggers its dissociation.[11,111,136] The underlying mechanism by which Rho dissociates the paused EC remains unconfirmed, but Rho must approach the exit channel to exert force on RNA or RNAP.

its ability to remove any obstacle, including RNAP, from RNA. Various inconsistencies with experimental findings and explanatory limitations of this model have led to the development of an alternative model that focuses on the dynamic properties of RNAP instead.[108] Several studies in the last year have provided the first direct evidence in favor of this model and new insights into the complex cellular regulation of Rho-dependent termination.

5.4.1 Kinetic Coupling: an RNA-centric, Steric Model

The first identified Rho targets, λ *cro* and the *trp* operon, contain strong *rut* elements. Binding of Rho to these targets and to some other RNAs not bound by RNAP with high (sub-nanomolar) affinity[34] indicated that Rho first binds to a *rut* site on a nascent RNA transcript and then engages in a race with

RNAP described in the "kinetic coupling" model.[109] At some point within the termination window, RNAP pauses, at which point Rho catches up, rear-ends RNAP, and causes EC dissociation and RNA release.

Rho is a powerful molecular motor expected to exert forces in excess of 200 pN, estimated on the basis of its ability to break biotin–streptavidin interactions.[11] Several variations of this model have been proposed, in which Rho uses this considerable force to either *push* RNAP forward without RNA extension;[110] *pull out* the nascent RNA, thus disrupting the RNA–DNA hybrid;[34] or *invade* RNAP and directly unwind the RNA–DNA hybrid.[108] In all three scenarios, Rho would use its well-documented translocase and helicase activities to trigger RNA release.

Although Rho is expected to contact RNAP during the final steps of termination, these interactions do not need to be specific, since Rho can terminate transcription by phylogenetically and structurally distinct phage T7 RNAP and eukaryotic RNAP II.[111,112] Such non-specificity is indeed completely consistent with a purely steric model in which Rho functions by brute force.

5.4.2 Trap-and-release: an RNAP-centric, Allosteric Model

However, the *rut*-centric steric model is at odds with the existence of numerous Rho targets that appear to lack recognizable *rut* sites[12] and fails to explain how Rho discriminates in favor of nascent RNAs and against transcripts that have been released, either prematurely or upon completion, to avoid futile cycles. A model in which Rho is first recruited to the EC and only subsequently binds the nascent RNA is supported by direct[108] and NusA-mediated[97] binding of Rho to RNAP *in vitro* and by the overlapping genome-wide distributions of RNAP, NusA, and Rho in *E. coli* inferred from chromatin immunoprecipitation (ChIP)-chip experiments.[71]

By "hitchhiking" on RNAP rather than running along in its wake, Rho can play its role as sentinel by surveying RNAs as they are made and acting quickly to terminate harmful or useless transcripts. Such a spring into action requires the ability to trigger the dissociation of an exceptionally stable EC and to do so rapidly and irreversibly, in response to RNA and/or other physiological signals. These criteria, along with the exclusion of Rho from released transcripts and imperfect correlations among the helicase, translocase, and termination competencies of Rho variants, indicate an allosteric model of action, with specific Rho–RNAP contacts playing a key role in driving the refolding of the EC into a metastable state in which dissociation is inevitable.

Nudler and colleagues[108] postulated that Rho acts in two steps, first inactivating the EC by trapping it in a moribund state and then slowly releasing the nascent RNA. Specific interactions between Rho and the β′ lid domain, which is positioned at the upstream boundary of the transcription bubble and cooperates with NusG to stabilize the upstream DNA duplex,[79,113] are proposed to induce changes in the catalytic β′ trigger loop (TL) that weaken RNAP–nucleic acid contacts, inactivating the EC and making it an easier target for Rho.[108]

5.5 It Wants to be Found: Catching Rho in the Act with Structures of Pre-termination ECs

The overwhelming strength of the translocase and helicase activities of Rho, the efficient unwinding of RNA–DNA hybrids by Rho in the absence of RNAP, and the lack of genetic evidence in support of direct Rho–RNAP contacts led to broad acceptance of the RNA-centric model of Rho-dependent termination.[31] Direct evidence in support of the allosteric mechanism of termination came from single-particle cryo-EM studies. In 2020, two groups succeeded in obtaining structures of pre-termination *E. coli* ECs in which Rho is recruited to RNAP bound to NusA and NusG.[114,115] These structures revealed an intricate and dynamic network of interactions among the Rho hexamer, RNAP subunits, nucleic acids, and Nus factors, which are rearranged in a stepwise fashion to eventually inactivate the transcription complex (Figure 5.4). At least seven distinct intermediates that punctuate this pathway can be identified.[115]

During *engagement*, Rho is recruited to the EC *via* extensive contacts to RNAP, NusA, and NusG, but initially makes no contacts to the long *rut*-bearing nascent RNA; in fact, Rho forms a very similar complex with an EC in which the RNA is too short to reach Rho.[114] NusA is required for the formation of a stable Rho–RNAP complex,[114,115] as first shown in 1987,[95] whereas Rho contacts with all subunits of RNAP contribute to binding and/or function; for example, Rho interactions with βIS2, an insertion into the β flap domain, modulate termination *in vitro* and *in vivo.*[114,115] Following recruitment,

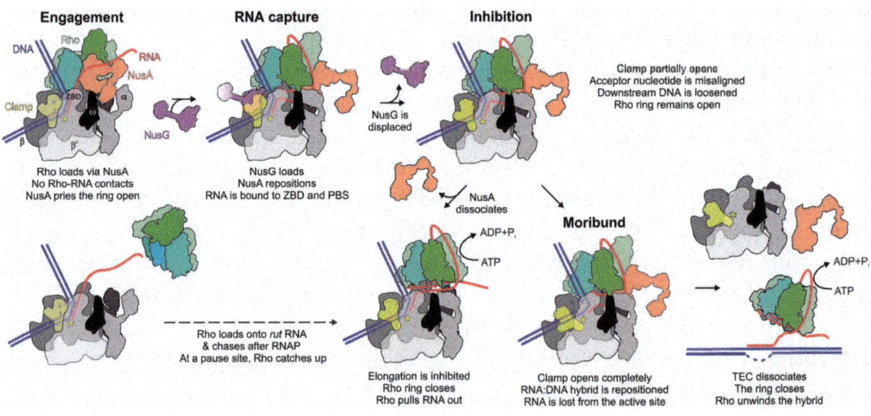

Figure 5.4 A hypothetical EC-dependent Rho-mediated termination pathway. Four key intermediates among seven captured by cryo-EM analysis are depicted.[115] RNAP is shown in shades of gray, with the clamp domain in yellow; the active site is indicated by a star. The Rho hexamer is green–cyan, NusA is orange, and NusG is purple; both NusG domains are shown even though the CTD is not visible in the structures. The alternative pathway (*via* RNA) of Rho recruitment to and dissociation of the NusA-less EC is also shown.

structural changes in the Rho hexamer, RNAP, NusA, and DNA duplex set up a stage for *RNA capture* by Rho, which is aided by the β′ subunit zinc-binding domain (ZBD). Once the RNA is engaged, Rho induces rearrangements of several RNAP domains, triggering the displacement of NusG and partial opening of the RNAP clamp, thereby *inactivating* the EC. Finally, a *moribund* complex in which the clamp is wide open, RNAP contacts to the RNA–DNA hybrid and the downstream DNA are lost, and the RNA 3′ end is dislodged from the active site, is formed.[115] The key contributions of Rho–EC contacts observed in the structures are supported by genetic and biochemical data.[114,115]

Unexpectedly, although *rut* RNA, NusG, and ADP-BeF$_3$, which promote ring closure,[35,36] were present, the Rho ring remained open throughout the entire pathway, even in the moribund complex, preventing Rho from translocating the RNA. The open state is locked by numerous Rho–protein contacts, particularly by NusA, which is lodged between two Rho protomers (Figure 5.4). When, therefore, does Rho close and unwind? We envision several possibilities (Figure 5.4). *First*, the open-clamp, moribund EC is likely to be unstable, and may release the nucleic acids with Rho still bound to the RNA, followed by unwinding of the RNA–DNA hybrid. *Second*, the inhibited EC may release NusA, facilitating ring closure and subsequent unwinding. *Third*, Rho may bind to a high-affinity *rut* element instead of the EC, particularly if RNAP fails to load Nus factors, and engage in closed-ring ATP-powered translocation toward RNAP, catching and releasing the paused EC. The ability of a heterologous phage T4 RNA–DNA helicase ultraviolet susceptibility W (UvsW) to substitute for otherwise essential *E. coli* Rho,[27] the scarcity of *rut* elements in *E. coli*,[116] and efficient termination of *rut*-less transcripts[12] indicate that the latter, RNA-centric mode of recruitment and termination is not the predominant one.

5.6 Regulating Rho: Timing Isn't Everything

The persistent "hitchhiking" of Rho with the EC involved in the trap-and-release model readily explains how Rho can selectively target RNAs in the making, ensuring that it does not terminate transcription too late. However, it also raises a challenge: how can regulators of Rho ensure that it doesn't do so too early, and that transcripts are guarded from premature release? Numerous gene-specific mechanisms of Rho downregulation have been described in the literature; here we will discuss general mechanisms, interpreting findings in the light of models of both Rho recruitment and RNA release.

5.6.1 NusA and NusG: Dueling Roles for a Duo of General Factors

NusA, which is critical for Rho recruitment,[114,115] may offer a first line of defense. *First*, NusA appears to direct the nascent RNA away from Rho; indeed, mutations in the S1 and KH1 domains that strengthen NusA–RNA contacts display enhanced anti-Rho activity *in vitro* and *in vivo*.[92] However, *in vivo* observations that transcripts affected by BCM do not significantly overlap with those affected by a deletion of all three RNA-binding domains of NusA[12] indicate that competition for RNA is not the major mechanism by which NusA inhibits Rho-dependent termination. *Second*, by modifying the RNA exit channel and altering RNAP conformation,[94,117] NusA could make it resistant to termination. *Third*, NusA prevents ring closure, which is a committed step toward EC dissociation.

NusG also appears to play conflicting roles in termination. Despite the well-documented enhancement of Rho termination by NusG–CTD,[36,81] this domain does not make any contacts to the Rho–EC,[114,115] and holding the ring open may inhibit stable NusG binding. Furthermore, NusG–NTD binding to the β′ clamp interferes with Rho-mediated EC inactivation (Figure 5.4), and NusG release is a prerequisite for clamp opening. Additional experiments are needed to reveal if, once released, NusG engages Rho and promotes subsequent ring closure.

In addition to modulating termination through contacts to Rho and RNAP, NusA and NusG contribute to antitermination by serving as adaptors between the RNAP and other proteins. NusA and NusG are integral components of nucleoprotein ATCs[69,70] and can couple RNAP to the trailing ribosome.[77,78] Once formed, the ATC and the expressome physically block Rho access to the EC, precluding termination.

5.6.2 Proteins that Bind to and Inhibit Rho

Among proteins that inhibit termination *via* direct contacts to Rho, bacteriophage P4 capsid protein polarity suppression (Psu) has been studied the most extensively. *In silico* docking indicates that a Psu dimer binds to two Rho monomers across the central pore, capping the SBS and blocking translocation, a conclusion supported by the results of functional analyses.[48,118] Mapping of a Psu-binding site to Rho residues 139–153[48] and demonstration that Psu inhibits Rho proteins in diverse Gram-negative and Gram-positive pathogens[119] highlight the functional importance of the disordered connector region.

E. coli proteins host factor for bacteriophage Q (Hfq) and YaeO[120,121] have also been shown to inhibit Rho. A mechanism in which YaeO blocks RNA–PBS interactions has been proposed based on NMR studies of YaeO alone and in complex with the Rho NTD.[122] Six YaeO molecules were modeled to bind to the Rho hexamer, explaining why YaeO overexpression was required to observe antitermination.[121,122] A search for proteins with structures similar to YaeO identified RNA-binding domains of several Sm-fold ribonucleoproteins, including the bacterial RNA chaperone Hfq.[122] *E. coli* Hfq is a hexameric protein that shares no common sequence motifs with YaeO, but is also known to inhibit Rho,[120] albeit not in all contexts.[20] Hfq acts in concert with several small RNAs (sRNAs) thought to inhibit Rho through competition for the PBS.[20] The exact molecular mechanism by which Hfq inhibits Rho remains to be determined, but one possibility is that Hfq may deliver inhibitory sRNAs to the Rho NTD and then leave, as it does after promoting annealing between an sRNA and its mRNA target.[123] A recent crystal structure of *Vibrio cholerae* YaeO,[124] whose sequence is quite distinct from its *E. coli* homolog, revealed similarities to *E. coli* YaeO and Hfq. In contrast to both proteins, *V. cholerae* YaeO is proposed to act as a dimer that directly binds to, and disrupts, the Rho hexamer and may also compete with Rho NTD for binding to the PBS RNA.[124]

5.6.3 sRNAs Protect Leader Regions from Premature Termination

A large fraction of coding RNAs have 5′ untranslated leader regions (UTRs) that could be easy targets for Rho. Indeed, Sedlyarova *et al.* found that hundreds of *E. coli* genes with UTRs that could recruit Rho were upregulated by BCM.[20] Identification of the RNA polymerase, sigma S (*rpoS*) leader among these UTRs prompted the authors to investigate whether sRNAs, known to control expression of the stationary phase σ factor RpoS,[125] could confer protection from premature termination. Bacterial sRNAs bind to a significant fraction of cellular mRNAs to modulate their translation or decay.[126] Aided by Hfq, sRNAs Aerobic respiration control Z (ArcZ), downstream of regulator of capsular synthesis A (DsrA), and RpoS regulator RNA (RprA) pair with the *rpoS* leader to disrupt an inhibitory hairpin and positively regulate translation of RpoS.[125]

Sedlyarova *et al.* showed that ArcZ, DsrA, and RprA suppressed Rho termination of *rpoS in vivo* and *in vitro* and that Hfq stimulated antitermination. Moreover, they identified other likely targets of these sRNAs whose expression was also insulated from Rho action in an Hfq-dependent manner. The role of Hfq therein is likely to be limited to its ability to promote the sRNA–mRNA duplex formation, because Hfq rapidly dissociates from the ternary complex[123] and because Hfq did not induce *rpoS* antitermination *in vitro* when present alone.[20]

The abundance of sRNA targets indicate that many physiological responses could be mediated by an interplay between sRNAs and Rho. The expression of *rho* itself is controlled by *Salmonella* small RNA L (SraL) sRNA, which binds to the *rho* leader in an Hfq-dependent manner to relieve autoinhibition.[127] Although some sRNAs could stimulate both mRNA synthesis and translation, as is observed in *rpoS*, others may exert an opposite effect by inhibiting translation and promoting Rho-dependent termination, a mechanism exemplified by *Salmonella* chitoporin X regulator (ChiX) sRNA.[128] Additional regulatory mechanisms may have evolved to curtail widespread sRNA-mediated antitermination. For example, bacteriophages have a high density of sRNAs, some of which mimic the mRNA target regions of bacterial sRNAs and can act as sponges to diminish the effects of host sRNAs.[129]

5.6.4 Riboswitching of Rho

Among UTRs affected by Rho, several are associated with well-known riboswitches,[20] regulatory elements that respond to signal molecules from ions to tRNAs[130,131] to regulate a wide range of cellular processes. Riboswitches typically fold into complex structures that, upon binding of a cognate ligand, rearrange their structures to control transcription, mRNA decay, or translation. Most riboswitches are located in the UTR of the operon they control, although examples of *trans*-acting riboswitches have been described.[132,133] Elaborate secondary structures of riboswitches indicate that they should be refractory to Rho.[34] However, this rule is not iron-clad, as many examples demonstrate. Among them are λ *cro* tR1 terminator, in which an N-utilization (*nut*) hairpin is positioned next to the *rut* site;[134] and the *Salmonella mgtA* leader, in which the primary *rut* site is wedged in between hairpins.[19] Consistent with a lack of significant hairpin penalty, systematic evolution of ligands by exponential enrichment (SELEX) for Rho ligands identified two equally represented classes of C-rich RNAs, with and without an adjacent 6-bp hairpin.[135] Once Rho loads onto RNA, it is able to unwind RNA duplexes *in vitro*[11,136] and *in vivo.*[55]

In contrast to many *B. subtilis* riboswitches that rely on intrinsic termination to control transcription, known *E. coli* riboswitches lack such signals and are translationally controlled; well-documented Rho–ribosome genetic switches[128,137] make Rho termination within riboswitches an attractive regulatory strategy. Indeed, a pioneering study by Groisman and colleagues revealed that Rho induces termination in a Mg-sensing magnesium transporter A (*mgtA*) riboswitch from *Salmonella* and a flavin mononucleotide (FMN)-sensing *ribB* riboswitch from *E. coli* in the presence of their cognate ligands.[19] The ligand-induced conformational changes allow Rho to bind to *rut* sites otherwise sequestered in secondary structures. A conceptually similar mechanism is widely used to modulate ribosome access to the ribosome binding sites and the start codon,[138] indicating that Rho-controlled riboswitches could also be common.

This hypothesis was confirmed by Lafontaine and colleagues,[17] who found that four *E. coli* riboswitches, thiamine pyrophosphate (TPP)-sensing *thiB, thiC*, and *thiM* and lysine-sensing *lysC*, were also controlled by Rho. Although the ligand binding to the *thiC* riboswitch unmasks the *rut* site, allowing Rho to terminate transcription,[18] ligand-dependent Rho activation in the *thiM* riboswitch is achieved through inhibition of translation.[17] This mechanism is thus more similar to other examples of Rho–ribosome competition than to a classical ligand-switching between alternative states that controls Rho binding directly.[18,19] It remains to be determined how other riboswitches integrate Rho into their functional outputs.

5.6.5 *Cis*-acting RNAP Aptamers

RNAP makes largely non-specific interactions with the nucleic acids in the EC but can also bind to isolated RNAs with high affinity,[139] and examples of nascent RNA structures that inhibit termination, presumably through binding to surface sites on RNAP, have been reported.[140,141] Results from a recent genome-wide study in *E. coli* indicated that RNAP-binding aptamers (RAPs) are common, enriched in antisense transcripts, and are near translation start sites.[142] RAPs appear to be particularly important for promoting Rho silencing of antisense transcripts,[22] but have also been shown to control the expression of coding genes. A RAP in the *nadD* ORF appears to act by uncoupling transcription and translation, thereby exposing RNA to Rho;[142] a ribosome pause observed just upstream of *nadD* RAP[143] is consistent with a loss of coupling.[144] It is conceivable that binding of the aptamer near the RNAP exit channel could disrupt coupling by interfering with direct interactions between RNAP and 30S;[145–147] however, the uncoupling effect alone cannot explain increased termination in RAP-containing antisense transcripts.[142] Aptamer binding near the exit channel could reposition NusA,[94] relieving inhibition of Rho. RAPs could also favor termination kinetically: the *nadD* RAP has been shown to stimulate RNAP pausing *in vitro*[142] by a mechanism likely to be distinct from those employed by nascent RNA structures that form within the RNA exit channel and induce conformational changes that stabilize the paused EC.[94,148,149] Effects of RAPs on transcription could be as diverse as their structures, awaiting new studies to reveal the details of RAP actions, but the available evidence implicates them as RNA-based regulators of gene expression.

5.7 Regulatory Routes and Checkpoints in Rho Termination

Although Rho terminates transcription consistently at the end of some genes,[55] the majority of its action appears to be context-dependent. This demands that Rho termination is modulated to meet regulatory needs. The new Rho–EC structures[114,115] reveal many intrinsic checkpoints for regulation along the route followed in the trap-and-release model. Similar extents of Rho–EC contacts in consecutive states (Figure 5.4) indicate that the hypothetical termination pathway could be readily reversible and sensitive to the nascent RNA sequence. If no *rut* site is available, the pathway may be halted prior to RNA capture, whereas a poor *rut* may support inefficient termination. NusG can dissociate from the EC during elongation,[79] and it is likely that so can NusA; different conformations of the EC and the nascent RNA could affect their association with RNAP. The various states will have different propensities to termination, leading to a broad variation of Rho termination windows, in some cases reflecting several attempts by Rho at termination.

The pathway can be influenced by extrinsic inhibitors and activators of Rho-dependent termination acting at several points (Figure 5.5). Controlling RNA accessibility is the simplest and apparently most widely used mechanism. Control of ring closure also emerges as a common post-recruitment checkpoint, as evidenced by effects of BCM, NusG, and RNA ligands.[16,35,36] Some accessory factors could merely shift the termination window, but more potent inhibitory effects are feasible. For example, a Rho-antagonizing RNA element (RARE) in the *Salmonella mgtCBR* leader inhibits Rho *in vitro* and *in vivo*, but does not block *rut* interactions;[150] RARE may interfere with ring closure by binding to a second RNA site on the NTD,[44] or alter NusA position.

Step	Termination	Readthrough
RNAP binding	NusA Poor translation	σ initiation factor ATC Expressome
RNA capture	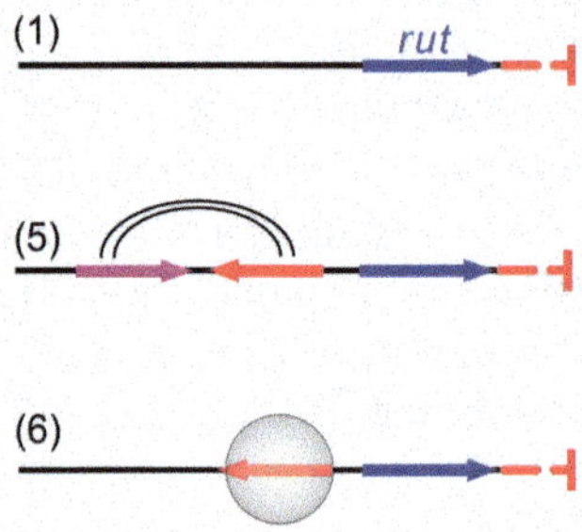 NusG	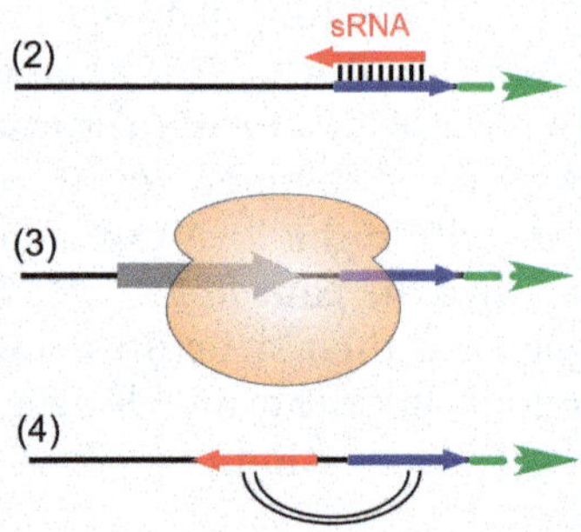 Hfq
EC inactivation & release	NusG Bridged NAP filaments RNAP pausing	RARE, *put* RNA BCM ATFs: Psu, YaeO, RfaH

Figure 5.5 Factors that modulate Rho recruitment to RNAP, RNA capture, and termination, categorized by which favor termination or readthrough at each step. RNAP binding: Rho binds to RNAP with help from NusA,[97,114,115] which requires dissociation of the σ initiation factor. A coupled expressome[77,78] or multicomponent ATCs[69,70] sterically block Rho interaction with the EC. RNA capture: Once loaded on the EC, Rho captures RNA (1) unless the *rut* element is masked by (2) a sRNA,[20] or (3) a stalled ribosome,[137] or (4) is base-paired to a complementary anti-*rut* RNA;[19] formation of this inhibitory hairpin can be blocked by (5) an alternative structure[61] or (6) an RNA-binding protein.[178] Regulation is sensitive to metabolites, such as Trp[137] or FMN.[61] EC inactivation and release: Following productive Rho recruitment and RNA capture, diverse factors could affect termination, including RNAP pausing and Rho ring closure. Other known Rho effectors thought to act during these steps are shown, but their exact mechanisms remain to be determined; ATF, antitermination factor.

Rho binding to RNAP can be directly hindered by ribosomes or ATCs.[69,70,77,78] Alternatively, RNAP could be modified into a state that resists Rho-induced conformational changes, *e.g.* by *cis*-acting RNAs that remodel the RNA exit channel[141] or by specialized NusG paralogs that stabilize the clamp in a closed state.[79] It seems likely that more factors that regulate Rho *via* these and other mechanisms will be discovered in the future.

In contrast, the route involved in the kinetic-coupling model offers essentially only a single regulatory checkpoint: a sufficiently long pause by RNAP within the termination window to allow Rho to catch up, since it is precisely the RNAP speed that is coupled to the Rho translocase speed. Early reports that Rho release sites coincided with pause sites[151–153] and that pause-prone RNAP variants increased termination[109] indicated that a reduction in RNAP speed is critical to enable Rho to catch up.

This is not only contradicted by the trap-and-release model, in which Rho hitchhikes with RNAP first and translocates later, completely decoupling Rho translocase speed from RNAP speed (Figure 5.4). Statistical evidence has also been provided by the sum total of RNAP variants with altered pausing propensities that have been characterized since the 1980s, which have revealed a poor correlation between RNA synthesis rate and pausing.[154] Revisiting this point, Said *et al.* showed that reducing RNAP speed by limiting substrate NTP concentration did not promote termination.[115] In contrast, V550A RNAP, which is resistant to pausing despite being only marginally faster than the wild-type,[154] was also resistant to Rho.[115] If, as the latest genetic, biochemical, and structural evidence have indicated, the kinetic-coupling model is the road less traveled, we can now understand why the transcribing speed of RNAP has little effect on its sensitivity to Rho: it is instead the propensity of RNAP to undergo the conformational changes associated with pausing that will make the difference.

5.8 The Importance of Pausing in Rho-dependent Termination

The role of RNAP pausing in Rho-dependent termination has been studied in detail at two riboswitch-regulated transcriptional attenuators. Riboswitches' functions depend on their ability to form complex secondary structures, and pausing is critical for co-transcriptional folding of diverse riboswitches.[149,155–157] In the *Salmonella* leader region of *mgtA*, which encodes a Mg transporter, Rho terminates at a site that corresponds to a long-lived pause observed *in vitro* and *in vivo*.[158] This pause is required to stimulate Rho-dependent termination in response to Mg^{2+} availability, and mutations in the pause element reduce termination as much as those in the *rut* site.[19] The *mgtA* pause shares key features with the canonical hairpin-stabilized *his* pause[159] but has a longer half-life that is further increased in the presence of high concentrations of Mg^{2+}. This hyper-stabilization effect has been attributed to a hairpin (located upstream from the pause hairpin), which could make interactions with RNAP (or an elongation factor *in vivo*). Pausing also appears to control termination in the TPP-responsive *thiC E. coli* riboswitch.[18] RNAP pauses near the start codon, downstream from a *rut* element that becomes unmasked in the presence of TPP. *In vitro*, TPP stimulates pausing and weak Rho-dependent termination, which occurs exactly at the pause site; NusG strongly increases termination. Termination is inhibited by mutations that reduce pausing and is stimulated by TPP-mediated increase in pause time.[18] Unlike *mgtA*, the *thiC* transcript does not encode an RNA hairpin positioned 11–12 nt upstream from the pause site, spacing that supports hairpin stabilization in *E. coli.*[160] Intriguingly, results from a recent bioinformatic analysis indicate that Rho targets are devoid of elemental pauses[116] and may instead utilize distinct pause signals, a provocative hypothesis that will require validation by mapping of the genomic pause and termination sites at single-nucleotide resolution.

Findings with *mgtA* and *thiC* demonstrate that ligand-stabilized pausing stimulates termination and defines its position. Although the details of pause stabilization may be different, the contribution of the cognate ligand is a key regulatory feature of both mechanisms. In striking contrast to a typical broad termination zone, Rho terminates transcription exactly at the pause site. This unusual pattern may be explained by a strong pause that occurs just as the termination window opens, 93 and 88 nucleotides from a hypothetical *rut* in *mgtA* and *thiC*, respectively. The simplest explanation, consistent with the kinetic-coupling model,[109] is that RNAP waits for Rho arrival, although it is possible that pausing-associated changes in the EC[148] also facilitate release. Is ligand-stimulated-pausing termination a universal feature of Rho-controlled riboswitches? Observation of a strong FMN-enhanced pause in *E. coli ribB*[18] supports this idea. Is termination at a precise position particularly important in riboswitches because the regulatory window is narrow, or does it serve to produce a *trans*-acting RNA?[158] Why do *mgtA* and *his* pause hairpins promote termination *in vivo*[158] but the *hisP* signal resists Rho *in vitro*?[161] This difference could be due to NusA, which stabilizes pausing at *hisP* through a network of protein–RNA interactions,[87,94] an effect that could be mimicked by a combination of high Mg^{2+} and hairpin B in *mgtA*.[158] Answers to these questions await future investigation.

5.9 R-loops and Genome Stability

Termination is critical for proper control of gene expression, and the identification of numerous Rho sites genome-wide indicates that termination constitutes the essential (in some species) function of Rho.

Surprisingly, it is the resolution of R-loops, *via* its helicase activity, that makes Rho indispensable, at least in *E. coli.*[27] R-loops form when the nascent RNA anneals to the template DNA strand outside the transcription bubble, displacing the non-template DNA strand and forming an extended RNA–DNA hybrid.[162,163] R-loops accumulate in *E. coli* strains with defects in RNase HI, which cleaves the RNA strand in R-loops; RecG helicase, which unwinds R-loops; or topoisomerase I, which removes negative supercoils and thus prevents R-loop formation.[163] Studies of plasmid runaway replication in *E. coli*[164] implicated Rho in R-loop regulation. During replication of ColE1 plasmids, RNase HI cleaves an R-loop intermediate to generate the 3′ end needed for initiation of plasmid replication. Excessive accumulation of R-loops in *rnhA* (which encodes RNase HI) or *recG* mutants dramatically increases the plasmid copy number, killing the cell. Termination-defective alleles of *nusG* and *rho* did not survive transformation with ColEI plasmids and were synthetically lethal with *rnhA* and *recG* mutants,[164] prompting a model in which defects in Rho function trigger widespread R-loop accumulation on the chromosome, titrating a host factor away from the plasmid to allow its over-replication. This model was validated in a follow-up study, where Gowrishankar and colleagues showed that R-loops occur at a basal level in wild-type *E. coli* and become more frequent in a *rho* mutant.[27] Strikingly, both *rho* and *nusG* could be deleted in strains expressing UvsW, an R-loop helicase from phage T4.[27]

The resolution of R-loops underlies the ability of Rho to minimize the deleterious effects of conflicts between replication and transcription. Collisions between the EC and the replisome are unavoidable because these complexes move on the same DNA track at vastly different rates (reviewed by Dutta *et al.*[165]). Co-directional conflicts arise in genes encoded on the leading strand and trigger replisome dissociation and restart. Their consequences are generally mild but significantly exacerbated when the replication fork encounters a backtracked RNAP.[166] Head-on conflicts occur when a gene is encoded on the lagging strand and have more severe consequences, including double-stranded DNA breaks (DSBs). R-loops trigger DSBs and genome instability in bacteria and eukaryotes and have been observed during head-on[167] and co-directional[166] collisions in bacteria.

Cells use diverse mechanisms to reduce collisions and to repair collapsed forks before the next round of cell division, including helicases and R-loop processing enzymes. Rho, which can dissociate ECs and unwind RNA–DNA hybrids, is particularly well suited to lessen the burden of conflicts. Consistently, Washburn and Gottesman reported that inhibition of Rho by BCM or mutations in *nusA* or *nusG* induced DSBs in *E. coli.*[25] Conversely, an *rpoB**35 (βH1244Q) RNAP mutation that had been proposed to destabilize the EC[168] reduced the dependence on Rho[25] and suppressed defects of *rho* and *nusG* mutants.[27] RNAP variants with properties similar to RpoB*35 were found to tolerate toxic effects of head-on collisions.[169] However, a moving RNAP may not trigger replication-fork collapse, unlike an irreversibly stalled EC. Dutta *et al.* showed that ECs arrested by Nun induce DSBs at the site of arrest during co-directional collisions.[166] Active Rho and *rpoB**35 corrected the problem, as did RNase HI, implying the existence of an R-loop intermediate. However, the results of the *in vitro* analysis indicated that, instead of triggering EC dissociation, *rpoB**35 mutation reduced its propensity to backtrack,[166] suggesting a model in which backtracking triggers R-loop formation and DSBs.[166] This view is consistent with recent studies in bacterial and human cells indicating that R-loops, and not the EC, are responsible for the severity of head-on conflicts.[167,170] Extensively backtracked ECs appear to be resistant to Rho, at least *in vitro*,[161] indicating that Rho may reduce backtracking indirectly, by releasing the RNA that is not translated. In addition, Rho could directly inhibit R-loop formation.

Many physiological roles of R-loops and outstanding questions about their formation, resolution, and functions have been extensively covered in recent reviews,[163,171,172] but several questions centered around Rho are worth noting. *First*, does Rho prevent R-loop formation, resolve them afterwards, or both? R-loops form on ribosome-free and unstructured nascent RNAs,[164] preferred targets for Rho, and Rho would probably terminate transcription before R-loops form. If the displacement of an extensively backtracked RNAP by the co-directional replisome leaves a long RNA–DNA hybrid with an RNA tail behind,[166] this is a perfect substrate for Rho helicase unless all cellular Rho is bound to RNAP. *Second*, does NusG play a role in R-loop formation and resolution? In addition to supporting Rho activity, NusG could inhibit R-loop formation independently, through stabilization of the upstream DNA junction[113] and perhaps by altering the nascent RNA path.[79] *Third*, why is Rho dispensable in some bacteria and completely absent from some phyla?[41] Perhaps these bacteria use hyperactive enzymes for R-loop removal[163] or encode alternative termination factors; it is also possible that their RNAPs are intrinsically resistant to backtracking.[173] *Finally*, are R-loops always deleterious? Results from a recent study indicated that cells adapt to conditions of Rho limitation by using R-loops, which form on antisense RNAs when Rho is inhibited, to block subsequent antisense transcription at numerous *E. coli* chromosomal loci.[174]

5.10 Conclusions

In the wake of the semi-centennial anniversary of its discovery, Rho is emerging as a general transcription factor in bacteria with exceptionally diverse functions. Rho terminates RNA synthesis at the end of genes, dampens transcriptional noise, ensures that stalled RNAPs do not interfere with genome maintenance and that translation occurs only under favorable conditions, and enables cells to safely explore paths for adaptation and evolution. In recent years, a prevailing steric, RNA-centric model of Rho-dependent termination developed over decades has been increasingly challenged by an allosteric, RNAP-centric model that provides additional regulatory avenues and roles for the myriad known Rho effectors and cellular partners. The lessons learned from Rho are likely to translate across all domains of life, since a eukaryotic termination factor, Senataxin, resolves R-loops,[175] inhibits antisense RNA synthesis,[176] and shares many regulatory features with Rho despite the lack of homology,[177] indicating the ubiquity of allosteric mechanisms of transcription termination and stressing the importance of RNA surveillance by helicases in all living cells.

References

1. J. W. Roberts, *Nature*, 1969, **224**, 1168–1174.
2. L. P. Guarente, D. H. Mitchell and J. Beckwith, *J. Mol. Biol.*, 1977, **112**, 423–436.
3. D. Ratner, *Nature*, 1976, **259**, 151–153.

4. Y. Matsumoto, K. Shigesada, M. Hirano and M. Imai, *J. Bacteriol.*, 1986, **166**, 945–958.
5. P. Gollnick and C. Yanofsky, *J. Bacteriol.*, 1990, **172**, 3100–3107.
6. J. M. Peters, A. D. Vangeloff and R. Landick, *J. Mol. Biol.*, 2011, **412**, 793–813.
7. C. Lowery and J. P. Richardson, *J. Biol. Chem.*, 1977, **252**, 1381–1385.
8. M. Rabhi, V. Gocheva, F. Jacquinot, A. Lee, E. Margeat and M. Boudvillain, *J. Mol. Biol.*, 2011, **405**, 497–518.
9. C. A. Brennan, E. J. Steinmetz, P. Spear and T. Platt, *J. Biol. Chem.*, 1990, **265**, 5440–5447.
10. E. J. Steinmetz, C. A. Brennan and T. Platt, *J. Biol. Chem.*, 1990, **265**, 18408–18413.
11. A. Schwartz, E. Margeat, A. R. Rahmouni and M. Boudvillain, *J. Biol. Chem.*, 2007, **282**, 31469–31476.
12. J. M. Peters, R. A. Mooney, J. A. Grass, E. D. Jessen, F. Tran and R. Landick, *Genes Dev.*, 2012, **26**, 2621–2633.
13. L. Botella, J. Vaubourgeix, J. Livny and D. Schnappinger, *Nat. Commun.*, 2017, **8**, 14731.
14. P. Nicolas, U. Mader, E. Dervyn, T. Rochat, A. Leduc, N. Pigeonneau, E. Bidnenko, E. Marchadier, M. Hoebeke, S. Aymerich, D. Becher, P. Bisicchia, E. Botella, O. Delumeau, G. Doherty, E. L. Denham, M. J. Fogg, V. Fromion, A. Goelzer, A. Hansen, E. Hartig, C. R. Harwood, G. Homuth, H. Jarmer, M. Jules, E. Klipp, L. Le Chat, F. Lecointe, P. Lewis, W. Liebermeister, A. March, R. A. Mars, P. Nannapaneni, D. Noone, S. Pohl, B. Rinn, F. Rugheimer, P. K. Sappa, F. Samson, M. Schaffer, B. Schwikowski, L. Steil, J. Stulke, T. Wiegert, K. M. Devine, A. J. Wilkinson, J. M. van Dijl, M. Hecker, U. Volker, P. Bessieres and P. Noirot, *Science*, 2012, **335**, 1103–1106.
15. U. Mader, P. Nicolas, M. Depke, J. Pane-Farre, M. Debarbouille, M. M. van der Kooi-Pol, C. Guerin, S. Derozier, A. Hiron, H. Jarmer, A. Leduc, S. Michalik, E. Reilman, M. Schaffer, F. Schmidt, P. Bessieres, P. Noirot, M. Hecker, T. Msadek, U. Volker and J. M. van Dijl, *PLoS Genet.*, 2016, **12**, e1005962.
16. V. Valabhoju, S. Agrawal and R. Sen, *J. Biol. Chem.*, 2016, **291**, 22386–22403.
17. L. Bastet, A. Chauvier, N. Singh, A. Lussier, A. M. Lamontagne, K. Prevost, E. Masse, J. T. Wade and D. A. Lafontaine, *Nucleic Acids Res.*, 2017, **45**, 7474–7486.
18. A. Chauvier, F. Picard-Jean, J. C. Berger-Dancause, L. Bastet, M. R. Naghdi, A. Dube, P. Turcotte, J. Perreault and D. A. Lafontaine, *Nat. Commun.*, 2017, **8**, 13892.
19. K. Hollands, S. Proshkin, S. Sklyarova, V. Epshtein, A. Mironov, E. Nudler and E. A. Groisman, *Proc. Natl. Acad. Sci. U. S. A.*, 2012, **109**, 5376–5381.
20. N. Sedlyarova, I. Shamovsky, B. K. Bharati, V. Epshtein, J. Chen, S. Gottesman, R. Schroeder and E. Nudler, *Cell*, 2016, **167**, 111–121 e113.

21. P. Lin, Q. Pu, Q. Wu, C. Zhou, B. Wang, J. Schettler, Z. Wang, S. Qin, P. Gao, R. Li, G. Li, Z. Cheng, L. Lan, J. Jiang and M. Wu, *Nat. Commun.*, 2019, **10**, 3728.
22. A. Magan, F. Amman, F. El-Isa, N. Hartl, I. Shamovsky, E. Nudler, R. Schroeder and N. Sedlyarova, *Nucleic Acids Res.*, 2019, **47**, 10894–10905.
23. A. M. Stringer, G. Baniulyte, E. Lasek-Nesselquist, K. D. Seed and J. T. Wade, *Elife*, 2020, **9**, e58182.
24. R. Menouni, S. Champ, L. Espinosa, M. Boudvillain and M. Ansaldi, *Proc. Natl. Acad. Sci. U. S. A.*, 2013, **110**, 14414–14419.
25. R. S. Washburn and M. E. Gottesman, *Proc. Natl. Acad. Sci. U. S. A.*, 2011, **108**, 792–797.
26. S. Jain, R. Gupta and R. Sen, *Nat. Commun.*, 2019, **10**, 1207.
27. J. K. Leela, A. H. Syeda, K. Anupama and J. Gowrishankar, *Proc. Natl. Acad. Sci. U. S. A.*, 2013, **110**, 258–263.
28. P. L. Freddolino, H. Goodarzi and S. Tavazoie, *PLoS Genet.*, 2012, **8**, e1002744.
29. P. Mitra, G. Ghosh, M. Hafeezunnisa and R. Sen, *Annu. Rev. Microbiol.*, 2017, **71**, 687–709.
30. A. Ray-Soni, M. J. Bellecourt and R. Landick, *Annu. Rev. Biochem.*, 2016, **85**, 319–347.
31. M. Boudvillain, N. Figueroa-Bossi and L. Bossi, *Curr. Opin. Microbiol.*, 2013, **16**, 118–124.
32. E. Bidnenko and V. Bidnenko, *Curr. Genet.*, 2018, **64**, 541–546.
33. E. P. Gogol, S. E. Seifried and P. H. von Hippel, *J. Mol. Biol.*, 1991, **221**, 1127–1138.
34. J. P. Richardson, *Biochim. Biophys. Acta*, 2002, **1577**, 251–260.
35. M. R. Lawson, K. Dyer and J. M. Berger, *Proc. Natl. Acad. Sci. U. S. A.*, 2016, **113**, 13714–13719.
36. M. R. Lawson, W. Ma, M. J. Bellecourt, I. Artsimovitch, A. Martin, R. Landick, K. Schulten and J. M. Berger, *Mol. Cell*, 2018, **71**, 911–922 e914.
37. E. Skordalakes and J. M. Berger, *Cell*, 2003, **114**, 135–146.
38. E. Skordalakes and J. M. Berger, *Cell*, 2006, **127**, 553–564.
39. N. D. Thomsen and J. M. Berger, *Cell*, 2009, **139**, 523–534.
40. N. D. Thomsen, M. R. Lawson, L. B. Witkowsky, S. Qu and J. M. Berger, *Proc. Natl. Acad. Sci. U. S. A.*, 2016, **113**, E7691–E7700.
41. F. D'Heygere, M. Rabhi and M. Boudvillain, *Microbiology*, 2013, **159**, 1423–1436.
42. C. E. Bogden, D. Fass, N. Bergman, M. D. Nichols and J. M. Berger, *Mol. Cell*, 1999, **3**, 487–493.
43. J. P. Richardson, *J. Biol. Chem.*, 1982, **257**, 5760–5766.
44. T. K. Hitchens, Y. Zhan, L. V. Richardson, J. P. Richardson and G. S. Rule, *J. Biol. Chem.*, 2006, **281**, 33697–33703.
45. A. Canals, I. Uson and M. Coll, *J. Mol. Biol.*, 2010, **400**, 16–23.
46. W. L. Nowatzke, C. M. Burns and J. P. Richardson, *J. Biol. Chem.*, 1997, **272**, 2207–2211.
47. A. H. Yuan and A. Hochschild, *Science*, 2017, **355**, 198–201.

48. A. Ranjan, S. Sharma, R. Banerjee, U. Sen and R. Sen, *Nucleic Acids Res.*, 2013, **41**, 6839–6856.
49. K. Hu and I. Artsimovitch, *mBio*, 2017, **8**, e00753-17.
50. D. E. Kim and S. S. Patel, *J. Biol. Chem.*, 2001, **276**, 13902–13910.
51. G. R. Galluppi and J. P. Richardson, *J. Mol. Biol.*, 1980, **138**, 513–539.
52. A. Schwartz, M. Rabhi, F. Jacquinot, E. Margeat, A. R. Rahmouni and M. Boudvillain, *Nat. Struct. Mol. Biol.*, 2009, **16**, 1309–1316.
53. I. Faus and J. P. Richardson, *J. Mol. Biol.*, 1990, **212**, 53–66.
54. S. Pereira and T. Platt, *J. Biol. Chem.*, 1995, **270**, 30401–30407.
55. J. M. Peters, R. A. Mooney, P. F. Kuan, J. L. Rowland, S. Keles and R. Landick, *Proc. Natl. Acad. Sci. U. S. A.*, 2009, **106**, 15406–15411.
56. K. M. Walstrom, J. M. Dozono, S. Robic and P. H. von Hippel, *Biochemistry*, 1997, **36**, 7980–7992.
57. E. J. Steinmetz and T. Platt, *Proc. Natl. Acad. Sci. U. S. A.*, 1994, **91**, 1401–1405.
58. D. J. Koslover, F. M. Fazal, R. A. Mooney, R. Landick and S. M. Block, *J. Mol. Biol.*, 2012, **423**, 664–676.
59. B. L. Stitt, *J. Biol. Chem.*, 1988, **263**, 11130–11137.
60. H. Noji, R. Yasuda, M. Yoshida and K. Kinosita Jr, *Nature*, 1997, **386**, 299–302.
61. J. Geiselmann, Y. Wang, S. E. Seifried and P. H. von Hippel, *Proc. Natl. Acad. Sci. U. S. A.*, 1993, **90**, 7754–7758.
62. Y. Wang and P. H. von Hippel, *J. Biol. Chem.*, 1993, **268**, 13947–13955.
63. S. Pereira and T. Platt, *J. Mol. Biol.*, 1995, **251**, 30–40.
64. J. Chalissery, S. Banerjee, I. Bandey and R. Sen, *J. Mol. Biol.*, 2007, **371**, 855–872.
65. H. Mori, M. Imai and K. Shigesada, *J. Mol. Biol.*, 1989, **210**, 39–49.
66. R. S. Washburn and B. L. Stitt, *J. Mol. Biol.*, 1996, **260**, 332–346.
67. Y. Miwa, T. Horiguchi and K. Shigesada, *J. Mol. Biol.*, 1995, **254**, 815–837.
68. Y. Xu, H. Kohn and W. R. Widger, *J. Biol. Chem.*, 2002, **277**, 30023–30030.
69. Y. H. Huang, T. Hilal, B. Loll, J. Burger, T. Mielke, C. Bottcher, N. Said and M. C. Wahl, *Mol. Cell*, 2020, **79**, 1024–1036 e1025.
70. F. Krupp, N. Said, Y. H. Huang, B. Loll, J. Burger, T. Mielke, C. M. T. Spahn and M. C. Wahl, *Mol. Cell*, 2019, **74**, 143–157 e145.
71. R. A. Mooney, S. E. Davis, J. M. Peters, J. L. Rowland, A. Z. Ansari and R. Landick, *Mol. Cell*, 2009, **33**, 97–108.
72. B. A. Boudreau, D. R. Hron, L. Qin, R. A. van der Valk, M. V. Kotlajich, R. T. Dame and R. Landick, *Nucleic Acids Res.*, 2018, **46**, 5525–5546.
73. M. V. Kotlajich, D. R. Hron, B. A. Boudreau, Z. Sun, Y. L. Lyubchenko and R. Landick, *Elife*, 2015, **4**, e04970.
74. S. L. Sullivan and M. E. Gottesman, *Cell*, 1992, **68**, 989–994.
75. J. Li, S. W. Mason and J. Greenblatt, *Genes Dev.*, 1993, **7**, 161–172.
76. S. Saxena, K. K. Myka, R. Washburn, N. Costantino, D. L. Court and M. E. Gottesman, *Mol. Microbiol.*, 2018, **108**, 495–504.

77. C. Wang, V. Molodtsov, E. Firlar, J. T. Kaelber, G. Blaha, M. Su and R. H. Ebright, *Science*, 2020, **369**, 1359–1365.
78. M. W. Webster, M. Takacs, C. Zhu, V. Vidmar, A. Eduljee, M. Abdelkareem and A. Weixlbaumer, *Science*, 2020, **369**, 1355–1359.
79. J. Y. Kang, R. A. Mooney, Y. Nedialkov, J. Saba, T. V. Mishanina, I. Artsimovitch, R. Landick and S. A. Darst, *Cell*, 2018, **173**, 1650–1662 e1614.
80. B. M. Burmann, K. Schweimer, X. Luo, M. C. Wahl, B. L. Stitt, M. E. Gottesman and P. Rosch, *Science*, 2010, **328**, 501–504.
81. R. A. Mooney, K. Schweimer, P. Rosch, M. Gottesman and R. Landick, *J. Mol. Biol.*, 2009, **391**, 341–358.
82. K. W. Nehrke, F. Zalatan and T. Platt, *Gene Expression*, 1993, **3**, 119–133.
83. R. Shashni, M. Z. Qayyum, V. Vishalini, D. Dey and R. Sen, *Nucleic Acids Res.*, 2014, **42**, 9677–9690.
84. L. Bossi, M. Ratel, C. Laurent, P. Kerboriou, A. Camilli, E. Eveno, M. Boudvillain and N. Figueroa-Bossi, *PLoS Genet.*, 2019, **15**, e1008425.
85. C. M. Burns and J. P. Richardson, *Proc. Natl. Acad. Sci. U. S. A.*, 1995, **92**, 4738–4742.
86. A. Sevostyanova, G. A. Belogurov, R. A. Mooney, R. Landick and I. Artsimovitch, *Mol. Cell*, 2011, **43**, 253–262.
87. K. S. Ha, I. Toulokhonov, D. G. Vassylyev and R. Landick, *J. Mol. Biol.*, 2010, **401**, 708–725.
88. C. D. Sigmund and E. A. Morgan, *Biochemistry*, 1988, **27**, 5622–5627.
89. C. L. Chan and R. Landick, *J. Biol. Chem.*, 1989, **264**, 20796–20804.
90. L. F. Lau, J. W. Roberts and R. Wu, *Proc. Natl. Acad. Sci. U. S. A.*, 1982, **79**, 6171–6175.
91. C. M. Burns, L. V. Richardson and J. P. Richardson, *J. Mol. Biol.*, 1998, **278**, 307–316.
92. M. Z. Qayyum, D. Dey and R. Sen, *J. Biol. Chem.*, 2016, **291**, 8090–8108.
93. T. Pan, I. Artsimovitch, X. W. Fang, R. Landick and T. R. Sosnick, *Proc. Natl. Acad. Sci. U. S. A.*, 1999, **96**, 9545–9550.
94. X. Guo, A. Myasnikov, J. Chen, C. Crucifix, G. Papai, M. Takacs, P. Schultz and A. Weixlbaumer, *Mol. Cell*, 2018, **69**, 816–827.
95. M. C. Schmidt and M. J. Chamberlin, *J. Mol. Biol.*, 1987, **195**, 809–818.
96. S. Mondal, A. V. Yakhnin, A. Sebastian, I. Albert and P. Babitzke, *Nat. Microbiol.*, 2016, **1**, 15007.
97. M. C. Schmidt and M. J. Chamberlin, *J. Biol. Chem.*, 1984, **259**, 15000–15002.
98. G. Muteeb, D. Dey, S. Mishra and R. Sen, *Nucleic Acids Res.*, 2012, **40**, 11213–11228.
99. G. Ralling and T. Linn, *J. Bacteriol.*, 1987, **169**, 2277–2280.
100. S. Saxena and J. Gowrishankar, *J. Bacteriol.*, 2011, **193**, 3842–3850.
101. D. F. Ward and M. E. Gottesman, *Nature*, 1981, **292**, 212–215.
102. K. Singh, J. N. Milstein and W. W. Navarre, *Annu. Rev. Microbiol.*, 2016, **70**, 199–213.
103. M. V. Kotlajich, D. R. Hron, B. A. Boudreau, Z. Sun, Y. L. Lyubchenko and R. Landick, *Elife*, 2015, **4**, e04970.

104. S. S. Ali, J. C. Whitney, J. Stevenson, H. Robinson, P. L. Howell and W. W. Navarre, *J. Biol. Chem.*, 2013, **288**, 13356–13369.
105. D. Chandraprakash and A. S. Seshasayee, *J. Biosci.*, 2014, **39**, 53–61.
106. L. Tran, J. A. van Baarsel, R. S. Washburn, M. E. Gottesman and J. H. Miller, *J. Bacteriol.*, 2011, **193**, 2229–2235.
107. S. Saxena and J. Gowrishankar, *J. Bacteriol.*, 2011, **193**, 3832–3841.
108. V. Epshtein, D. Dutta, J. Wade and E. Nudler, *Nature*, 2010, **463**, 245–249.
109. D. J. Jin, R. R. Burgess, J. P. Richardson and C. A. Gross, *Proc. Natl. Acad. Sci. U. S. A.*, 1992, **89**, 1453–1457.
110. J. S. Park and J. W. Roberts, *Proc. Natl. Acad. Sci. U. S. A.*, 2006, **103**, 4870–4875.
111. W. H. Lang, T. Platt and R. H. Reeder, *Proc. Natl. Acad. Sci. U. S. A.*, 1998, **95**, 4900–4905.
112. Z. Pasman and P. H. von Hippel, *Biochemistry*, 2000, **39**, 5573–5585.
113. M. Turtola and G. A. Belogurov, *Elife*, 2016, **5**, e18096.
114. Z. Hao, V. Epshtein, K. H. Kim, S. Proshkin, V. Svetlov, V. Kamarthapu, B. Bharati, A. Mironov, T. Walz and E. Nudler, *Mol. Cell*, 2021, **81**, 281–292 e288.
115. N. Said, T. Hilal, N. D. Sunday, A. Khatri, J. Burger, T. Mielke, G. A. Belogurov, B. Loll, R. Sen, I. Artsimovitch and M. C. Wahl, *Science*, 2021, **371**, eabd1673.
116. C. Nadiras, E. Eveno, A. Schwartz, N. Figueroa-Bossi and M. Boudvillain, *Nucleic Acids Res.*, 2018, **46**, 8245–8260.
117. N. Said, F. Krupp, E. Anedchenko, K. F. Santos, O. Dybkov, Y. H. Huang, C. T. Lee, B. Loll, E. Behrmann, J. Burger, T. Mielke, J. Loerke, H. Urlaub, C. M. T. Spahn, G. Weber and M. C. Wahl, *Nat. Microbiol.*, 2017, **2**, 17062.
118. B. Pani, S. Banerjee, J. Chalissery, A. Muralimohan, R. M. Loganathan, R. B. Suganthan and R. Sen, *J. Biol. Chem.*, 2006, **281**, 26491–26500.
119. G. Ghosh, J. Reddy, S. Sambhare and R. Sen, *J. Bacteriol.*, 2018, **200**, e00380-17.
120. M. Rabhi, O. Espeli, A. Schwartz, B. Cayrol, A. R. Rahmouni, V. Arluison and M. Boudvillain, *EMBO J.*, 2011, **30**, 2805–2816.
121. S. Pichoff, L. Alibaud, A. Guedant, M. P. Castanie and J. P. Bouche, *Mol. Microbiol.*, 1998, **29**, 859–869.
122. P. Gutierrez, G. Kozlov, L. Gabrielli, D. Elias, M. J. Osborne, I. E. Gallouzi and K. Gehring, *J. Biol. Chem.*, 2007, **282**, 23348–23353.
123. J. F. Hopkins, S. Panja and S. A. Woodson, *Nucleic Acids Res.*, 2011, **39**, 5193–5202.
124. K. Pal, M. Yadav, S. Jain, B. Ghosh, R. Sen and U. Sen, *J. Mol. Biol.*, 2019, **431**, 4749–4766.
125. T. Soper, P. Mandin, N. Majdalani, S. Gottesman and S. A. Woodson, *Proc. Natl. Acad. Sci. U. S. A.*, 2010, **107**, 9602–9607.
126. M. C. Carrier, D. Lalaouna and E. Masse, *Annu. Rev. Microbiol.*, 2018, **72**, 141–161.
127. I. J. Silva, S. Barahona, A. Eyraud, D. Lalaouna, N. Figueroa-Bossi, E. Masse and C. M. Arraiano, *Proc. Natl. Acad. Sci. U. S. A.*, 2019, **116**, 3042–3051.

128. L. Bossi, A. Schwartz, B. Guillemardet, M. Boudvillain and N. Figueroa-Bossi, *Genes Dev.*, 2012, **26**, 1864–1873.
129. J. J. Tree, S. Granneman, S. P. McAteer, D. Tollervey and D. L. Gally, *Mol. Cell*, 2014, **55**, 199–213.
130. R. R. Breaker, *Mol. Cell*, 2011, **43**, 867–879.
131. A. V. Sherwood and T. M. Henkin, *Annu. Rev. Microbiol.*, 2016, **70**, 361–374.
132. S. DebRoy, M. Gebbie, A. Ramesh, J. R. Goodson, M. R. Cruz, A. van Hoof, W. C. Winkler and D. A. Garsin, *Science*, 2014, **345**, 937–940.
133. E. Loh, O. Dussurget, J. Gripenland, K. Vaitkevicius, T. Tiensuu, P. Mandin, F. Repoila, C. Buchrieser, P. Cossart and J. Johansson, *Cell*, 2009, **139**, 770–779.
134. J. E. Graham and J. P. Richardson, *J. Biol. Chem.*, 1998, **273**, 20764–20769.
135. D. Schneider, L. Gold and T. Platt, *FASEB J.*, 1993, **7**, 201–207.
136. C. A. Brennan, A. J. Dombroski and T. Platt, *Cell*, 1987, **48**, 945–952.
137. K. V. Konan and C. Yanofsky, *J. Bacteriol.*, 2000, **182**, 3981–3988.
138. R. R. Breaker, *Cold Spring Harbor Perspect. Biol.*, 2018, **10**, a032797.
139. N. Windbichler, F. von Pelchrzim, O. Mayer, E. Csaszar and R. Schroeder, *RNA Biol.*, 2008, **5**, 30–40.
140. I. Irnov and W. C. Winkler, *Mol. Microbiol.*, 2010, **76**, 559–575.
141. N. Komissarova, T. Velikodvorskaya, R. Sen, R. A. King, S. Banik-Maiti and R. A. Weisberg, *Mol. Cell*, 2008, **31**, 683–694.
142. N. Sedlyarova, P. Rescheneder, A. Magan, N. Popitsch, N. Rziha, I. Bilusic, V. Epshtein, B. Zimmermann, M. Lybecker, V. Sedlyarov, R. Schroeder and E. Nudler, *Mol. Cell*, 2017, **67**, 30–43 e36.
143. G. W. Li, E. Oh and J. S. Weissman, *Nature*, 2012, **484**, 538–541.
144. S. Elgamal, I. Artsimovitch and M. Ibba, *mBio*, 2016, **7**, e01373-16.
145. G. Demo, A. Rasouly, N. Vasilyev, V. Svetlov, A. B. Loveland, R. Diaz-Avalos, N. Grigorieff, E. Nudler and A. A. Korostelev, *Elife*, 2017, **6**, e28560.
146. H. Fan, A. B. Conn, P. B. Williams, S. Diggs, J. Hahm, H. B. Gamper Jr, Y. M. Hou, S. E. O'Leary, Y. Wang and G. M. Blaha, *Nucleic Acids Res.*, 2017, **45**, 11043–11055.
147. R. Kohler, R. A. Mooney, D. J. Mills, R. Landick and P. Cramer, *Science*, 2017, **356**, 194–197.
148. J. Y. Kang, T. V. Mishanina, M. J. Bellecourt, R. A. Mooney, S. A. Darst and R. Landick, *Mol. Cell*, 2018, **69**, 802–815.
149. J. R. Widom, Y. A. Nedialkov, V. Rai, R. L. Hayes, C. L. Brooks 3rd, I. Artsimovitch and N. G. Walter, *Mol. Cell*, 2018, **72**, 541–552 e546.
150. A. Sevostyanova and E. A. Groisman, *Proc. Natl. Acad. Sci. U. S. A.*, 2015, **112**, E6835–E6843.
151. W. D. Morgan, D. G. Bear and P. H. von Hippel, *J. Biol. Chem.*, 1983, **258**, 9553–9564.
152. L. F. Lau and J. W. Roberts, *J. Biol. Chem.*, 1985, **260**, 574–584.
153. G. A. Kassavetis and M. J. Chamberlin, *J. Biol. Chem.*, 1981, **256**, 2777–2786.

154. A. M. Malinen, M. Nandymazumdar, M. Turtola, H. Malmi, T. Grocholski, I. Artsimovitch and G. A. Belogurov, *Nat. Commun.*, 2014, **5**, 3408.
155. S. Gong, Y. Wang and W. Zhang, *J. Chem. Phys.*, 2015, **143**, 045103.
156. G. A. Perdrizet 2nd, I. Artsimovitch, R. Furman, T. R. Sosnick and T. Pan, *Proc. Natl. Acad. Sci. U. S. A.*, 2012, **109**, 3323–3328.
157. H. Steinert, F. Sochor, A. Wacker, J. Buck, C. Helmling, F. Hiller, S. Keyhani, J. Noeske, S. Grimm, M. M. Rudolph, H. Keller, R. A. Mooney, R. Landick, B. Suess, B. Furtig, J. Wohnert and H. Schwalbe, , *Elife*, 2017, **6**, e21297.
158. K. Hollands, A. Sevostiyanova and E. A. Groisman, *Proc. Natl. Acad. Sci. U. S. A.*, 2014, **111**, E1999–2007.
159. C. L. Chan, D. Wang and R. Landick, *J. Mol. Biol.*, 1997, **268**, 54–68.
160. I. Toulokhonov, I. Artsimovitch and R. Landick, *Science*, 2001, **292**, 730–733.
161. D. Dutta, J. Chalissery and R. Sen, *J. Biol. Chem.*, 2008, **283**, 20243–20251.
162. J. Sollier and K. A. Cimprich, *Trends Cell Biol.*, 2015, **25**, 514–522.
163. J. Gowrishankar, J. K. Leela and K. Anupama, *Transcription*, 2013, **4**, 153–157.
164. R. Harinarayanan and J. Gowrishankar, *J. Mol. Biol.*, 2003, **332**, 31–46.
165. K. S. Lang and H. Merrikh, *Annu. Rev. Microbiol.*, 2018, **72**, 71–88.
166. D. Dutta, K. Shatalin, V. Epshtein, M. E. Gottesman and E. Nudler, *Cell*, 2011, **146**, 533–543.
167. K. S. Lang, A. N. Hall, C. N. Merrikh, M. Ragheb, H. Tabakh, A. J. Pollock, J. J. Woodward, J. E. Dreifus and H. Merrikh, *Cell*, 2017, **170**, 787–799 e718.
168. B. W. Trautinger and R. G. Lloyd, *EMBO J.*, 2002, **21**, 6944–6953.
169. H. Boubakri, A. L. de Septenville, E. Viguera and B. Michel, *EMBO J.*, 2010, **29**, 145–157.
170. S. Hamperl, M. J. Bocek, J. C. Saldivar, T. Swigut and K. A. Cimprich, *Cell*, 2017, **170**, 774–786 e719.
171. C. H. Freudenreich, *Curr. Genet.*, 2018, **64**, 789–794.
172. A. Kuzminov, *Trends Genet.*, 2018, **34**, 111–120.
173. I. Artsimovitch, V. Svetlov, L. Anthony, R. R. Burgess and R. Landick, *J. Bacteriol.*, 2000, **182**, 6027–6035.
174. N. Raghunathan, R. M. Kapshikar, J. K. Leela, J. Mallikarjun, P. Bouloc and J. Gowrishankar, *Nucleic Acids Res.*, 2018, **46**, 3400–3411.
175. M. Groh, L. O. Albulescu, A. Cristini and N. Gromak, *J. Mol. Biol.*, 2017, **429**, 3181–3195.
176. M. Castelnuovo, J. B. Zaugg, E. Guffanti, A. Maffioletti, J. Camblong, Z. Xu, S. Clauder-Munster, L. M. Steinmetz, N. M. Luscombe and F. Stutz, *Nucleic Acids Res.*, 2014, **42**, 4348–4362.
177. Z. Han and O. Porrua, *Transcription*, 2018, **9**, 152–158.
178. N. Figueroa-Bossi, A. Schwartz, B. Guillemardet, F. D'Heygere, L. Bossi and M. Boudvillain, *Genes Dev.*, 2014, **28**, 1239–1251.

CHAPTER 6

Global Regulation of Transcription by Nucleotides and (p)ppGpp

LUCAS M. ONDER[a], JUE D. WANG*[a], JONATHAN JAGODNIK[a], WILMA ROSS[a] AND RICHARD L. GOURSE*[a]

[a]University of Wisconsin–Madison, Department of Bacteriology, 1550 Linden Drive, Madison, WI 53706, USA
*E-mail: wang@bact.wisc.edu, rgourse@bact.wisc.edu

6.1 Introduction

The activity of the RNA polymerase (RNAP) molecular machine is highly regulated in organisms from humans to bacteria. Gene expression is regulated in response to the external environment and to internal cues in order to mediate feedback and homeostatic controls, to mount stress responses, and to initiate developmental pathways. Most attention surrounding gene expression control has focused on protein-based transcription factors, receptor-mediated protein modifications, and ligands that allosterically regulate repressors, activators, riboswitches, or alternative sigma factors. In this chapter, we focus on less appreciated regulators of transcription, notably ribonucleotides themselves and ribonucleotide derivatives that act as signaling molecules to control the synthesis or degradation of the substrates of the transcription machinery or the activity of the core transcription machinery itself. We focus on more recent aspects of the voluminous literature on these subjects.

Chemical Biology No. 19
RNA Polymerases as Molecular Motors: On the Road, 2nd Edition
Edited by Robert Landick, Terence Strick and Jue Wang

Published by the Royal Society of Chemistry, www.rsc.org

Nucleotide derivatives can serve as signals of the nutritional status of the cell and/or of the external environment, and therefore they are sometimes referred to as alarmones.[1] Perhaps the most well-characterized of these alarmones are guanosine tetraphosphate (ppGpp) and guanosine pentaphosphate (pppGpp), collectively called (p)ppGpp. Although these alarmones are produced throughout the bacterial domain of life, as well as in chloroplasts,[2] there is significant evolutionary diversity in how they are utilized. Here we will focus primarily on the action of these alarmones in the model Gram-negative bacterium *Escherichia coli* and in the model Gram-positive bacterium *Bacillus subtilis*, but we emphasize that there are additional mechanisms for (p)ppGpp utilization that are not represented by the mechanisms observed in these two species.

6.2 Changing Nucleotide Triphosphate (NTP) Concentrations Can Regulate Transcription Initiation and Elongation

Adenosine triphosphate (ATP), guanosine triphosphate (GTP), cytosine triphosphate (CTP), and uridine triphosphate (UTP) concentrations range from low micromolar to millimolar in *E. coli* and *B. subtilis*. Fluctuations in the environment, including nutrient starvation and entrance into stationary phase, result in changes in nucleotide levels. The efficiency of transcription initiation is sensitive to the concentrations of the NTP substrates, as the Michaelis–Menten constant (K_m) for the first NTP is much higher than that for the elongating NTPs.[3] In both *E. coli* and *B. subtilis*, the K_m for the initiating NTP for rRNA promoters is unusually high (weaker affinity) compared with standard promoters.[4–8] As a result, fluctuations in NTP concentrations have stronger effects on rRNA promoters than on most other promoters. The effects of NTP concentration on rRNA promoters results primarily from the instability of the open complex, which therefore has a tendency to fall back to earlier transcription initiation intermediates in the absence of NTP substrates. When NTP pools are high, however, they drive transcription initiation forward by mass action. Although the promoter sequences that determine open complex lifetime are complex, there is an almost perfect correlation between regulation of rRNA promoters by NTP concentration and open complex lifetime.[9] A decline in NTP levels during nutrient depletion and as cells enter stationary phase thereby contributes to the decrease in rRNA transcription at that time, and when cells go deeper into stationary phase, further NTP depletion may be the major contributor to reduction in rRNA synthesis.[4,8]

Specific DNA sequences also differentially affect transcription elongation. Although the affinity of the transcription complex for the elongating substrates is generally much higher and more uniform than that for the initial substrate, each succeeding NTP has its own characteristic K_m depending on the identity of the template base and the surrounding sequences.[10] RNAP pauses at positions where the affinities are weak, depending on the availability of the substrate NTP (see also Belogurov and Artsimovitch[11]). An evolutionarily conserved "universal pause sequence" has been identified that increases the propensity for pausing.[12–14] Low cellular NTP concentrations increase the magnitude of the effects of DNA sequences and elongation and termination factors. Thus, changes in NTP concentration can change the transcription start site or cause backtracking, transcription slippage, reiterative transcription, and even transcription attenuation, and thereby alter transcription elongation, pausing, and termination (reviewed by Turnbough and Switzer[15]). Such effects of NTP concentrations were documented long ago in several pyrimidine biosynthetic operons, where high UTP concentrations feedback regulate gene expression by a variety of different mechanisms. For example, in the cytosine deaminase *codBA* operon, high intracellular levels of UTP favor transcriptional initiation at a position that results in non-productive, reiterative transcription, whereas low UTP concentration leads to initiation at the adjacent position and productive elongation.[16] In the pyrimidine biosynthetic *pyrBI* operon, high UTP concentrations lead to transcriptional attenuation, whereas low concentrations of UTP lead to productive elongation.[17] A long A+T tract in the initial transcribed region of the uracil phosphoribosyltransferase (*upp*) operon leads to reiterative transcription, sensitizing the operon to the effects of high UTP concentrations.[18]

More generally, results from recent studies have indicated that NTP concentrations can affect the rate of promoter clearance. Specifically, high NTP concentrations promote the transition from the initiation to the elongation complex by helping RNAP escape from a scrunched, backtracked, paused intermediate in which σ region 3.2 competes with the growing RNA chain for the RNA exit channel (Lerner *et al.*;[19] see also Winkelman *et al.*[20]).

6.3 The Nucleotide Derivatives pppGpp and ppGpp Control Transcription by Multiple Mechanisms

6.3.1 Historical Background

ppGpp and pppGpp were first discovered as atypical "spots" on thin-layer chromatography plates from ^{32}P-radiolabelled cellular extracts prepared from amino acid-starved *E. coli* cells.[21] Because the appearance of these spots correlated with the decrease in stable RNA synthesis (Cashel and Gallant,[21] Fiil *et al.*[22]), ppGpp and pppGpp were proposed as potential regulators of ribosome synthesis and referred to as 'magic spots'.[23] ppGpp and pppGpp are produced from GDP or GTP, respectively, by multidomain (p)ppGpp synthetases RSH (RelA and SpoT homologs), and by small alarmone synthetases SAS[24] such as YwaC and YjbM (also called SasA and SasB) in *B. subtilis*,[24,25] but ppGpp results primarily from conversion of the pentaphosphate to the tetraphosphate *in vivo* because of the enzyme guanosine-5′-triphosphate, 3′-diphosphate pyrophosphatase (GppA).[23] The efficiency of this conversion may vary with specific nutrient conditions and with the particular bacterium.

(p)ppGpp act as signaling molecules that adapt the cell to many environmental stresses, not just amino acid starvation.[23,26–28] The cell regulates its rate of ribosome synthesis to match the need for protein synthesis, a correlation referred to as growth rate dependent regulation of ribosome synthesis.[29,30] Generally, mutations that disturb the rate of ribosome synthesis decrease the growth rate, presumably as a result of misappropriation of cellular resources between the synthesis of the translation apparatus *versus* other biosynthetic processes. Because the (p)ppGpp concentration under most conditions contributes to regulation of ribosome synthesis and therefore correlates inversely with the cellular growth rate,[31] and mutations that alter (p)ppGpp production reduce the growth rate, it is sometimes assumed that (p)ppGpp is responsible for determining the cellular growth rate. However, there are many contributors to the cellular growth rate, and (p)ppGpp concentration should not be assumed to be uniquely responsible.

6.3.2 Regulation of Transcription by (p)ppGpp Binding to Enzymes Responsible for the Synthesis of NTPs

(p)ppGpp production consumes GTP. Therefore, (p)ppGpp synthesis reduces purine nucleotide pools in both proteobacteria like *E. coli* and in firmicutes like *B. subtilis*. This depletion is most apparent in Gram-positives, as reported in early work from Freese and colleagues[32,33] showing that (p)ppGpp promotes sporulation in *B. subtilis*, largely as a result of the reduction of the GTP concentration. (p)ppGpp also inhibits GTP synthesis in *B. subtilis* by binding directly to, and inhibiting, enzymes needed for GTP synthesis (Figure 6.1). This includes guanylate kinase, the enzyme that produces guanosine diphosphate (GDP) from guanosine monophosphate (GMP)[34,35] and hypoxanthine–guanine phosphoribosyltransferase (HprT), and xanthine phosphoribosyltransferase (XpT),[35–37] purine salvage enzymes that convert guanine, hypoxanthine and xanthine to GMP, inosine monophosphate (IMP) and xanthine monophosphate (XMP), respectively. While (p)ppGpp does not inhibit guanylate kinase in proteobacteria like *E. coli*, it has been shown recently that (p)ppGpp also reduces GTP synthesis directly by binding to phosphoribosylpyrophosphate amidotransferase (PurF), HprT, and guanosine kinase (Gsk) enzymes in *E. coli* needed for *de novo* and salvage purine biosynthesis.[38–40]

The reduction of GTP concentration by (p)ppGpp has promoter-specific effects on transcription initiation. Transcription initiates with GTP from both the *rrn* P1 and P2 promoters in all 10 rRNA operons in *B. subtilis*. Because these promoters form short-lived open complexes (as they do in *E. coli*), rRNA transcription initiation is strongly dependent on high GTP concentrations.[8] Since (p)ppGpp induction reduces the GTP concentration by 60–90% in *B. subtilis*, rRNA synthesis is reduced accordingly. This appears to be the major mechanism for control of rRNA synthesis in firmicutes, as the rRNA response to (p)ppGpp after amino acid starvation is defective in *relA* mutants.[8] In contrast, the primary mechanism by which (p)ppGpp regulates

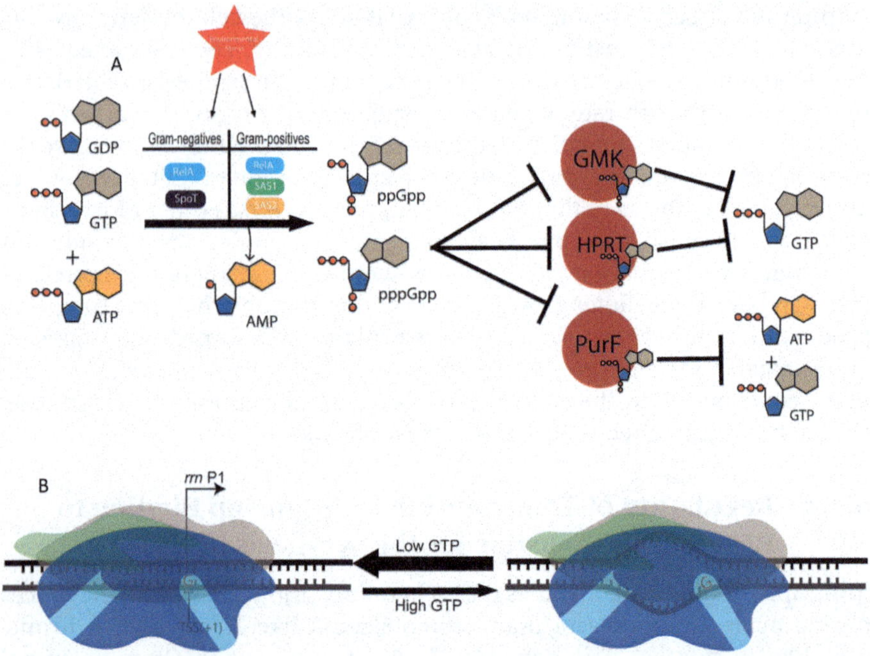

Figure 6.1 **ppGpp, *de novo* and salvage purine synthesis pathways, and effects of NTP concentrations on rRNA transcription initiation.** Top. After an environmental stressor such as nutrient limitation (p)ppGpp is synthesized from GDP or GTP by the addition of a pyrophosphate from ATP to the 3′-hydroxyl group resulting in ppGpp or pppGpp, respectively. In *E. coli*, production of ppGpp is by RelA and SpoT only, whereas production of ppGpp in Gram-positives is by RelA, SAS1, and SAS2. Direct inhibition of guanylate kinase (Gmk), HprT, and PurF by ppGpp results in a negative feedback loop, resulting in repression of GTP and ATP levels. PurF is inhibited by ppGpp only in *E. coli*, whereas Gmk is inhibited by ppGpp only in *B. subtilis*. HprT is inhibited by ppGpp in both organisms. Bottom. Reduction of NTP levels results in inhibition of promoters that require high NTP concentrations in order to initiate transcription. For example, all 10 rRNA operons in *B. subtilis* initiate with GTP, and low levels of GTP result in inhibition of rRNA in *B. subtilis* due to the instability of the open complex formed by rRNA promoters with RNAP. In *E. coli*, six of the seven *rrn* P1 promoters initiate with ATP and are inhibited by low ATP concentrations, the other *rrn* P1 promoter starts with GTP and is inhibited by low GTP concentrations, and the 7 *rrn* P2 promoters start with CTP and are inhibited by low CTP concentrations.

rRNA transcription in proteobacteria is by binding directly to RNAP[41,42] (see later in this chapter).

Whereas (p)ppGpp negatively regulates rRNA promoters by decreasing GTP pools in *B. subtilis*, (p)ppGpp positively regulates transcription from promoters for branched chain amino acids (BCAA). The best characterized is the *ilv* operon encoding enzymes for synthesis of leucine and isoleucine. During amino acid starvation, *ilv* operon expression is upregulated more than 30-fold.[43] This results in large part from an increase in transcription from promoters controlled by cytosine deaminase operon protein Y (CodY), a repressor for which binding to DNA is dependent on GTP as a cofactor, *i.e.* when (p)ppGpp reduces the GTP concentration, this derepresses promoters inhibited by CodY–GTP.[44,45] Consistent with this model, deletion of the genes encoding the three (p)ppGpp synthetases, RelA, YwaC and YjbM, results in a loss of *ilv* transcription.[43]

It has also been reported that changes in ATP concentrations have a CodY-independent effect on some promoters.[46,47] In this case, it was shown that (p)ppGpp accumulation results in a decrease in the intracellular GTP concentration and a reciprocal increase in the ATP concentration, stimulating transcription initiation from genes whose initiating nucleotide is ATP, including the *ilvB* operon. In support of this model, maximum transcription initiation from these promoters requires a much higher concentration of the initiating nucleotide (ATP), and switching the initiating nucleotide from A to G disrupts this regulation.

To examine the effects of (p)ppGpp on the transcriptome of *B. subtilis* comprehensively, microarray gene expression profiling was performed using wild type and (p)ppGpp^0 strains. To delineate the downstream effects of (p)ppGpp, these gene expression profiles were compared with that of a (p)ppGpp^0 Δ*codY* strain and a (p)ppGpp^0 *guaB*^{down} strain in which GTP synthesis is reduced by lowering expression of *guaB* which encodes IMP dehydrogenase, an enzyme involved in GTP biosynthesis.[48] Their transcriptomes were characterized after guanosine addition (to increase GTP levels) or by addition of arginine hydroxamate (to induce amino acid starvation), allowing for a comprehensive snapshot of the transcriptome in various backgrounds. Analysis of expression profiles of the *ilvB* operon, including the *ilvBHC* and *leuABCD* genes, confirmed that they are indeed induced in wild type cells upon (p)ppGpp induction. In (p)ppGpp^0 cells, expression is strongly inhibited even during amino acid starvation, indicating that amino acid biosynthesis in (p)ppGpp^0 cells is extremely misregulated. This misregulation is responsible for the branched-chain amino acid auxotrophy observed in (p)ppGpp^0 cells. Genes in the *ilv* operon are highly expressed in a (p)ppGpp^0 Δ*codY* strain, revealing that CodY-dependent regulation is mostly responsible for the repression of *ilv* under normal growth conditions and its activation by (p)ppGpp.

The genome-wide transcription profiling also revealed profound (p)ppGpp-dependent upregulation of other components of the CodY regulon.[43,48] These include other BCAA genes including *ilvD* and *ybgE*, the *app* and *opp* oligopeptide ABC transporter operons, and the *dpp* dipeptide ABC transporter operon. CodY functions as a transcription repressor by binding to these promoters when GTP levels are high. Production of (p)ppGpp reduces GTP levels, thus derepressing CodY and inducing gene expression.

Some of the genes whose transcription is strongly activated upon (p)ppGpp accumulation, such as 5-methyltetrahydropteroyltriglutamate--homocysteine methyltransferase (*metE*) and the homoserine dehydrogenase (*hom*) operon including the threonine synthesis *thrBC* genes, are not identified as members of the CodY regulon. Their upregulation can be explained, at least in part, by the changes in GTP and/or ATP levels that affect transcription independently of CodY.[43]

Finally, in addition to the GTP and ATP mediated transcriptional effects, there is a strong general decrease in expression of the *de novo* purine biosynthesis genes (the *purEKBCSQLFMNHD* operon) upon (p)ppGpp induction. This is due to direct binding of (p)ppGpp to the transcription regulator PurR. Thus (p)ppGpp, in addition to regulating gene expression indirectly *via* its effects on NTP levels, also directly regulates gene expression of the PurR regulon.[49]

6.3.3 The Nucleotide Derivatives pppGpp and ppGpp Control Transcription Directly by Multiple Mechanisms

6.3.3.1 Effects on Transcription Initiation

In *B. subtilis*, characterization of specific promoters provided strong evidence for both positive and negative stringent regulation through changes in GTP concentrations.[8,43,49] However, in proteobacteria, (p)ppGpp exerts its effects on transcription primarily by binding directly to RNAP. We focus first on the effect of (p)ppGpp on transcription initiation in *E. coli*, followed by briefly mentioning its reported effects on transcription elongation.

Full regulation of transcription by (p)ppGpp in *E. coli* requires the small protein DksA.[50] It was initially assumed that DksA prevented a protein folding defect since it was identified originally as a multicopy suppressor of a *dnaK* mutant.[51] However, it was later shown that DksA is a transcription factor that binds directly to the RNAP secondary channel[50,52] and that the suppression of a *dnaK* mutant was dependent on non-physiologically high levels of expression of DksA.[53] The β′ rim helices, the binding site for DksA, are also the binding site for the transcription factors A and B (GreA and GreB).[54–56] However, DksA shares little sequence homology with the Gre factors and primarily affects transcription initiation, whereas the Gre factors primarily affect elongation.[26,50,57–59]

The discovery that DksA is a transcription factor that works synergistically with (p)ppGpp was critical for understanding the stringent response. Together, DksA and (p)ppGpp reduce transcription from rRNA promoters *in vitro* approximately 20-fold compared with the approximately threefold reduction observed with (p)ppGpp alone.[41,50,60] (p)ppGpp and DksA also activate transcription of many amino acid biosynthesis genes.[42,58] Consistent with the effects of DksA *in vitro*, deletion of the *dksA* gene results in defects that share similarities with cells incapable of producing (p)ppGpp, including polyauxotrophy.[50,58,61,62]

6.3.3.2 (p)ppGpp Binding Sites on RNAP

In *E. coli*, (p)ppGpp interacts directly with RNAP to regulate transcription initiation at specific promoters. Since (p)ppGpp lacks the appropriate 3′OH groups for attack by the incoming NTP, (p)ppGpp should not be able to be utilized as a potential substrate for RNAP chain elongation. Therefore, identification of the (p)ppGpp binding site(s) on RNAP was a crucial step in understanding its mechanism of action. Although an early cocrystal of ppGpp with *Thermus thermophilus* RNAP indicated that (p)ppGpp might bind near the catalytic site,[63] the results of further investigation indicated that potential binding of (p)ppGpp to this site in *E. coli* RNAP was not responsible for effects of (p)ppGpp on transcription by *E. coli* RNAP,[64] and, in fact, (p)ppGpp binding to that site on *T. thermophilus* RNAP was not likely to be responsible for effects of (p)ppGpp on *T. thermophilus* transcription either.[65] In retrospect, it appears that binding of ppGpp in the *T. thermophilus* co-crystal might have derived from artifactual binding of the alarmone to the so-called E-site of RNAP, a binding site favored by NTPs on their approach to the active site.[66] Consistent with this hypothesis, when the (p)ppGpp binding site(s) were finally determined and confirmed genetically, it was found that it bound to two distinct sites (which were then dubbed Sites 1 and 2), neither of which corresponded to the site identified in the *T. thermophilus* co-crystal.

Site 1 was discovered by using ^{32}P-6-thio-ppGpp to produce a UV-dependent crosslink within RNAP. After finding that 6-thio-ppGpp crosslinked to one of the two large subunits of RNAP, the subunit harboring the crosslink was identified by fusing β′ to green fluorescent protein (GFP) to distinguish its migration from that of β on polyacrylamide gels. Single thrombin sites were engineered at a series of positions in β′, and then a crosslink was mapped to an interval of 36 amino acids within the β′ subunits using cleavage with thrombin. Substitutions were created in RNAP near the site of the crosslink, and RNAPs containing these substitutions were then tested for their responses to (p)ppGpp by *in vitro* transcription.[67] RNAPs containing a deletion of the N-terminal amino acids in ω (ωΔ2–5) or multiple substitutions in β′ were insensitive to inhibition by (p)ppGpp *in vitro*, indicating that a pocket was formed by two sections of β′ and the N-terminal residues in the ω subunit. X-ray structures of co-crystals of *E. coli* RNAP holoenzyme containing ppGpp confirmed the position of the binding site and extended the identification of the liganding residues responsible for forming the binding pocket[68,69] [see Figure 6.2 and legend for positions of substitutions that reduced or eliminated (p)ppGpp function]. However, major conformational change(s) that could account for the effect on transcription were not observed by comparison of the crystal structures with and without ppGpp.

Mutations in the *rpoB* or *rpoZ* genes that resulted in RNAPs that failed to respond to ppGpp *in vitro* were introduced into the bacterial chromosome, and the resulting strains were tested *in vivo*.[67] Surprisingly, the mutant strains lacking Site 1 exhibited only mild defects in responding to amino acid starvation. Furthermore, RNAPs lacking the ppGpp binding site partially recovered a response to ppGpp when DksA was included in the *in vitro* transcription reaction. These results led to the hypothesis that there might be another ppGpp binding site on RNAP (now called Site 2). To identify the putative binding site, RNAP lacking Site 1 was incubated with DksA and ^{32}P-6-thio-ppGpp, and crosslinks were identified to DksA itself. The crosslink to DksA was completely dependent on inclusion of RNAP.[41]

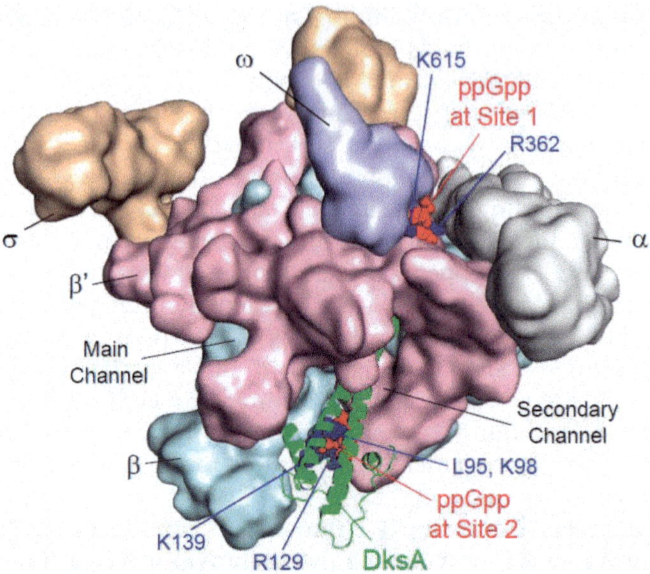

Figure 6.2 **Surface view of ppGpp binding sites 1 and 2 on *E. coli* RNAP.** Site 1 is at an interface between the β′ and ω subunits, and Site 2 is at an interface between the β′ subunit and DksA.[41,67–69,71] DksA binds to the RNAP secondary channel. The RNAP surface representation was created in Pymol using PDB4JKR, with the β′ subunit in pink, β in cyan, α in grey, σ in wheat, and ω in light blue. DksA, in green, is shown as a cartoon adapted from PDB 1TJL. ppGpp is shown as red spheres. Residues indicated as blue spheres in Sites 1 and 2 are required for ppGpp binding and function[41,67] including β′R362 and β′K615 in Site 1 and DksA L95, K98, R129, and K139 in Site 2. The locations of the main and secondary channels in RNAP are also indicated.

ppGpp binding to RNAP was tested using direct binding assays [differential radial capillary action of ligand assay (DRaCALA)].[41,70] These data supported a model in which RNAP contains a ppGpp binding site at an interface of the rim-helices of β′ and DksA, as well as the one at the interface of β′ and ω (Site 1). Substitutions for specific DksA residues and for β′ rim-helix residues reduced effects of ppGpp on inhibition and activation without affecting DksA binding to RNAP, identifying the crucial residues contributing to the ppGpp Site 2 binding pocket (see Figure 6.2). Whereas Sites 1 and 2 were both required for full inhibition of transcription from an rRNA promoter, only Site 2 was required for full activation of amino acid biosynthetic promoters.[41]

A crystal structure containing *E. coli* RNAP, DksA, and (p)ppGpp confirmed the positions of the (p)ppGpp binding sites[71] and displayed a small rigid body rotation centered around (p)ppGpp Site 2. However, crystal packing constraints made the significance of this rotation unclear and furthermore prevented identification of major conformational changes resulting from binding of DksA and (p)ppGpp. Recent single-particle cryo-electron microscopy (cryo-EM) structures, in conjunction with the results of biochemical and mutational studies,[72] indicate that DksA and (p)ppGpp exert their effects on transcription by adjusting the conformations of multiple dynamic regions of RNAP, altering the rates of multiple steps in the transcription initiation mechanism (see Section 6.3.3.4).

Although the structural analyses confirmed the positions of the (p)ppGpp binding sites and permitted determination of the identities of all of the amino acids contacted by (p)ppGpp, they did not provide information about the relative affinities of the two sites for (p)ppGpp. This question has finally been answered: the two sites on RNAP bind ppGpp with similar affinity when omega and DksA are saturating.[73]

6.3.3.3 Genome-wide Effects of (p)ppGpp Binding to RNAP

The effects of (p)ppGpp on transcription are global and pervasive. The development of functional genomic tools has facilitated a view of the cellular response at a genomic level, complementing the results from mechanistic studies. Early studies of the genome-wide effects of (p)ppGpp in *E. coli* utilized expression microarrays on cells treated with serine hydroxamate, an inhibitor of serine tRNA aminoacylation, to induce the stringent response.[74] Additional expression microarray studies compared transcripts produced in wild-type and Δ*relA* Δ*spoT* strains following isoleucine-exhaustion.[75,76] A fourth study compared steady-state expression in wild-type, Δ*dksA*, and Δ*relA* Δ*spoT* strains.[77] Although each of the studies identified a large number of transcripts that changed with (p)ppGpp levels, there were major differences in the identities of the affected transcripts in the different studies, perhaps not surprising since amino acid limitation itself has profound (p)ppGpp-independent effects on cellular metabolism. Furthermore, the stringent response is a rapid, transient response, dramatically altering transcription within a few minutes of (p)ppGpp induction, and the response has profound consequences for subsequent gene expression. Measurements conducted after the first few minutes of (p)ppGpp induction therefore probably include indirect effects resulting from the initial direct effects of (p)ppGpp on transcription.

Sanchez-Vazquez *et al.* therefore attempted to identify the (p)ppGpp regulon using RNA sequencing (RNA-seq) of transcripts 5 or 10 minutes after ectopic expression of *relA*.[42] (p)ppGpp was induced without concurrent amino acid limitation. Expression of 757 genes changed at least two-fold within 5 minutes of induction of (p)ppGpp in the strain containing the wild-type (p)ppGpp binding sites on RNAP, whereas there was little or no response in a strain in which (p)ppGpp was produced but the two binding sites on RNAP were eliminated by mutation. Thus, in *E. coli* at least, (p)ppGpp does not affect transcription directly by binding to transcription factors other than RNAP. In large part because of the increased sensitivity and reliability of RNA-seq *versus* expression microarrays, more than 75% of the promoter targets reported in this study had not been identified in the previous microarray studies.[74,76] *In vitro* transcription analysis of effects of (p)ppGpp–DksA on a large number of the promoters identified in the RNA-seq experiment confirmed that virtually all the effects on transcription identified in the RNA-seq experiment were direct.

The regulon members identified in the RNA-seq study confirmed previous conclusions that (p)ppGpp inhibits synthesis of multiple components of the translation apparatus, but the number of genes in the translation category was much greater than expected; more than 100 different transcripts related to translation were regulated by (p)ppGpp. Not only were ribosomal RNA and ribosomal protein genes regulated by (p)ppGpp, but so were many of the genes coding for rRNA and tRNA processing enzymes, translation factors, the GTPases that aid in ribosome assembly and maturation, and genes coding for enzymes that modify rRNAs, tRNAs, and r-proteins. Similarly, transcription of genes responsible for biosynthesis of 16 of the 20 amino acids was activated by (p)ppGpp, many more than predicted previously. The regulon also included many unexpected categories of genes, including a large number of genes related to the cell exterior.

6.3.3.4 Promoter Specificity and the Mechanism of Regulation by ppGpp–DksA and TraR

Unlike canonical factors that regulate transcription initiation by binding to specific DNA sequences adjacent to, or overlapping, the binding site for RNAP, promoter specificity for members of the ppGpp–DksA regulon is not determined by a binding sequence for the factor on DNA. Rather, (p)ppGpp and DksA bind directly to RNAP whether or not the output from the promoter is regulated. Specificity, whether the promoter is activated, inhibited, or unaffected, is determined by the intrinsic kinetic properties of the promoter.[78]

Transcription initiation is a multistep process in which there are a series of conformational changes of the complex. Interfaces established in earlier intermediates trigger subsequent rearrangements to form later intermediates, ultimately leading to DNA opening and alignment of the start site base on the template strand with the catalytic site of the enzyme. The whole process is driven by binding free energy, not by hydrolysis of ATP or other high energy molecules.[79]

Studies on the effects of (p)ppGpp and DksA on transcription have recently been supplemented by studies on transcriptional activator R (TraR), a distant DksA homolog that is produced by the proteobacterial F element. TraR is only half the size of DksA, but it binds in the secondary channel of RNAP like DksA, and it activates or inhibits roughly the same set of promoters as (p)ppGpp and DksA.[80,81] Recently, structural studies using single particle cryo-EM on TraR–RNAP complexes and ppGpp–DksA–RNAP complexes have identified multiple conformational changes in RNAP that occur in response to the regulatory factors.[72,82,83] The structures indicate that (p)ppGpp–DksA and TraR cause conformational changes in RNAP that were not apparent from X-ray structures of RNAP containing (p)ppGpp–DksA or TraR because of crystal packing constraints.[71] The results of the cryo-EM studies, together with previous mechanistic studies that indicated that (p)ppGpp–DksA and TraR alter "isomerization" steps, *i.e.* steps in transcription initiation after formation of the closed complex and prior to the NTP incorporation step,[50,58,62,81,84] are providing models for how the factor-induced conformational changes affect specific steps in open complex formation.

The prototypic (p)ppGpp-inhibited promoter, the rRNA promoter *rrnB* P1, associates with RNAP very rapidly to form a closed complex, but the open complex is very short-lived.[84,85] As proposed by Travers almost 40 years ago, a G+C-rich region between the −10 hexamer and the transcription start site "discriminates" promoters that are negatively regulated by (p)ppGpp from those that are activated or unaffected.[86] Results of more recent analyses indicate that the discriminator region is actually a composite of two determinants.[42,87] In general, negatively-regulated promoters are characterized by the absence of a non-template strand G two positions downstream of the −10 element (which interacts with σ region 1.2 in unregulated or activated promoters) and by the presence of a G+C-rich region just upstream of the transcription start site (that favors reannealing of the transcription bubble). These two sequence determinants conspire to create a very short-lived open complex. Inhibited promoters also tend to have −10 hexamers that are closer to the consensus at positions between the highly conserved −11A and −7T positions that flip into pockets in the housekeeping sigma factor during open complex formation.[42,88]

rRNA promoters are among the strongest in *E. coli*, largely because of the presence of UP elements and near consensus −10 and −35 hexamers that together result in a rapid rate of promoter binding by RNAP. The high activity of rRNA promoters is also attributable to sequences that lead to a rapid rate of promoter escape. rRNA promoters form a "scrunched open complex" that allows for more efficient promoter escape, bypassing the requirement for creating strain to break contacts with promoter DNA during early RNA synthesis.[20] A C residue at position −1 in the non-template strand is conserved in >85% of (p)ppGpp–DksA inhibited promoters, including rRNA promoters.[42] This strong preference for the non-template strand C at −1 has been proposed to reflect stabilization of the incoming rNTP at +1 (most often a purine) by base stacking with the purine at −1 on the template strand.[89] Thus, the conservation of the C at −1 in negatively regulated promoters may reflect more the requirement for rapid promoter escape than regulation by ppGpp–DksA.

Positively regulated promoters (such as the promoter for the anti-adapter protein *iraP* and a variety of amino acid biosynthesis promoters, including *argI*) isomerize to the open complex very slowly in the absence of (p)ppGpp–DksA or TraR, but they make very stable open complexes.[41,58,72,84] (p)ppGpp–DksA or TraR activated promoters tend to lack consensus bases in the −10 element at −10, −9, and −8 and have A+T-rich discriminator regions. Consistent with the importance of a non-consensus −10 hexamer for activation, mutations in the −10 element that increase the similarity to the consensus sequence bypass the requirement for (p)ppGpp.[90,91]

The conformational changes in RNAP caused by ppGpp–DksA or TraR binding have been identified in single-particle cryo-EM studies.[72,83,92] These studies have identified a few general principles for positive and negative control of transcription initiation, depending on the kinetic properties of the promoter. As predicted, (p)ppGpp–DksA and TraR affect multiple steps in the mechanism.[93,94] They activate transcription of positively regulated promoters by increasing the rates of formation of specific intermediates subsequent to closed complex formation, including capture of the −10 hexamer and displacement of sigma region 1.1, facilitating entry of promoter DNA into the main DNA channel of the enzyme.[72,83,92] It has also been proposed that (p)ppGpp–DksA could increase transcription from the universal stress protein A (*uspA*) promoter by facilitating promoter clearance.[95]

In contrast, the results of cryo-EM studies indicate that (p)ppGpp–DksA and TraR inhibit transcription initiation of negatively regulated promoters by at least three different mechanisms. They stabilize an intermediate subsequent to nucleation of strand opening, thereby impeding progression to later intermediates,[83,92] they alter interactions between RNAP and the downstream DNA duplex, thereby inhibiting productive open complex formation,[50,62,72,96] and, on the fraction of the promoter population that escapes the inhibitory effects in earlier intermediates, they sterically interfere with incorporation of the initial NTP.[71] The fraction of the overall effect attributable to each mechanism would depend on which step(s) in the mechanism are rate limiting for the specific promoter.

Recent single-molecule imaging studies indicate that secondary channel binding factors (SCFs), including GreB and DksA, do not bind to preformed open complexes but instead associate with promoters only as a complex with RNAP, and that they dissociate relatively quickly from the promoter complex.[97] These data are consistent with those described in a previous report indicating that DksA has a tenfold higher affinity for free RNAP than for RNAP in a preformed open complex.[98] On negatively-regulated promoters like *rrnB* P1 where RNAP dissociates quickly from the DNA, the SCF remains with the complex only as long as RNAP is bound, dissociating from the promoter together with RNAP. Consistent with this model, no SCFs were observed on rRNA–promoter complexes that had initiated transcription. Thus, on strongly negatively-regulated promoters like *rrnB* P1, DksA–ppGpp or TraR must exert most of their inhibitory effects before the NTP incorporation step. On weakly inhibited promoters (that form somewhat longer-lived promoter complexes), the SCFs might sterically inhibit NTP incorporation (in agreement with the model proposed by Molodtsov *et al.*[71]).

In contrast, promoters positively regulated by DksA–ppGpp or TraR form long-lived open complexes. DksA–ppGpp or TraR would be expected to dissociate from RNAP before RNAP dissociates from the promoter. Dissociation of the factor while RNAP is still bound to the promoter would allow completion of the transcription initiation program and nucleotide incorporation subsequent to the activated steps that occur earlier in the kinetic pathway.

6.3.3.5 *Effects of (p)ppGpp and/or DksA on Transcription Elongation, RNA Fidelity, and Genome Integrity*

Effects of (p)ppGpp and DksA on transcription are not limited to initiation. They also have been reported to affect transcription elongation, which, in turn, can affect other processes that share the DNA template, most importantly, DNA replication.

Although elongation by RNAP is processive, it is interrupted by pauses whose durations vary widely. RNAP pauses frequently at defined sequences, referred to as elemental pauses, that are relatively short-lived, and from which RNAP generally recovers with high efficiency.[12] Most pauses that mediate gene regulation are thought to originate from elemental paused complexes that interrupt the nucleotide addition cycle,[99] but *cis*- or *trans*-acting factors can exacerbate the elemental paused state and drive the complex into a backtracked state that requires rescue by Gre factors. Since (p)ppGpp stimulates pausing[100] and promotes backtracking,[101] (p)ppGpp could stimulate pausing by stabilizing the backtracked elongation complex.

(p)ppGpp can reduce the chain elongation rate of *E. coli* RNAP *in vitro*.[100,102] Since these early studies were done before the discovery of DksA, the effects must have resulted from (p)ppGpp binding to Site 1. However, there are also reports that DksA amplifies the effects of ppGpp on elongation.[103]

Older reports had described the finding that amino acid limitation induces (p)ppGpp production, decreasing the mRNA chain elongation rate *in vivo*. The effects of (p)ppGpp were only observed on translated genes *in vivo*, and not on rRNA operons.[104] Furthermore, an antitermination sequence determinant in the rRNA precursors, boxA, was required for reducing the inhibitory effects of (p)ppGpp on elongation,[105] indicating that the rRNA antitermination system counteracts the effects of (p)ppGpp on stable RNA genes (which are not translated). In contrast, (p)ppGpp dependent pausing of RNAP on translated genes helps facilitate coupling of transcription and translation and thus it was proposed that it prevents premature termination of mRNA synthesis.

RNAP variants have been identified that partially mimic the effects of (p)ppGpp by selecting for growth of Δ*relA* Δ*spoT* or Δ*dksA* mutant cells on media without amino acids,[61,62,106] or by suppression of the loss of viability caused by various mutations needed for DNA repair.[107,108] These RNAPs all made open complexes with reduced open complex lifetimes[62,84,106] and/or made shorter arrays of RNAPs on DNA templates, indicating that they made less stable elongation complexes.[109] DNA damage induces (p)ppGpp synthesis, and it was proposed that (p)ppGpp works synergistically with UvrD to facilitate backtracking of RNAP away from the site of DNA lesions or adducts to promote DNA repair.[101]

The results of chromatin immunoprecipitation (ChIP) experiments examining genome-wide RNAP occupancy indicated that RNAP processivity is decreased on protein-encoding genes in Δ*dksA* cells compared with wild type cells: RNAP signals are enriched closer to the promoters and fewer signals were found near the 3′ end of the transcribed sequences in Δ*dksA* cells.[110] Thus, it was proposed that DksA prevents loss of cell viability because it prevents stalling of transcription elongation complexes on protein-encoding genes.[110] When coupled with a mutation that slows the translation rate, cells lacking *dksA* lost viability, indicating a genetic interaction between *dksA* and the ribosome.

The results of these *in vivo* studies also supported a role for DksA in facilitating transcription processivity under circumstances where translation was inhibited, such as during amino acid starvation.[110] DksA was found with RNAP both at promoters and throughout transcribed sequences.[110] In a strain lacking *dksA*, RNAP was more prone to pause at promoter proximal sites, and this effect was amplified when cells were starved for amino acids, especially in protein coding genes. How DksA and (p)ppGpp are able to decrease elongation rates and promote transcription termination *in vitro*[103] while DksA is able to increase processivity *in vivo*[110] is a paradox that has not yet been resolved.

In bacteria, transcription errors occur about 10 000-fold more frequently than replication errors.[111,112] There have been conflicting reports on the effect of DksA and (p)ppGpp on transcription fidelity.[113] Using two different kinds of *in vivo* reporter assays, it was reported that Δ*dksA* cells showed a higher level of transcription errors than wild-type cells,[113,114] but it was recently reported that only GreA, and not GreB or DksA, acted as a transcription fidelity factor using RNA-seq measurements on a genome wide scale during log-phase growth.[115] One possibility is that GreA does not directly increase transcription fidelity *per se*, but instead results in an increase of fidelity *in vivo* because of its ability to restart backtracked RNAP complexes.[115] Measurements of transcription error rates were not performed under (p)ppGpp-inducing conditions.

Finally, DksA has been implicated in prevention of replication–transcription conflicts by reducing replication arrest during amino acid starvation.[116] In the presence of DksA, RNAP allowed completion of DNA replication following amino acid starvation, presumably because DksA reduces RNA synthesis at the levels of initiation and prevents stalling of the transcription elongation complex and its subsequent blocking of replication fork progression. In the absence of DksA, however, replication arrest was only rescued by the addition of rifampicin, which blocks RNA chain growth beyond the first two or three nucleotides and thus ultimately reduces elongation complexes interfering with replication.[116]

6.3.3.6 Regulation of Transcription by ppGpp Binding to RNA

Riboswitches are transcripts whose expression is regulated by ligands, usually small molecules that bind to the transcript with high-affinity.[117,118] The riboswitch is composed of two RNA domains, a ligand-binding motif, sometimes referred to as the aptamer, and a domain that responds to a conformational change in the aptamer domain upon ligand binding and changes gene expression, sometimes referred to as the expression platform.[118] Riboswitches can alleviate or provoke premature termination of transcription, or they can alter ribosome binding to the mRNA, increasing or decreasing translation. Riboswitches act in *cis*, in contrast to *trans*-acting small RNAs that regulate gene expression by base-pairing with target transcripts. Thus, riboswitches couple sensing and regulatory functions, providing fast responses to changing ligand concentrations.

Recently, (p)ppGpp-binding riboswitches from several related firmicutes have been described.[119] Among these, the (p)ppGpp-binding motif of the *ilvE* mRNA from *Thermosediminibacter oceani*, encoding a branched chain amino acid transferase, has a remarkably high affinity for both ppGpp and pppGpp, with a dissociation constant of about 10 nM. The (p)ppGpp-binding motifs constitute a subclass of a widespread riboswitch family in mRNAs involved in branched chain amino acid metabolism, identified by *in silico* methods.[120] Another subclass of the same riboswitch family was identified as a guanidine-binding motif, and a third subclass was characterized as having a phosphoribosyl pyrophosphate (PRPP) binding motif.[121] The structures of one of the PRPP binding aptamers and one of the (p)ppGpp aptamers[122] were solved with the bound ligand. Strikingly, point mutations in the PRPP and (p)ppGpp binding motifs switch the specificities of the two aptamers.[119,122]

These results highlight the ability of RNA aptamers to acquire new binding specificities for related ligands. Presumably this has happened many times during the course of evolution. In any case, these results indicate that (p)ppGpp-binding riboswitches, along with other base- and nucleotide-derived riboswitches, may have been important regulatory mechanisms in primordial living cells, perhaps even preceding the evolution of protein-binding transcription factors.[117]

Acknowledgements

Work in the Gourse lab is supported by R01 GM37048 from the National Institutes of Health. J Jagodnik was supported in part by a long-term EMBO fellowship (ALTF 1119-2018). Work in the Wang lab is supported by R35 GM127088 from the National Institutes of Health and an HHMI Faculty Scholar Program. L Onder was supported in part by an NIH predoctoral training grant T32 GM007133.

References

1. K. Potrykus and M. Cashel, *Annu. Rev. Microbiol.*, 2008, **62**, 35.
2. B. Field, *J. Exp. Bot.*, 2018, **69**, 2797.
3. W. C. Nierman and M. J. Chamberlin, *J. Biol. Chem.*, 1980, **254**(16), 7921.
4. L. Krásny and R. L. Gourse, *EMBO J.*, 2004, **23**(22), 4473.
5. H. D. Murray, D. A. Schneider and R. L. Gourse, *Mol. Cell*, 2003, **12**(1), 125.

6. L. Sojka, T. Kouba, I. Barvíkš, H. Šanderová, Z. Maderová, J. Jonák and L. Krásný, *Nucleic Acids Res.*, 2011, **39**(11), 4598.
7. T. Gaal, M. S. Bartlett, W. Ross, C. L. Turnbough and R. L. Gourse, *Science*, 1997, **278**(5346), 2092.
8. C. M. Lew and J. D. Gralla, *J. Biol. Chem.*, 2004, **279**(19), 19481.
9. M. M. Barker and R. L. Gourse, *J. Bacteriol.*, 2001, **183**(21), 6315.
10. N. V. Vo, L. M. Hsu, C. M. Kane and M. J. Chamberlin, *Biochemistry*, 2003, **42**(13), 3798.
11. G. A. Belogurov and I. Artsimovitch, *J. Mol. Biol.*, 2019, **431**(20), 3975.
12. M. H. Larson, R. A. Mooney, J. M. Peters, T. Windgassen, D. Nayak, C. A. Gross, S. M. Block, W. J. Greenleaf, R. Landick and J. S. Weissman, *Science*, 2014, **344**(6187), 1042.
13. I. O. Vvedenskaya, H. Vahedian-Movahed, J. G. Bird, J. G. Knoblauch, S. R. Goldman, Y. Zhang, R. H. Ebright and B. E. Nickels, *Science*, 2014, **344**(6189), 1285.
14. J. T. Winkelman, C. Pukhrambam, I. O. Vvedenskaya, Y. Zhang, D. M. Taylor, P. Shah, R. H. Ebright and B. E. Nickels, *Mol. Cell*, 2020, **79**(5), 797.
15. C. L. Turnbough and R. L. Switzer, *Microbiol. Mol. Biol. Rev.*, 2008, **72**(2), 266.
16. F. Qi and C. L. Turnbough, *J. Mol. Biol.*, 1995, **254**(4), 552.
17. C. Liu, L. S. Heath and C. L. Turnbough, *Genes Dev.*, 1994, **8**(23), 2904.
18. Y. Cheng, S. M. Dylla and C. L. Turnbough, *J. Bacteriol.*, 2001, **183**(1), 221.
19. E. Lerner, S. Chung, B. L. Allen, S. Wang, J. Lee, S. W. Lu, L. W. Grimaud, A. Ingargiola, X. Michalet, Y. Alhadid, S. Borukhov, T. R. Strick, D. J. Taatjes and S. Weiss, *Proc. Natl. Acad. Sci. U. S. A.*, 2016, **113**(43), E6562–E6571.
20. J. T. Winkelman, P. Chandrangsu, W. Ross and R. L. Gourse, *Proc. Natl. Acad. Sci. U. S. A.*, 2016, **113**(13), E1787–E1795.
21. M. Cashel and J. Gallant, *Nature*, 1969, **221**(5183), 838.
22. N. Fiil, K. von Meyenburg and J. D. Friesen, *J. Mol. Biol.*, 1972, **71**(3), 769.
23. M. Cashel, D. Gentry, V. J. Hernandez and D. Vinella, in *Escherichia coli and Salmonella: Cellular and Molecular Biology*, ed. F. C. Neidhardt, ASM Press, Washington, DC, 1996, vol. 1, p. 1458.
24. V. Hauryliuk, G. C. Atkinson, K. S. Murakami, T. Tenson and K. Gerdes, *Nat. Rev. Microbiol.*, 2015, **13**(5), 298.
25. A. Srivatsan, Y. Han, J. Peng, A. K. Tehranchi, R. Gibbs, J. D. Wang and R. Chen, *PLoS Genet.*, 2008, **4**(8), e1000139.
26. D. Lee, G. Feng and R. Landick, *J. Biol. Chem.*, 1994, **269**(35), 22295.
27. S. Iyer, D. Le, B. R. Park and M. Kim, *Nat. Microbiol.*, 2018, **3**(6), 741.
28. A. Battesti and E. Bouveret, *Mol. Microbiol.*, 2006, **62**(4), 1048.
29. M. Nomura, R. Gourse and G. Baughman, *Annu. Rev. Biochem.*, 1984, **53**, 75.
30. M. Schaechter, O. Maaloe and N. O. Kjeldgaard, *J. Gen. Microbiol.*, 1958, **19**(3), 592.

31. J. Ryals, R. Little and H. Bremer, *J. Bacteriol.*, 1982, **151**(3), 1261.
32. J. M. Lopez, A. Dromerick and E. Freese, *J. Bacteriol.*, 1981, **146**(2), 605.
33. K. Ochi, J. Kandala and E. Freese, *J. Bacteriol.*, 1982, **151**(2), 1062.
34. K. Liu, A. R. Myers, J. L. Keck, J. D. Wang, K. Liu, A. R. Myers, T. Pisithkul, K. R. Claas, K. A. Satyshur and D. Amador-Noguez, *Mol. Cell*, 2015, **57**(4), 735.
35. A. Kriel, A. N. Bittner, S. H. Kim, K. Liu, A. K. Tehranchi, W. Y. Zou, S. Rendon, R. Chen, B. P. Tu and J. D. Wang, *Mol. Cell*, 2012, **48**(2), 231.
36. B. W. Anderson, A. Hao, K. A. Satyshur, J. L. Keck and J. D. Wang, *J. Mol. Biol.*, 2020, **432**(14), 4108.
37. B. W. Anderson, K. Liu, C. Wolak, K. Dubiel, F. She, K. A. Satyshur, J. L. Keck and J. D. Wang, *eLife*, 2019, **8**, e47534.
38. B. Wang, R. A. Grant, M. T. Laub, B. Wang, R. A. Grant and M. T. Laub, *Mol. Cell*, 2020, **80**(1), 29.
39. B. Wang, P. Dai, D. Ding, A. del Rosario, R. A. Grant, B. L. Pentelute and M. T. Laub, *Nat. Chem. Biol.*, 2019, **15**(2), 141.
40. Y. Zhang, E. Zborníková, D. Rejman and K. Gerdes, *mBio*, 2018, **9**, e02188–e02217.
41. W. Ross, P. Sanchez-Vazquez, A. Y. Chen, J. H. Lee, H. L. Burgos and R. L. Gourse, *Mol. Cell*, 2016, **62**, 811.
42. P. Sanchez-Vazquez, C. N. Dewey, N. Kitten, W. Ross and R. L. Gourse, *Proc. Natl. Acad. Sci. U. S. A.*, 2019, **116**(17), 8310.
43. A. Kriel, S. R. Brinsmade, J. L. Tse, A. K. Tehranchi, A. N. Bittner, A. L. Sonenshein and J. D. Wang, *J. Bacteriol.*, 2014, **196**(1), 189.
44. V. M. Levdikov, E. Blagova, P. Joseph, A. L. Sonenshein and A. J. Wilkinson, *J. Biol. Chem.*, 2006, **281**(16), 11366.
45. S. R. Brinsmade and A. L. Sonenshein, *J. Bacteriol.*, 2011, **193**(20), 5637.
46. L. Krásný, H. Tišerová, J. Jonák, D. Rejman and H. Šanderová, *Mol. Microbiol.*, 2008, **69**(1), 42.
47. S. Tojo, T. Satomura, K. Kumamoto, K. Hirooka and Y. Fujita, *J. Bacteriol.*, 2008, **190**(18), 6134.
48. V. Molle, Y. Nakaura, R. P. Shivers, H. Yamaguchi, R. Losick, Y. Fujita and A. L. Sonenshein, *J. Bacteriol.*, 2003, **185**(6), 1911.
49. B. W. Anderson, M. A. Schumacher, J. Yang, A. Turdiev, H. Turdiev, Q. He, V. T. Lee, R. G. Brennan and J. D. Wang, *bioRxiv*, DOI: 10.1101/2020.12.02.409011.
50. B. J. Paul, M. M. Barker, W. Ross, D. A. Schneider, C. Webb, J. W. Foster and R. L. Gourse, *Cell*, 2004, **118**(3), 311.
51. P. J. Kang and E. A. Craig, *J. Bacteriol.*, 1990, **172**, 2055–2064.
52. A. Perederina, V. Svetlov, M. N. Vassylyeva, T. H. Tahirov, S. Yokoyama, I. Artsimovitch and D. G. Vassylyev, *Cell*, 2004, **118**, 297.
53. P. Chandrangsu, L. Wang, S. H. Choi and R. L. Gourse, *J. Bacteriol.*, 2012, **194**(6), 1437.
54. S. Borukhov, V. Sagitov and A. Goldfarb, *Cell*, 1993, **72**(3), 459.
55. N. Opalka, M. Chlenov, P. Chacon, W. J. Rice, W. Wriggers and S. A. Darst, *Cell*, 2003, **114**(3), 335.

56. J. H. Lee, C. W. Lennon, W. Ross and R. L. Gourse, *J. Mol. Biol.*, 2012, **416**(4), 503.
57. G. H. Feng, D. N. Lee, D. Wang, C. L. Chan and R. Landick, *J. Biol. Chem.*, 1994, **269**(35), 22282.
58. B. J. Paul, M. B. Berkmen and R. L. Gourse, *Proc. Natl. Acad. Sci. U. S. A.*, 2015, **102**(22), 7823.
59. S. T. Rutherford, J. J. Lemke, C. E. Vrentas, T. Gaal, W. Ross and R. L. Gourse, *J. Mol. Biol.*, 2007, **366**(4), 1243.
60. C. E. Vrentas, T. Gaal, W. Ross, R. H. Ebright and R. L. Gourse, *Genes Dev.*, 2005, **19**(19), 2378.
61. H. Murphy and M. Cashel, *Methods Enzymol.*, 2003, **371**, 596.
62. S. T. Rutherford, C. L. Villers, J. H. Lee, W. Ross and R. L. Gourse, *Genes Dev.*, 2009, **23**(2), 236.
63. I. Artsimovitch, V. Patlan, S. Sekine, M. N. Vassylyeva, T. Hosaka, K. Ochi, S. Yokoyama and D. G. Vassylyev, *Cell*, 2004, **117**(3), 299.
64. C. E. Vrentas, T. Gaal, M. B. Berkmen, S. T. Rutherford, S. P. Haugen, D. G. Vassylyev, W. Ross and R. L. Gourse, *J. Mol. Biol.*, 2008, **377**(2), 551.
65. K. Kasai, T. Nishizawa, K. Takahashi, T. Hosaka, H. Aoki and K. Ochi, *J. Bacteriol.*, 2006, **188**(20), 7111.
66. K. D. Westover, D. A. Bushnell and R. D. Kornberg, *Cell*, 2004, **119**(4), 481.
67. W. Ross, C. E. Vrentas, P. Sanchez-Vazquez, T. Gaal and R. L. Gourse, *Mol. Cell*, 2013, **50**(3), 420.
68. Y. Zuo, Y. Wang and T. A. Steitz, *Mol. Cell*, 2013, **50**(3), 430.
69. U. Mechold, K. Potrykus, H. Murphy, K. S. Murakami and M. Cashel, *Nucleic Acids Res.*, 2013, **41**(12), 6175.
70. K. G. Roelofs, J. Wang, H. O. Sintim and V. T. Lee, *Proc. Natl. Acad. Sci. U. S. A.*, 2011, **108**(37), 15528.
71. V. Molodtsov, E. Sineva, L. Zhang, X. Huang, M. Cashel, S. E. Ades and K. S. Murakami, *Mol. Cell*, 2018, **69**(5), 828.
72. A. Y. Chen, A. Feklistov, J. Chen, S. Gopalkrishnan, Z. Zhang, C. S. Potter, B. Carragher, R. L. Gourse, S. A. Darst and W. Ross, unpublished.
73. A. R. Myers, D. P. Thistle, W. Ross and R. L. Gourse, *Front. Microbiol.*, 2020, **11**, 587098.
74. T. Durfee, A. M. Hansen, H. Zhi, F. R. Blattner and J. J. Ding, *J. Bacteriol.*, 2008, **190**(3), 1084.
75. M. F. Traxler, V. M. Zacharia, S. Marquardt, S. M. Summers, H. T. Nguyen, S. E. Stark and T. Conway, *Mol. Microbiol.*, 2011, **79**(4), 830.
76. M. F. Traxler, S. M. Summers, H. T. Nguyen, V. M. Zacharia, G. A. Hightower, J. T. Smith and T. Conway, *Mol. Microbiol.*, 2008, **68**(5), 1128.
77. A. Åberg, J. Fernández-Vázquez, J. D. Cabrer-Panes, A. Sánchez and C. Balsalobre, *J. Bacteriol.*, 2009, **191**(10), 3226.
78. S. P. Haugen, W. Ross and R. L. Gourse, *Nat. Rev. Microbiol.*, 2008, **6**, 507.
79. R. M. Saecker, M. T. Record Jr and P. L. Dehaseth, *J. Mol. Biol.*, 2011, **412**(5), 754.

80. M. D. Blankschien, K. Potrykus, E. Grace, A. Choudhary, D. Vinella, M. Cashel and C. Herman, *PLoS Genet.*, 2009, **5**(1), e1000345.
81. S. Gopalkrishnan, W. Ross, A. Y. Chen and R. L. Gourse, *Proc. Natl. Acad. Sci. U. S. A.*, 2017, **114**(28), E5539.
82. J. Chen, H. Boyaci and E. A. Campbell, *Nat. Rev. Microbiol.*, 2021, **19**, 95.
83. J. Chen, C. Chiu, S. Gopalkrishnan, A. Y. Chen, P. D. B. Olinares, R. M. Saecker, J. T. Winkelman, M. F. Maloney, B. T. Chait, W. Ross, R. L. Gourse, E. A. Campbell and S. A. Darst, *Mol. Cell*, 2020, **78**, 275.
84. M. M. Barker, T. Gaal, C. A. Josaitis and R. L. Gourse, *J. Mol. Biol.*, 2001, **305**, 673.
85. L. Rao, W. Ross, J. Appleman, T. Gaal, S. Leirmo, P. J. Schlax, M. T. Record and R. L. Gourse, *J. Mol. Biol.*, 1994, **235**, 1421.
86. A. A. Travers, *J. Bacteriol.*, 1980, **141**, 973.
87. S. P. Haugen, M. B. Berkmen, W. Ross, T. Gaal, C. Ward and R. L. Gourse, *Cell*, 2006, **125**, 1069.
88. A. Feklistov and S. A. Darst, *Cell*, 2011, **147**, 1257.
89. M. L. Gleghorn, E. K. Davydova, R. Basu, L. B. Rothman-Denes and K. S. Murakami, *Proc. Natl. Acad. Sci. U. S. A.*, 2011, **108**, 3566.
90. X. J. da Costa and S. W. Artz, *J. Bacteriol.*, 1997, **179**, 5211.
91. D. L. Riggs, R. D. Mueller, H. S. Kwan and S. W. Artz, *Proc. Natl. Acad. Sci. U. S. A.*, 1986, **83**, 9333.
92. J. Chen, S. Gopalkrishnan, C. Chiu, A. Y. Chen, E. A. Campbell, R. L. Gourse, W. Ross and S. A. Darst, *eLife*, 2019, **8**, e49375.
93. R. L. Gourse, A. Y. Chen, S. Gopalkrishnan, P. Sanchez-Vazquez, A. Myers and W. Ross, *Annu. Rev. Microbiol.*, 2018, **72**, 163.
94. E. A. Galburt, *Proc. Natl. Acad. Sci. U. S. A.*, 2018, **115**(50), E11604.
95. B. Gummesson, M. Lovmar and T. Nyström, *J. Biol. Chem.*, 2013, **288**(29), 21055.
96. V. Mekler, L. Minakhin, S. Borukhov, A. Mustaev and K. Severinov, *J. Mol. Biol.*, 2014, **426**, 3973.
97. S. K. Stumper, H. Ravi, L. J. Friedman, R. A. Mooney, I. R. Corrêa, A. Gershenson, R. Landick and J. Gelles, *eLife*, 2019, **8**, e40576.
98. C. W. Lennon, T. Gaal, W. Ross and R. L. Gourse, *J. Bacteriol.*, 2009, **191**, 5854.
99. J. Saba, X. Y. Chua, T. V. Mishanina, D. Nayak, T. A. Windgassen, R. A. Mooney and R. Landick, *eLife*, 2019, **8**, e40981.
100. R. E. Kingston, W. C. Nierman and M. J. Chamberlin, *J. Biol. Chem.*, 1981, **256**, 2787.
101. V. Kamarthapu, V. Epshtein, B. Benjamin, S. Proshkin, A. Mironov, M. Cashel and E. Nudler, *Science*, 2016, **352**(6288), 993.
102. R. E. Kingston and M. J. Chamberlin, *Cell*, 1981, **27**, 523.
103. R. Furman, A. Sevostyanova and I. Artsimovitch, *Nucleic Acids Res.*, 2012, **40**(8), 3392.
104. U. Vogel and K. F. Jensen, *J. Biol. Chem.*, 1994, **269**, 16236.
105. U. Vogel and K. F. Jensen, *J. Biol. Chem.*, 1995, **270**(31), 18335.

106. M. S. Bartlett, T. Gaal, W. Ross and R. L. Gourse, *J. Mol. Biol.*, 1998, **279**(2), 331.
107. P. McGlynn and R. G. Lloyd, *Cell*, 2000, **101**(1), 35.
108. B. W. Trautinger and R. G. Lloyd, *EMBO J.*, 2002, **21**, 6944.
109. B. W. Trautinger, R. P. Jaktaji, E. Rusakova and R. G. Lloyd, *Mol. Cell*, 2005, **19**(2), 247–258.
110. Y. Zhang, R. Mooney, J. Grass, P. Sivaramakrishnan, C. Herman, R. Landick and J. Wang, *Mol. Cell*, 2014, **53**(5), 766.
111. C. C. Traverse and H. Ochman, *Proc. Natl. Acad. Sci. U. S. A.*, 2016, **113**, 3311.
112. M. Imashimizu, T. Oshima, L. Lubkowska and M. Kashlev, *Nucleic Acids Res.*, 2013, **41**, 9090.
113. M. Roghanian, N. Zenkin and Y. Yuzenkova, *Nucleic Acids Res.*, 2015, **43**, 1529.
114. D. Satory, A. J. Gordon, M. Wang, J. A. Halliday, I. Golding and C. Herman, *Nucleic Acids Res.*, 2015, **43**(21), 10190.
115. C. C. Traverse and H. Ochman, *G3: Genes, Genomes, Genet.*, 2018, **8**(7), 2257.
116. A. K. Tehranchi, M. D. Blankschien, Y. Zhang, J. A. Halliday, A. Srivatsan, J. Peng, C. Herman and J. D. Wang, *Cell*, 2010, **141**(4), 595.
117. A. Serganov and E. Nudler, *Cell*, 2013, **152**, 17.
118. W. Winkler, A. Nahvi and R. R. Breaker, *Nature*, 2002, **419**, 952.
119. M. E. Sherlock, N. Sudarsan and R. R. Breaker, *Proc. Natl. Acad. Sci. U. S. A.*, 2018, **115**, 6052.
120. J. W. Nelson, R. M. Atilho, M. E. Sherlock, R. B. Stockbridge and R. R. Breaker, *Mol. Cell*, 2017, **65**, 220.
121. M. E. Sherlock, N. Sudarsan, S. Stav and R. R. Breaker, *eLife*, 2018, **7**, e33908.
122. A. J. Knappenberger, C. W. Reiss and S. A. Strobel, *eLife*, 2018, **7**, e36381.

CHAPTER 7

Helicases and the Obstructive RNAP

JEFFREY W. ROBERTS[a] AND TERENCE R. STRICK*[b]

[a]Department of Molecular Biology and Genetics, Cornell University, 349 Biotechnology Building, Ithaca, NY 14850, USA; [b]Institut de Biologie, Ecole normale supérieure, Université Paris Sciences et Lettres, 46 rue d'Ulm, 75005 Paris, France
*E-mail: strick@bio.ens.psl.eu

7.1 Introduction

The high stability of transcription complexes ensures the completion of RNA synthesis, but also obstructs other essential DNA transactions. Primary among these is replication, which occurs at the same time and place as transcription in bacterial cells. Since the replication fork moves along DNA about ten times as fast as RNA polymerase (RNAP), there must be a mechanism to clear RNAP from DNA for replication; from what we know about the forces stabilizing the transcription complex and its structure, any notion that it might be stable to passing replication is implausible. An early and dramatic demonstration of the encounter of replication and transcription is the electron microscopic analysis of French,[1] who examined the interaction of the replication fork with ribosomal RNA genes, which during rapid cell growth contain a train of essentially close packed RNAP – of the order of a hundred per *rRNA* operon. The natural directions in *Escherichia coli* of replication and rRNA synthesis are co-linear, and in this configuration the ribosomal genes appear to provide no barrier to replication. However, they do in an artificially

Chemical Biology No. 19
RNA Polymerases as Molecular Motors: On the Road, 2nd Edition
Edited by Robert Landick, Terence Strick and Jue Wang

Published by the Royal Society of Chemistry, www.rsc.org

reversed rRNA operon, although replication eventually passes.[1,2] These results are indicative of a specific mechanism coordinated with movement of the replication fork that disrupts co-linear transcription complexes, as well as mechanisms to deal with barriers like a converging RNAP.

Cellular pathways for repair of DNA damage also must deal with obstructive RNAP. For example, transcribing RNAP stalls at a pyrimidine dimer, preventing both gene expression of the damaged segment and access of the damage site to DNA repair enzymes.[3] Enzymatic processes can recognize and remove such stalled complexes; these include the helicase mutation frequency decline (Mfd),[3] the transcription termination and release RNA translocase Rho,[4–7] and potentially other helicases, such as ultraviolet resistance D (UvrD).[2,8–10] The direct involvement of RNAP in inhibiting DNA repair is shown by the ability of mutationally altered RNAP to facilitate repair of DNA damage, presumably by destabilizing and removing transcription complexes, although the precise biochemical mechanism of this phenotype is not clear.[11–13]

7.2 Destabilization of Transcription Complexes

The major known agents of transcription complex destabilization or deactivation are the well-characterized helicases discussed in the following sections (and elsewhere), especially UvrD, replication (Rep), Mfd, and Rho, which interact with RNAP or other barriers and transduce adenosine triphosphate (ATP) energy into dissociative forces. However, it is worth considering that dissociation of obstructing RNAP in replication might be caused by the replication fork directly. Thus, removing RNAP complexes might be an activity of *E. coli* DnaB, the ATP-dependent helicase that is the advance melting agent of the replication fork. Results from single-molecule experiments have indicated that unwinding forces imposed by the phage T7 replication complex either behind or ahead of an *E. coli* RNAP complex induce its dissociation.[14] The experiment hasn't been reported with the *E. coli* replication complexes, but it is clear that forces of the order of magnitude of those expected from the activity of a replicative helicase can disrupt transcription complexes.

7.2.1 UvrD

UvrD is an ATP-dependent 3′–5′ translocase that binds single stranded DNA and then unwinds duplex DNA that it encounters. It is well established that UvrD is required for two basic processes of DNA repair or error correction: DNA excision repair [as in repair of UV-induced damage (hence the Uvr name)] and methyl-directed DNA mismatch repair (MMR). In UvrABCD-dependent excision repair UvrD removes the 12-nucleotide excised fragment containing the lesion from its pairing with the complementary DNA strand. UvrD acts in MMR to facilitate the activity of exonucleases that degrade the mispaired segment from the newly synthesized DNA strand.

UvrD has a distinct role, shared with the Rep helicase, to remove proteins or protein complexes like RNAP that impede the replication fork progression from DNA.[2,15] An experiment to demonstrate this derives from the observation that the converging, but not the co-linear, configuration of ribosomal gene transcription and DNA replication causes delay in fork movement.[1,2] Boubrakri *et al.* examined the role of three helicases in cell survival, particularly in an artificially contrived case of the convergent configuration: UvrD; Rep, which is also a 3′–5′ DNA helicase; and the 5′–3′ helicase damage inducible G (DinG). In otherwise wild type cells the *uvrD dinG* and *rep dinG* double mutants are fully viable. However, the loss of both 3′–5′ helicases in a *uvrD rep* double mutant is lethal in rich medium (which presumably increases collisions relative to minimal medium). With a strain bearing convergent RNA and DNA synthesis, loss of any two helicases is lethal in rich medium, and the DinG mutant alone is very defective in rich medium. There is good evidence that rRNA gene transcription is responsible for these enhanced defects; an RNAP mutant selected to relieve DNA repair mutant defects also restores some growth of the multiple helicase mutants. Most directly, it was shown by gel analysis of replication intermediates that converging synthesis stalls the fork at the downstream end of the inverted ribosomal genes, and that helicase mutants exacerbate this effect.[2]

The helicases do more than remove interfering proteins: converging collision in particular can result in the deposition of an R-loop from the transcription complex, a barrier for the replication fork that is believed to be a major detrimental consequence of the collision.[16,17] The main role of the SOS-inducible DinG helicase is to remove R-loops;[2] there is evidence that overproducing RNase H relieves *dinG* lethality in the case of convergent complexes.[2] It is noteworthy that the RNAP mutations isolated to relieve DNA repair defects also suppress the putative R-loop-mediated lethality.[2] Assuming that these mutations destabilize complexes (which has not been definitively shown), one conjecture might be that R-loop formation is favored by stable stalled complexes, as opposed to released RNA, because trailing RNA is maintained in the vicinity of homologous segments of the template.

Hawkins *et al.*[9] have shown directly in reconstituted systems *in vitro* that *E. coli* transcription complexes block the *E. coli* replication apparatus, and that either the UvrD or Rep helicase relieves this block and allows replication to continue. DinG does not work, however, consistent with its having instead the primary role of removing R-loops. Results of other experiments have indicated that Rep helicase, which binds the replicative DnaB helicase and thus appears to travel with the replication fork, may be the principal agent to remove other interfering DNA binding proteins *in vivo*.[18] The replication fork is reported to move at only half the normal speed in a cell lacking Rep.[19]

UvrD has a domain [the C-terminal domain (CTD)] that binds RNA polymerase, as does the homologous helicase plasmid copy reduces A (PcrA) in *Bacillus subtilis*, indicating that their activity is directed particularly to RNAP.[10] *In vivo*, deletion of the CTD does not compromise excision repair, mismatch repair, or cellular ability to resolve transcription–replication conflicts, although it wasn't tested in the *in vitro* experiments of Hawkins *et al.* However, the CTD deletion does affect a distinct activity reported for UvrD, namely to induce backtracking of the RNAP complex. This stimulation of backtracking was proposed to underlie transcription-coupled repair and to be more important than function of the obvious candidate Mfd,[8] a notion that is controversial.[20] Backtracking would result from activity of a 3′–5′ helicase on the accessible non-template strand of a transcription bubble. The backtracking activity of UvrD was confirmed by Sanders *et al.*,[10] who also showed that deletion of the CTD partially but not completely degrades the ability of UvrD to stimulate backtracking. Whether stimulation of backtracking underlies the activity of UvrD to relieve collision interference is unknown. Increasing backtracking would help move RNAP out of the way and facilitate excision of a blocking lesion like a pyrimidine dimer, but this would not prevent collision with the replication complex. A second activity of UvrD on RNAP complexes reported by Sanders *et al.* is to *release* transcripts from the complex. In fact, complexes might be released if they were backtracked all the way to their 5′ ends, although such an activity has not been reported. Transcript release would be an entirely novel activity, perhaps akin to that of Rho; it awaits thorough investigation.

7.2.2 Rho

Rho is an ATP-dependent RNA translocase that binds accessible RNA and, in the prevalent model, pursues RNAP by moving along RNA in the 5′–3′ direction; Rho interaction with RNAP then leads to complex dissolution and RNA release.[5] (However, see Chapter 5 for a revised view of how Rho engages RNAP.) An important cofactor of Rho *in vivo* is the elongation factor NusG,[21–24] which stimulates Rho function by binding both it and RNAP. The molecular mechanism of Rho activity is still obscure, but must involve the translocase activity; one proposal is that the encounter with RNAP allows Rho to continue pulling on the RNA so as to extract it from the RNAP complex. Such a Rho–RNAP interaction could also stabilize a conformation that favors RNA release.[25] Functionally, Rho is required for terminating certain defined transcription units, and, particularly, for doing so in a way that is conditional upon the utilization of a transcript. Thus, Rho is the agent of genetic polarity, which is the termination of transcription within a multigene operon when a nonsense codon stops translation and leaves RNA emerging from RNAP empty of ribosomes and thus accessible to Rho.[26,27] Perhaps more importantly, Rho prevents a background of presumably adventitious transcription, called "pervasive transcription", which includes antisense transcription throughout the *E. coli* genome; most such transcripts would be non-utilized and thus their transcription would be accessible to Rho.[28,29] These transcripts give rise to R-loops, which then are believed to stop replication forks.[30,31] Rho in effect acts as a quality control device on the production of RNA.

Consistent with the notion that excess non-utilized RNA is dangerous and leads to interference with replication, diminished Rho function gives rise to double strand breaks in the chromosome.[6] Although total loss of Rho is lethal in *E. coli*, growth in a sublethal concentration of the Rho-inhibiting drug bicyclomycin reveals this damage. It is thought that the breaks result from disruption of replication complexes during active replication, and are substrates for recombination and replication fork restoration facilitated by the replication initiation factor primosomal protein A (PriA); consistently, *priA* and recombination B (*recB*) mutations increase sensitivity to bicyclomycin.[6] The actual obstructive agent could be both R-loops and the transcription complexes themselves. An RNA polymerase mutation that is thought to destabilize transcription complexes decreases sensitivity to bicyclomycin, indicating that some lethality is due to persistent transcription complexes themselves.[6]

Rho is important to the cellular response to exogenous DNA damage. It was an early observation that *rho* mutants are sensitive to ultraviolet light.[32] Since lesions like pyrimidine dimers in the template DNA strand stall transcription complexes,[4] they would produce a barrier to replication independent of whether the replication complex itself would stall at the lesion. (It presumably would not for a lesion in the lagging strand template, but instead would leave a gap in this strand.) Rho cooperates with Mfd in relieving transcription stalls induced by damage.[4] Loss of Mfd is not lethal or deleterious and, contrary to expectation, gives very modest sensitivity to DNA damage. However, impairment of Mfd and Rho together inhibits growth and enhances damage sensitivity, indicating that these are redundant agents to remediate the same problem.

7.2.3 Mfd

The activity of the Mfd protein was first observed genetically by Witkin for its ability to protect bacteria from the mutagenic effects of UV light.[33] UV light is known to convert adjacent pyrimidines into crosslinked photoproducts known as cyclopyrimidine dimers (CPDs) or 6–4 photoproducts. These photocrosslinked base pairs interfere with both RNA and DNA polymerases. They lead to mutations if not repaired prior to passage of the replication machinery as DNA polymerases are unfaithful in their replication of these damaged bases. They can also lead to mutations by irreversibly arresting the transcription elongation complex, which is a sufficiently stable roadblock to the replication fork that it leads to replication fork collapse and restart, increasing replicative stress and the likelihood of a range of mutational processes such as R-loop formation, single-strand breaks, and so on.[34] Results of genetic analysis indicated that the Mfd gene product reduced the mutational effects of UV light.

Separately, genetic and molecular biology work revealed that repair of UV lesions was not constant across the genome, but that it took place preferentially in transcribed regions, and furthermore that repair of lesions located on the non-coding (template) strand of genes was prioritized over repair of lesions located on the coding (non-template) strand of genes.

For this reason, the process was termed "transcription-coupled repair." Purification of the Mfd protein and its use in *in vitro* biochemical assays and *in vivo* complementation assays revealed that it was indeed the factor coupling transcription and repair, interacting with RNA polymerase to remove it from the DNA and make the lesion accessible to repair factors. Indeed, otherwise the RNAP protects the lesion from repair as has been shown biochemically[35] and structurally.[36] However, if removing RNAP from the lesion was Mfd's only activity, then template strand lesions would simply be repaired at the same rate as non-template strand lesions repair in the presence of Mfd, rather than prioritized.[20]

In fact, Mfd couples transcription and repair in a highly coordinated fashion, recruiting the downstream repair actors UvrA and UvrB (see Figure 7.1 for an overview). Although these components can act alone in repairing DNA in the process of nucleotide excision repair (NER), in such circumstances they are not able to prioritize repair of template over non-template strand.[37] As a result, the entire genome can be repaired by the UvrA–UvrB–UvrC–UvrD–polymerase–ligase machinery in a process termed global genome NER, or GG-NER, which is in contrast to TC-NER, for transcription-coupled NER.

In GG-NER, dimeric $UvrA_2$, or also possibly heterotetrameric $UvrA_2UvrB_2$, recognize damaged bases, such as UV lesions. Upon recognition, the UvrA species is released from the lesion while UvrB remains and recruits UvrC. UvrC contains two exonuclease sites which cleave 11 bases apart, generating a single-strand oligonucleotide bearing the lesion. This oligonucleotide is efficiently separated from DNA by the UvrD helicase, freeing a single-stranded gap which can be filled in and sealed by DNA polymerase and ligase. In TC-NER, RNAP is the initial damage sensor, which recruits Mfd to displace stalled RNAP from the DNA and reveal the lesion to DNA repair factors.

7.3 RNAP as a Lesion Sensor

There are several important features to the role of RNAP as a sensor for lesions on the transcribed strand of DNA. First of all, this prioritizes repair of transcribed genes in a manner which is proportional to their activity. Thus, heavily transcribed genes or operons, which are likely to be essential, will be prioritized for repair.

Second, RNAP can in fact patrol the entire genome for stalling lesions such as CPDs. Indeed, transcriptomics work over the last decade has made it abundantly clear that in reality the entire genome is continuously undergoing transcription.[38,39] In bacteria, where transcription proceeds unidirectionally from the asymmetric −10 element or Pribnow box, transcription termination is far from fail-proof, and termination read-through events lead to RNAP transcribing beyond its normal spatial bounds. As a result, RNAP can be found everywhere along both strands of the genome. This so-called "pervasive transcription" is also observed in eukaryotes, in no small part because eukaryotic promoters are in fact bidirectional and sustain both sense and antisense transcription.[39] Because RNAP remains indefinitely stalled at a UV

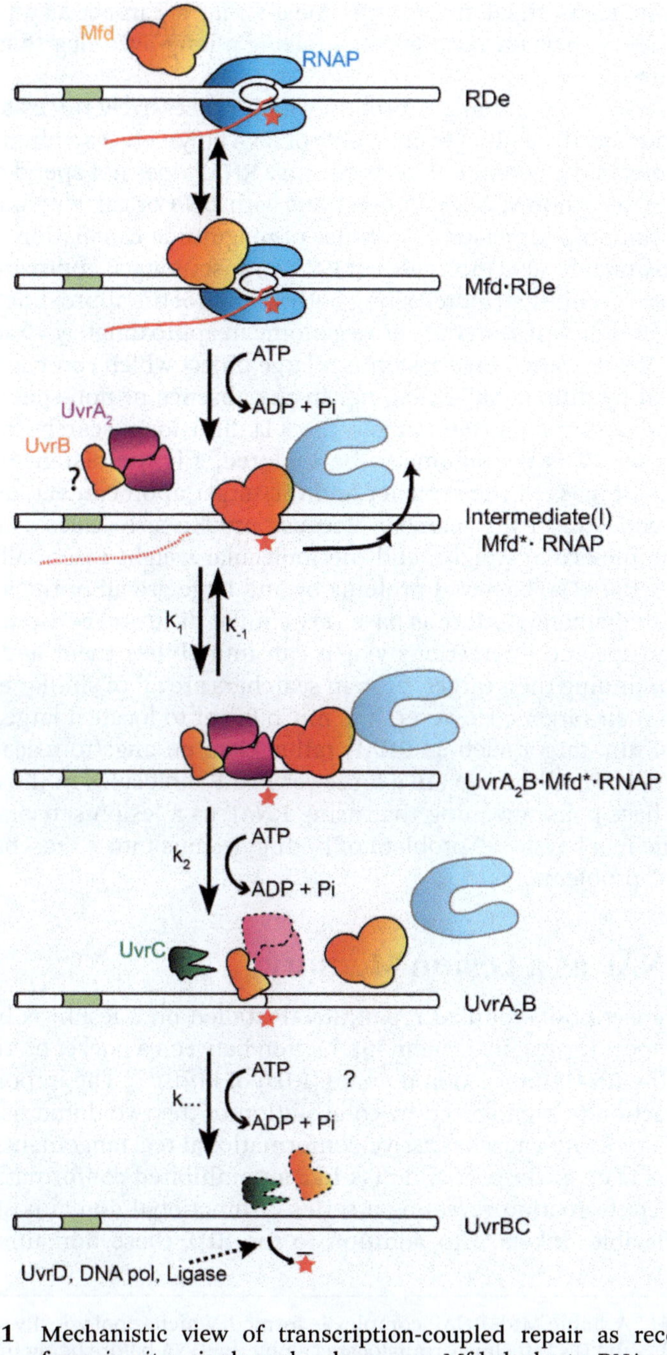

Figure 7.1 Mechanistic view of transcription-coupled repair as reconstituted from *in vitro* single-molecule assays. Mfd docks to RNA polymerase and becomes activated, tethering to DNA. Using its RecG-like ATPase–translocase domains it translocates towards RNAP. This torques and remodels RNA polymerase stalled at a lesion so that it releases RNA.

lesion on its transcribed strand, this means that it can act as an effective genome-wide sensor for such lesions, raising a long-lived flag that the cell will not miss.

Third, RNAP is an abundant protein which processively tracks along the genome, making the collective ensemble of RNA polymerases a rapid detector of UV lesions in the genome. This is because RNAP does not spend excessive time undergoing non-specific interactions with DNA or carrying out 3D diffusional searching for lesions. An order-of-magnitude calculation indicates that approximately 5000 molecules of RNAP transcribing at approximately 50 bases s^{-1} and evenly distributed along both strands of the approximately 5 Mb bacterial genome will scan the entire genome in approximately 40 seconds.

Fourth, RNAP stalled on a lesion is a large object which can itself be rapidly located by diffusional searching. In the absence of non-specific interactions and assuming a static target, the 3-D diffusional search time scales as $V/\varepsilon D$, where V is the volume to be explored, ε is the distance between the centers-of-mass of the searcher and the target upon contact, and D the diffusion constant of the searcher. Because protein size scales as the cube root of the molecular weight, and the molecular weight essentially ranges from 10 to 100 kDa, bacterial proteins by and large are all of the same size (5–10 nm in diameter) and thus have very similar diffusion constants. Thus, absent non-specific interactions which can immobilize them and prevent them from finding their target, protein searchers are all of similar efficiency in finding their targets. However, it is much faster to locate a large, approximately 10 nm, target such as RNAP, rather than an angstrom-scale defect in a base pair. Finally, there are a lot fewer RNAP molecules in the cell than there are base-pairs, meaning that using RNAP as a lesion sensor converts the "needle in a haystack" problem of finding lesions into a "baseball bat in a sandbox" problem.

7.4 RNAP as a Lesion Memory

During transcription-coupled repair, RNAP stalled on a lesion is bound by Mfd *via* a specific protein–protein interaction between a pocket on the β subunit and the RNAP-interaction domain (RID) of Mfd.[40–42] The importance of this interaction is highlighted by co-evolution of the two domains.[43] Docking of Mfd to RNAP causes massive conformational rearrangements in Mfd, which goes from a compact and disk-like autoinhibited conformation to an extended conformation revealing a series of functional domains separated by long flexible linkers.[44] In addition to the RID, these domains include

Figure 7.1 (*continued*) A stable Mfd–RNAP complex is formed, which topologically surrounds the DNA; it slowly translocates along the DNA before being intercepted by UvrA–UvrB. These arrest translocating Mfd and RNAP and release them from the DNA then carry out downstream steps of DNA damage recognition and recruitment of UvrC for DNA incision. Adapted from ref. 52 with permission from Springer Nature, Copyright 2015.

two RecG-like translocation modules which interact with DNA and use the energy of ATP hydrolysis to propel Mfd along DNA, as well as a UvrB homology module responsible for recruiting downstream repair components UvrA and UvrB.[40] The RecG-like domains of Mfd are inaccessible for DNA binding in the autoinhibited state, although there is evidence that in solution there are fluctuations between the autoinhibited closed and activated open states, allowing Mfd to attach to DNA even in the absence of contact with RNAP.[45,46] Similarly, the UvrB domain of Mfd is also inaccessible in the autoinhibited state.[42]

Single-molecule measurements using a magnetic trap (Figure 7.1) have shown that functional docking of Mfd to stalled RNAP is well-described by Michelis–Menten kinetics with a pseudo-first order binding rate of 2×10^5 M^{-1} s^{-1} and a spontaneous (*i.e.* futile) dissociation rate of 0.08 s^{-1}, slightly faster than the forward reaction rate for remodeling stalled RNAP of 0.06 s^{-1}.[47] The dissociation constant of Mfd from stalled RNAP is thus approximately 400 nM and the catalytic K_M for Mfd remodeling stalled RNAP is approximately 700 nM. Given an *in vivo* Mfd concentration in the range of 50–500 molecules per cell,[3,48,49] or 50–500 nM, these numbers also imply that up to about half of Mfd is associated with RNAP in the steady state. If it takes on average 10 seconds to probe an RNAP then it takes approximately 100–1000 seconds for Mfd to probe every one of the approximately 5000 RNAP in the cell. This level of surveillance of RNAP, roughly once per cell cycle, supports the view that Mfd is a background repair system, tasked with surveillance of low levels of UV damage which do not induce the SOS response. In this context it is noteworthy that Mfd level increases approximately twofold when bacteria are exposed to suprainhibitory (*i.e.* nonlethal yet effective) concentrations of antibiotics, such as the fluoroquinolone ciprofloxacin.[50] Similarly, processes which reduce transcriptional levels, such as the stringent stress response, should increase the ratio of patrolling Mfd to RNAP.

After docking to RNAP, the RecG motifs of Mfd bind to DNA and ATP, and using the energy of ATP hydrolysis translocate it towards RNAP situated just ahead. This applies force and/or torque to the stalled RNAP, either directly or *via* the DNA, resulting in its displacement from the DNA. Because Mfd and RNAP binding to DNA form a closed topological domain, translocation of Mfd towards RNAP could generate positive supercoiling between Mfd and RNAP, helping to dislodge the RNAP.[51] Single-molecule measurements indicate that stalled RNAP is successfully remodeled and cleared from the lesion at about 0.06 s^{-1}. Comparing this rate constant to the off-rate reveals that successful displacement of stalled RNAP is not only a slow process, but also that it is less likely than dissociation of Mfd from RNAP. Taken together, these two kinetic constants indicate that a kinetic proofreading mechanism prevents Mfd from displacing paused RNAP.[47] Indeed, only certain lesions, such as CPDs, can stall RNAP for a long enough time to ensure that, despite a slow catalytic step and a fast dissociation step, Mfd will successfully remodel it.

Intriguingly, remodeling of the stalled RNAP is not only slow, but results in formation of a long-lived intermediate on the DNA. Single-molecule experiments combining single-molecule nanomanipulation and single-molecule fluorescence by total internal fluorescence reflection (TIRF) provided new details into the fate of remodeled RNAP.[52] In TIRF the fluorescence excitation field decays exponentially as distance from the glass–buffer interface increases, and in these experiments the fluorescently-labelled RNAP was stalled while transcribing towards the surface. The magnetic trapping signal derived from this RNAP showed it to undergo stalling on a DNA lesion, followed by remodeling by Mfd. Simultaneous monitoring of the RNAP by fluorescence detection revealed, surprisingly, that after remodeling the RNAP remained visible in the fluorescence channel and hence associated with the DNA. Notably the fluorescence intensity of the RNAP, which had remained constant during its period of stalling, was observed to increase slowly after remodeling. As initial transcription was directed towards the surface where the TIRF field is stronger, this was consistent with slow, 4 nucleotides s^{-1} displacement of RNAP towards the surface after remodeling. Inverting the direction of initial transcription prior to stalling resulted in RNAP fluorescence decreasing with time after remodeling.

The fate of the nascent RNA in the stalled elongation complex was next monitored by hybridizing to it a complementary, fluorescent DNA oligonucleotide.[52] Simultaneous monitoring of the RNAP "mechanical" signal and the RNA fluorescence signal indicated that the RNA was present up until the moment where RNAP was remodeled by Mfd, but was lost at remodeling. This indicates that remodeled RNAP is no longer catalytically engaged with the DNA once remodeling has taken place. A third experiment using fluorescently-labeled Mfd showed it bound to RNAP just prior to remodeling and that it was present throughout the duration of the long-lived intermediate state.[52] As with the fluorescent RNAP, its fluorescence emission increased with time in the TIRF field. Altogether, these data indicated that an Mfd–RNAP complex was formed which displaced stalled RNAP from DNA by translocating along DNA, but then retained RNAP as a processivity factor as the Mfd–RNAP complex continued to slowly translocate along DNA.

In a fourth series of experiments, biotinylated RNAP was stalled as an elongation complex on DNA and then attached to a streptavidin-coated magnetic bead, and assembled under the magnetic trap (Figure 7.2). Addition of Mfd and ATP was sufficient to induce long-range translocation of the magnetic bead – here towards the surface – again colinear with the original direction of transcription of the stalled RNAP. It was observed that such complexes could translocate over thousands of base-pairs, with two-thirds of complexes displaying processivity greater than the 8 kbp DNA substrates used.[35,52] Under low extending force (approximately 1 pN) the motor reached a maximum velocity of 5 bp s^{-1} at saturating ATP, with a Michelis–Menten constant for ATP of only approximately 20 μM.[35]

The results of these experiments collectively indicated that RNAP remodeling by Mfd leads to formation of a robust new protein–DNA complex in which RNAP serves as a processivity factor for Mfd which, with RNAP, remains strongly bound to DNA.

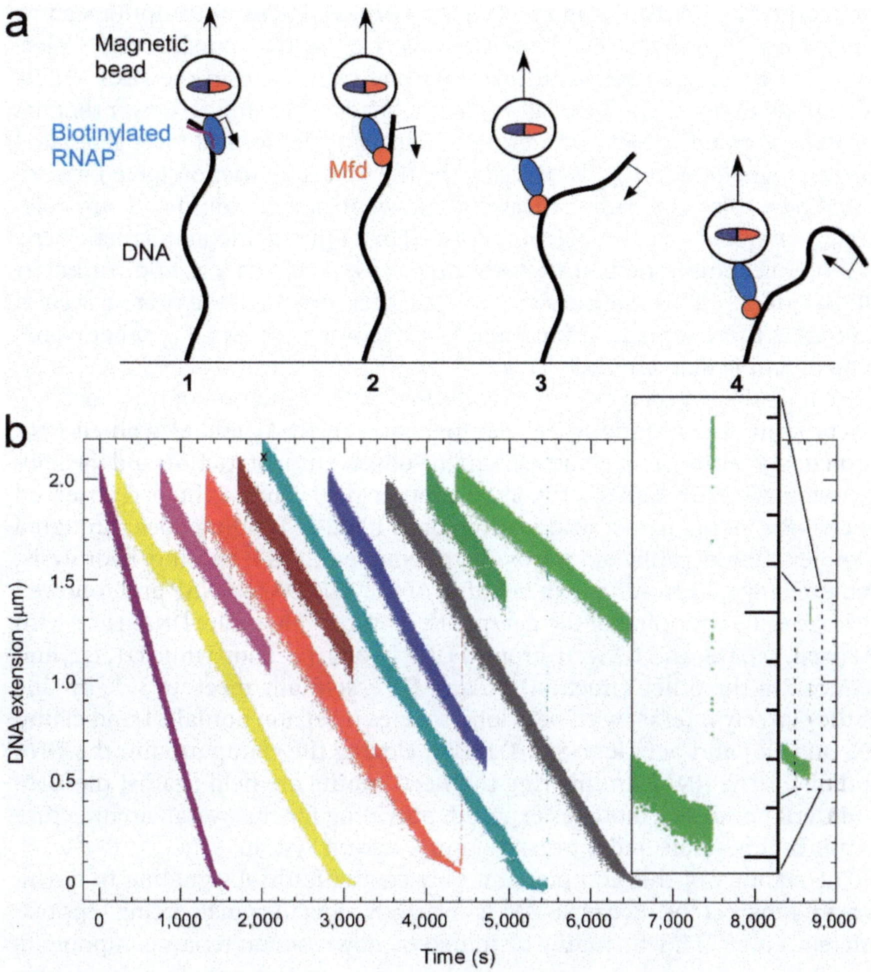

Figure 7.2 Real-time traces of Mfd–RNAP complexes translocating along DNA. (a) A biotinylated RNAP is stalled at one end of a DNA by withholding one of the four nucleotides. The RNAP is stalled transcribing towards the distal DNA end, which is digoxigenin-labelled. The RNAP–DNA complex is tethered to a streptavidin-coated magnetic bead and deposited onto an anti-digoxigenin surface. The DNA-RNAP-bead system is extended using a magnetic trap. (b) When Mfd and ATP are added bead motion towards the surface is observed. The inset shows an example of an event in which translocation is interrupted and the complex "slips" along DNA, but while maintaining sufficient contacts to cause frictional drag on the bead and even reattachment. The bead was tracked at 30 Hz, and the velocity in the rips was estimated to be on the order of 100 nm s^{-1}, far lower than the expected terminal velocity of approximately 1 µm s^{-1} for a freely traveling bead under the assay conditions. Reproduced from ref. 52 with permission from Springer Nature, Copyright 2015.

Some of the translocation trajectories published by Graves *et al.*[52] are interrupted before the bead reaches the surface. These interruptions show a rapid increase in the bead position, or "rip" in the translocation trajectory, as if the bead–RNAP–Mfd complex had torn off from the DNA. Oddly enough however, the velocity at which the bead rises is much slower than its expected terminal velocity. [At terminal velocity v_t the vertical force generated by the magnetic trap $\boldsymbol{F}_{\mathbf{mag}}$ is cancelled by the opposing friction force $\boldsymbol{f}$ experienced by a sphere of radius r travelling at velocity v_t in liquid with viscosity $\eta : F_{\text{mag}} = f = 6\pi\eta r v_t$. For a 1 pN force applied to a 1 μm diameter magnetic bead in buffer at room temperature with viscosity $\eta = 10^{-3}$ Pa s, v_t approximately 50–100 μm s^{-1}]. This indicates that despite detachment of the protein system from DNA there remains a frictional force between the two – in other words detachment is not complete.

We note that in certain events rebinding and re-initiation of translocation can be seen. A tight topological clamp formed by RNAP and Mfd about DNA upon RNAP remodeling could explain the observation of friction and the ability to rebind to DNA after a rip. Such a topological clamp is indeed observed in cryo-EM structures of Mfd loading onto RNAP.[53] Loading goes through a complex choreography of intermediates coupled to ATP binding, hydrolysis, and exchange, beginning with latching of the RID D4 to RNAP and followed by large-scale wrapping of the derepressed protein about the DNA. Here with D4 fixed, D5, D6, and D7 wrap around DNA in one direction while D1, D2, and D3 wrap in the other direction. D6 and D1 essentially meet up side-by-side as they largely interact with beta lobe-species insertion domain 1 and clamp (D5 and D6) and beta lobe-Si1 (D1–D3), closing the clamp around the DNA and distorting RNAP. In this ring the RecG motifs are held against the DNA and UvrB homology modules exposed, providing for the overall architecture of this highly stable and processive translocation system.

The stability of this complex can only ensure faithful signaling to downstream repair steps. Because Mfd is relatively slow in translocating (approximately 5 bp s^{-1}) it will readily be joined by downstream repair components UvrAB in the immediate vicinity of the lesion. In the translocation assays mentioned above UvrA alone and in complex with UvrB are observed to arrest the Mfd-RNAP complex and, using ATP hydrolysis, release at least RNAP from the DNA. In correlative single-molecule fluorescence and nanomanipulation fluorescence assays UvrAB is observed to release both fluorescent Mfd and fluorescent RNAP when it acts on the complex.[35] Kinetic analysis of the reaction using nanomanipulation assays indicate the association rate for UvrAB on Mfd-RNAP is on the order of 7×10^8 M^{-1} s^{-1} and the dissociation rate roughly 0.04 s^{-1}, and that catalytic removal of Mfd–RNAP takes about 40 seconds, for an overall reaction K_M in the 0.1 nM range,[35] nearly 100-fold greater than the affinity of UvrA and B for damaged DNA.

Recruitment of UvrA and B to Mfd–RNAP increases their concentration in the vicinity of the lesion. UvrA and B will then validate the lesion and recruit UvrC to it for double incision and handoff to UvrD, removing the damaged oligonucleotide from DNA. The presence of a UvrB homology module on Mfd indicates that the complex deposited on the DNA in this case of transcription-coupled repair could plausibly be in a $UvrA_2UvrB$ stoichiometry, maintaining asymmetry and so allowing the complex to continue to signal the damaged template strand to UvrC. This would potentially distinguish it from the symmetric $UvrA_2UvrB_2$ complex which would load independently of Mfd–RNAP, and as seen in so-called global genome nucleotide excision repair (GG-NER).[54]

These *in vitro* measurements are corroborated by the results of single-molecule fluorescence experiments *in vivo*. Fluorescent Mfd can be monitored in real-time interacting with RNAP *in vivo* in both wild-type and mutant backgrounds for UvrA or UvrB.[48,55] Single molecules of fluorescent Mfd are imaged as static fluorescent spots when Mfd is immobilized to the nucleoid *via* binding to RNAP actively engaged on DNA. Long, approximately 18 s, binding events are observed to elongating RNAP under normal growth conditions and in the absence of UV radiation. These events are not observed if Mfd carries RID mutation L499R, making it unable to interact with RNAP; their frequency decreases sevenfold upon application of rifampicin which prevents RNAP–promoter open complexes from initiating transcription; and their frequency increases twofold upon application of CBR703, a small molecule which induces stalling of elongating bacterial RNAP. The events observed thus correspond to specific interactions between Mfd and RNAP.

It is however difficult to know the exact state of the RNAP when these events initiate, namely paused, stalled, or actively elongating. Mfd notably does not interact with RNAP holoenzyme including σ^{70} as the latter occludes the site on the β subunit where Mfd binds.[3,42,51] UV irradiation of cells expressing fluorescent TCR components led to SOS induction, raising UvrA concentrations roughly threefold in the space of about 90 minutes and leading to an inversely proportional reduction of the static lifetime of fluorescent Mfd immobilized to DNA *via* RNAP.[56] UvrA and UvrB are typically present at approximately 100 copies per cell each prior to SOS induction.[56,57] Similarly, increasing UvrB concentration by overexpression also reduced the time Mfd was immobilized on the DNA *via* RNAP.[55] These results are consistent with the notion that the Mfd immobilized to DNA *via* RNAP must wait for UvrA or a complex including UvrA and UvrB to help release it. In the absence of UvrA, the binding duration increases to approximately 30 seconds,[48] and in the absence of UvrB the binding duration is on the order of 140 seconds.[55] It is noteworthy that in the absence of UvrB, fluorescent UvrA is seen to bind DNA for approximately 100 s, whereas in the presence of UvrB fluorescent UvrA exhibits second-timescale binding (80% of events) as well as longer approximately 10 s binding (20% of events)[56]—thus the absence of UvrB may indirectly reduce the pool of available UvrA. Over all these experiments reconstruct *in vivo* the process described *in vitro*, including formation of a stable Mfd–RNAP complex which can be very long-lived but which is turned over by UvrA and/or a UvrA–UvrB complex.

Single-molecule optical trapping experiments have also been carried out on the Mfd–RNAP complex.[46] In these experiments a dual-beam optical trap is used to pull on a long DNA hairpin by its stem, unzipping it. As the hairpin unzips the distance between the ends of the stem can be observed to increase in the optical trap. If a DNA-binding protein is located on the DNA it will transiently prevent unzipping and this can be detected in the optical trap. Because the distance signal can be converted into the number of base pairs that have been unzipped, this allows one to ascertain the position of the DNA-binding protein along the DNA. By carrying out these experiments with Mfd alone, it was shown that Mfd on its own can bind to DNA and translocate at the same approximately 5 bp s^{-1} velocity as observed in the magnetic trapping experiments. However, the average distance travelled by Mfd translocating alone on DNA is much shorter than that which is observed when it is coupled to RNAP – hundreds rather than thousands of base-pairs – indicating that different outcomes are obtained when Mfd binds first to DNA or first to RNAP.

In a second series of hairpin-unzipping experiments, an RNAP was loaded onto the long hairpin DNA but stalled as an elongation complex by withholding one of the four nucleotides. Mfd was then added to allow it to load onto RNAP. Restoring the full nucleotide complement restarted the RNAP before Mfd was able to remodel it. The unzipping assay then observed Mfd trailing behind the RNAP, consistent with the differential velocities of the two motors. RNAP could then be stalled again, at which point Mfd was observed to catch up to it, leading to a "release and catch-up model."

7.5 Mutagenic Activities Generated by Motor–Motor Interaction

In 2008 it was shown that *Campylobacter jejeuni* induces Mfd expression when subjected to the antimicrobial gyrase inhibitor ciprofloxacin.[50] Furthermore, Mfd overexpression increased the rate of appearance of resistant bacteria, while removal of Mfd reduced their rate of appearance. The accumulation of mutagenic events driven by Mfd collectively turned out to be responsible for about half of all point mutations and gene amplification events in the bacterial strain under study.[34] Importantly, loss of RNase HI further increased the rate of mutagenesis, but only if Mfd was present; in the absence of Mfd the removal of RNase HI no longer increased the rate of mutagenesis. Because RNase HI degrades the RNA in the RNA–DNA hybrids known as R-loops, it was proposed that Mfd was involved in R-loop biogenesis. The connection to RNA polymerase was highlighted by the fact that loss of the Mfd–RNAP interaction abrogated its ability to drive mutagenesis.[31] The implication of Mfd in the evolution of antibiotic resistance was further observed in multiple divergent bacteria subjected to a range of antibiotics with various mechanisms of action,[31] underscoring the generality of the phenomenon. Thus, although the historically-established function of Mfd is to displace RNAP stalled from a lesion and direct downstream GGR components UvrA and UvrB to the lesion for repair, a second function of Mfd also seems to be to drive R-loop formation in an RNAP-dependent manner, enhancing rates of bacterial evolution in conditions of stress.

Single-molecule observations have helped elucidate the mechanism whereby the Mfd–RNAP interaction drives R-loop formation.[58] The observations are technically involved and so are best explained by analogy with the classical Liu and Wang model for twin supercoiled domains resulting from transcription (Figure 7.3).[59] In this classical model if transcribing RNAP and DNA are not able to rotate about each other as RNAP tracks the DNA helix

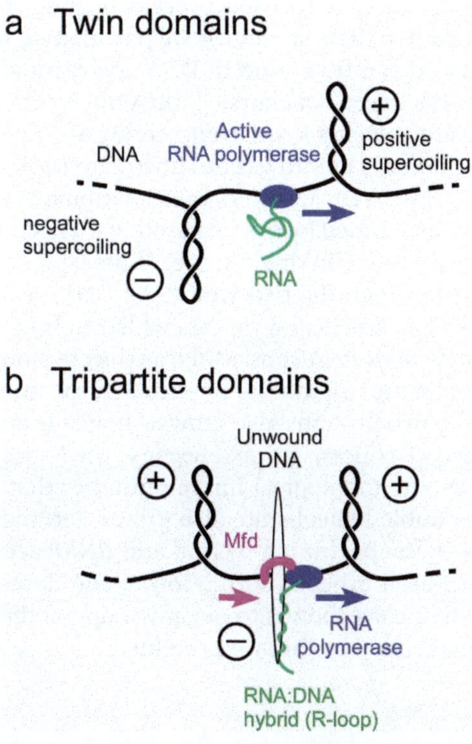

Figure 7.3 Twin and tripartite supercoiled domains. DNA is torsionally constrained at its extremities as in a topological chromosomal domain. (a) In the classical Liu and Wang model,[59] transcribing RNAP which is not free to rotate about DNA as it advances (*e.g.* because of frictional drag along the RNA and any associated ribosomes and polypeptides) will topologically accumulate positive supercoils in front and negative supercoils behind as it advances. Indeed, as transcription proceeds the number of base pairs in front of the polymerase decreases while the number of base pairs behind the polymerase increases, however the linking number of each domain remains fixed. An excess of linking number (*i.e.* positive supercoiling) accumulates in front of RNAP, and a deficit in linking number (negative supercoiling) accumulates behind RNAP. (b) When Mfd docks to elongating RNAP and then to DNA, the net differential velocity between the two motors means that the length of the DNA domain between the two proteins will grow, but not the linking number within this domain. As a result, this internal domain becomes highly negatively supercoiled and is susceptible to R-loop formation. Furthermore, positive supercoils now accumulate behind the complex as well as in front of it.

then positive supercoiling will form ahead of RNAP and negative supercoiling will form behind it (see Box 7.1). Rotation of RNAP relative to DNA can be hindered, for instance, due to drag against a nascent RNA undergoing translation. Supercoiling is likely to be, in part, dissipated into the DNA upstream and downstream of the RNA polymerase up until it reaches the boundary conditions at the end of chromosomal topological domains, and furthermore absorbed by the action of topoisomerases.

A similar yet more spatially focused process begins when Mfd interacts with transcribing, rather than stalled, RNA polymerase (Figure 7.3). As it becomes activated Mfd can then bind to DNA upstream of RNAP and begin to translocate as we have seen. Of course, if the velocity of Mfd is only 5 bp s^{-1} it will not succeed at displacing RNAP transcribing at a rate of 25 bases s^{-1} or more. However, the binding of Mfd to both upstream DNA and RNAP, and the further binding of RNAP to DNA, results in formation of a closed topological domain in the DNA bookended by the two proteins. This topological domain is formed with a small initial linking number (a measure of the total number of topological links between the two strands of DNA, see Box 7.1) which is fixed in size by the DNA-associated motors which tightly grip the ends of a short span of DNA. As both RNAP and Mfd translocate along DNA, the differential velocity between the two motors implies that the number of base-pairs of DNA spanning the two proteins also grows – typically at a rate of about 20 bp s^{-1} at saturating NTP concentrations. Because the length of DNA between the bookends grows, the topological linking number that would be needed to fully satisfy the double helical state also grows. Yet, the available linking number is fixed from the beginning by Mfd and RNAP acting as topological bookends bound to each other at formation of the domain. This growing linking number deficit corresponds to negative supercoiling which accumulates in the DNA bookended by the two proteins.

Box 7.1 DNA Linking number

The DNA linking number, Lk, is the sum of crossings between the two strands of the double helix and is the sum of two terms known as Twist, Tw, and Writhe, Wr: Lk = Tw + Wr. Twist counts the intramolecular crossings of the two strands; for torsionally-relaxed DNA in physiological conditions the natural twist Tw_0 of N base-pairs is N/h with pitch h of approximately 10.4 bp per turn for B-DNA. Writhe counts the number of times the two strands cross when the axis of the DNA molecule forms loops, or plectonemes, similar to those that form on a tangled phone cord; the natural writhe Wr_0 of DNA shorter than 50–100 kb is taken to be zero as DNA is stiff. The natural linking number for torsionally relaxed DNA is $Lk_0 = Tw_0 + Wr_0 = Tw_0$. If a DNA has a linking number Lk greater or smaller than Lk_0 it is positively or negatively supercoiled, respectively. The linking number of a torsionally-constrained domain of DNA is conserved as long as one of the strands is not broken. Supercoiling can redistribute as changes in DNA twist or changes in DNA writhe, provided the sum remains constant.

Conservation of the overall linking number of the nanomanipulated DNA within which this process occurs means that compensatory positive supercoiling accumulates in the DNA which is not included in this domain – *i.e.* in front of the RNAP and behind Mfd. If we now consider that, as in the original Liu and Wang model, the RNAP is rotationally immobilized and DNA rotates through it during elongation, then the domain in front of RNAP is topologically partitioned from the domain behind it. The domain behind it is itself partitioned into two domains, one between Mfd and RNAP and one behind Mfd, resulting in the formation of three, or tripartite, topological domains. The key difference between the twin and tripartite domain models is that in the tripartite domain model negative supercoiling generated just upstream of polymerase is focused into a small domain bookended by Mfd. As a result, the DNA between Mfd and RNAP is expected to be subjected to massive levels of unwinding approaching full topological unwinding of DNA. The minimal requirement of DNA upstream of RNAP for Mfd to bind as seen in footprinting assays is approximately 30 bp,[51] and recent cryo-EM structures[53] show interactions between Mfd and DNA from 14 to 34 basepairs during assembly upstream of RNAP. Thus, the topological domain at its inception is quite small, on the order of only tens of bases and only a few units of linking number Lk. As mentioned earlier, in one second RNAP will have advanced 25 bp and Mfd only 5 bp, meaning that the length of DNA between the two proteins will increase by approximately 20 bp – roughly doubling to 40 bp – while the DNA linking number will remain unchanged (Lk approximately 2). As a result, the DNA bookended by Mfd and RNAP will be unwound by roughly 50% a second after domain formation and this value tends with time to 100%; clearly this region rapidly undergoes massive negative supercoiling. Growth in the size of the negatively supercoiled domain can be quantified by monitoring the appearance of compensatory positive supercoiling in the magnetic trap. The rate of growth of the negatively supercoiled domain can be decreased by lowering the nucleotide concentration to a level low enough to slow RNAP but not Mfd (*e.g.* 50 μM), consistent with the model for differential motor velocity leading to DNA unwinding in a localized domain. We further note this process also takes place with Mfd deficient in its ability to translocate along DNA.

As a result of this immediate availability of highly unwound DNA, the nascent RNA being produced by the RNAP can invade this region and form an R-loop. Once RNAP terminates transcription, or in the unlikely event that Mfd dissociates spontaneously from DNA, the R-loop can be observed to remain associated with the DNA. The R-loop can then be detected and characterized in the magnetic trap thanks to the known supercoiling properties of such structures.[60] The structures detected in this manner are sensitive to RNase HI treatment; in the presence of RNase HI their lifetime drops from tens of thousands to tens of seconds, confirming their nature as RNA–DNA hybrids. These results are mechanistically consistent with genetic observations, and demonstrate that the interaction of Mfd with transcribing RNAP can result in co-transcriptional R-loop formation.

Although it remains to be seen structurally how this R-loop is formed, it is likely that the massive unwinding behind the polymerase tests the stability of DNA annealing at the upstream fork of the transcription bubble. Thus DNA unwinding in the negatively supercoiled domain may invade up to and including the transcription bubble. Now, it should be noted that there is no covalent polypeptide chain or domain physically barring the path for RNA present in the exit channel to collapse back onto its unwound DNA template. It is therefore likely that with time the R-loop will merge with the transcription bubble. Considering that RNAP is known to be stalled by negative torque behind it, formation of this negatively supercoiled domain behind it may cause it to stall, although *in vivo* growth regulator A and B (GreA and GreB) should be available to restart stalled RNAP if it comes to undergo backtracking from its stalled state. It is not known if Mfd has complete latitude to form a stable topological clamp on elongating RNAP, or if it perhaps forms a less stable complex more akin to the "release and catch up" models described from optical trapping experiments.[46]

An important question which arises from these experiments is how stress sets the balance between Mfd's pro-repair and pro-mutagenic activities, both of which target RNAP? First we note that the DNA repair activity should be basal: on stalled RNAP the only activity of Mfd is to displace it and initiate repair. Regarding the pro-mutagenic activity, as shown by Wimberly *et al.*,[34] the availability of ribosomes to transcribe the nascent RNA will reduce the ability of the RNA to invade the tripartite topological domain. Stress conditions, in which ribosome synthesis is abrogated and ppGpp induction limits transcriptional activity, could reduce the amount of ribosomes available to cover the RNA, raising the likelihood of R-loop formation. Stress conditions could also reduce the number of transcribing RNA polymerases while raising Mfd expression, raising coverage of remaining transcribing RNAPs by Mfd (in exponentially growing cells RNAP is present at about 5000 copies and Mfd at 100–500[3,49]).

According to this model, processes which remove Mfd from DNA should also limit the ability of the nascent RNA to form R-loops. The availability of UvrA and UvrB to remove Mfd–RNAP from the DNA is likely to be a factor in limiting formation of the tripartite topological domains. Because UvrA and UvrB are normally present at somewhat low concentrations in the absence of SOS induction (on the order of 50 dimers of each in the cell, with only 5–10 monomers of UvrC to turn over UvrB), any process which may solicit them for other purposes – such as low levels of DNA lesions – can reduce their availability for removing Mfd–RNAP complexes from DNA. Thus, for instance, low levels of replicative stress obtained using suprainhibitory concentrations of anti-gyrase antibiotics may lead to low-level accumulation of single stranded DNA (ssDNA) in the vicinity of the fork, which is known to result in DNA lesions in the exposed regions. This could titrate out available UvrA and therefore provide Mfd–RNAP complexes with more leeway to form R-loops. In this fashion replicative stress probably feeds back into transcriptional stress.

Very importantly, the fact that Mfd motor activity is not required for R-loop formation by Mfd as shown by Portman *et al.*[58] indicates that the topological process described can be carried out by any protein capable of simultaneously binding to elongating RNAP and upstream DNA. For instance, because transcription initiation is a stochastic process and promoter escape is likely to be independent of dissolution of contacts between RNAP and upstream DNA, it is reasonable to expect that this process may take place during promoter escape and thus be responsible for the overrepresentation of R-loops in promoter-proximal regions. The consequences of such processes for chromosome structure may not be trivial, as they have the potential of generating net positive supercoiling in topological chromosomal domains (*i.e.* in front of RNA polymerase and behind the protein anchored to DNA which partners with RNAP), at least until they are counteracted by topoisomerases.[61] Finally, it is interesting to note that the human transcription-coupled repair system is also implicated in tumor mutagenesis and evolution of resistance to chemotherapy.[62,63] It will be interesting to determine to what extent R-loop formation is at play here.

7.6 Conclusions

The three key proteins described are the best known of the proteins deployed by the cell which appear to minimize the obstructions caused by immobilized or misregulated RNA polymerase. Other similar helicases may still be waiting to be found in the trove of proteins with unknown functions, whose conditions for expression and activity have not yet been identified and studied. As described in Chapter 8, these regulatory helicases ultimately protect the cell from transcription–replication conflicts and the additional lesions which can act on RNAP and further solicit Mfd and Uvr components. The effects of the obstructive RNA polymerase on DNA damage and its repair, for instance, including single-strand breaks, double strand breaks, and the generation of oxidative damage in bases exposed in the non-template strand, is only beginning to be understood more completely at the bacterial level, and is certain to lead to exciting discoveries with medical repercussions in the coming years.

References

1. S. French, Consequences of replication fork movement through transcription units *in vivo*, *Science*, 1992, **258**, 1362–1365.
2. H. Boubakri, A. L. Septenville, E. Viguera and B. Michel, The helicases DinG, Rep and UvrD cooperate to promote replication across transcription units *in vivo*, *EMBO J.*, 2010, **29**, 145–157.
3. C. P. Selby and A. Sancar, Molecular mechanism of transcription-repair coupling, *Science*, 1993, **260**, 53–58.
4. S. Jain, R. Gupta and R. Sen, Rho-dependent transcription termination in bacteria recycles RNA polymerases stalled at DNA lesions, *Nat. Commun.*, 2019, **10**, 1207.

5. P. Mitra, G. Ghosh, M. Hafeezunnisa and R. Sen, Rho protein: roles and mechanisms, *Annu. Rev. Microbiol.*, 2017, **71**, 687–709.
6. R. S. Washburn and M. E. Gottesman, Transcription termination maintains chromosome integrity, *Proc. Natl. Acad. Sci. U. S. A.*, 2011, **108**, 792–797.
7. C. A. Brennan, A. J. Dombroski and T. Platt, Transcription-termination factor rho is an RNA-DNA helicase, *Cell*, 1987, **48**, 945–952.
8. V. Epshtein, V. Kamarthapu, K. McGary, V. Svetlov, B. Ueberheide, S. Proshkin, A. Mironov and E. Nudler, UvrD facilitates DNA repair by pulling RNA polymerase backwards, *Nature*, 2014, **505**, 372–377.
9. M. Hawkins, J. U. Dimude, J. A. L. Howard, A. J. Smith, M. S. Dillingham, N. J. Savery, C. J. Rudolph and P. McGlynn, Direct removal of RNA polymerase barriers to replication by accessory replicative helicases, *Nucleic Acids Res.*, 2019, **47**, 5100–5113.
10. K. Sanders, C.-L. Lin, A. J. Smith, N. Cronin, G. Fisher, V. Eftychidis, P. McGlynn, N. J. Savery, D. B. Wigley and M. S. Dillingham, The structure and function of an RNA polymerase interaction domain in the PcrA/UvrD helicase, *Nucleic Acids Res.*, 2017, **45**(7), 3875–3887.
11. P. McGlynn and R. G. Lloyd, Modulation of RNA polymerase by (p)ppGpp reveals a recG-dependent mechanism for replication fork progression, *Cell*, 2000, **101**, 35–45.
12. B. W. Trautinger, R. P. Jaktaji, E. Rusakova and R. G. Lloyd, RNA polymerase modulators and DNA repair activities resolve conflicts between DNA replication and transcription, *Mol. Cell*, 2005, **19**, 247–258.
13. Z. Baharoglu, R. Lestini, S. Duigou and B. Michel, RNA polymerase mutations that facilitate replication progression in the rep UvrD recF mutant lacking two accessory replicative helicases, *Mol. Microbiol.*, 2010, **77**, 324–336.
14. B. Sun, A. Singh, S. Sultana, J. T. Inman, S. S. Patel and M. D. Wang, Helicase promotes replication re-initiation from an RNA transcript, *Nat. Commun.*, 2018, **9**, 2306.
15. M. K. Gupta, C. P. Guy, J. T. Yeeles, J. Atkinson, H. Bell and R. G. Lloyd, Protein-DNA complexes are the primary source of replication fork pausing in *Escherichia coli*, *Proc. Natl. Acad. Sci. U. S. A.*, 2013, **110**, 7252–7257.
16. W. Gan, Z. Guan, J. Liu, T. Gui, K. Shen, J. L. Manley and X. Li, R-loop-mediated genomic instability is caused by impairment of replication fork progression, *Genes Dev.*, 2011, **25**, 2041–2056.
17. K. S. Lang, A. N. Hall, C. N. Merrikh, M. Ragheb, H. Tabakh, A. J. Pollock, J. J. Woodward, J. E. Dreifus and H. Merrikh, Replication-transcription conflicts generate R-loops that orchestrate bacterial stress survival and pathogenesis, *Cell*, 2017, **170**, 787–799.e18.
18. C. P. Guy, J. Atkinson, M. K. Gupta, A. A. Mahdi, E. J. Gwynn, C. J. Rudolph, P. B. Moon, I. C. van Knippenberg, C. J. Cadman, M. S. Dillingham, R. G. Lloyd and P. McGlynn, Rep provides a second motor at the replisome to promote duplication of protein-bound DNA, *Mol. Cell*, 2009, **36**, 654–666.
19. H. E. Lane and D. T. Denhardt, The rep mutation. IV. Slower movement of replication forks in *Escherichia coli* rep strains, *J. Mol. Biol.*, 1975, **97**, 99–112.

20. O. Adebali, A. Sancar and C. P. Selby, Mfd translocase is necessary and sufficient for transcription-coupled repair in *Escherichia coli*, *J. Biol. Chem.*, 2017, **292**, 18386–18391.
21. J. Y. Kang, R. A. Mooney, Y. Nedialkov, J. Saba, T. V. Mishanina, I. Artsimovitch, R. Landick and S. A. Darst, Structural basis for transcript elongation control by NusG family universal regulators, *Cell*, 2018, **173**, 1650–1662.e14.
22. J. Li, R. Horwitz, S. McCracken and J. Greenblatt, NusG, a new *Escherichia coli* elongation factor involved in transcriptional antitermination by the N protein of phage lambda, *J. Biol. Chem.*, 1992, **267**, 6012–6019.
23. S. Saxena, K. K. Myka, R. Washburn, N. Costantino, D. L. Court and M. E. Gottesman, *Escherichia coli* transcription factor NusG binds to 70S ribosomes, *Mol. Microbiol.*, 2018, **108**, 495–504.
24. S. L. Sullivan and M. E. Gottesman, Requirement for *E. coli* NusG protein in factor-dependent transcription termination, *Cell*, 1992, **68**, 989–994.
25. V. Epshtein, D. Dutta, J. Wade and E. Nudler, An allosteric mechanism of Rho-dependent transcription termination, *Nature*, 2010, **463**, 245–249.
26. S. Adhya, M. Gottesman and B. De Crombrugghe, Release of polarity in *Escherichia coli* by gene N of phage lambda: termination and antitermination of transcription, *Proc. Natl. Acad. Sci. U. S. A.*, 1974, **71**, 2534–2538.
27. N. C. Franklin, Altered reading of genetic signals fused to the N operon of bacteriophage lambda: genetic evidence for modification of polymerase by the protein product of the N gene, *J. Mol. Biol.*, 1974, **89**, 33–48.
28. J. M. Peters, R. A. Mooney, J. A. Grass, E. D. Jessen, F. Tran and R. Landick, Rho and NusG suppress pervasive antisense transcription in *Escherichia coli*, *Genes Dev.*, 2012, **26**, 2621–2633.
29. J. M. Peters, R. A. Mooney, P. F. Kuan, J. L. Rowland, S. Keles and R. Landick, Rho directs widespread termination of intragenic and stable RNA transcription, *Proc. Natl. Acad. Sci. U. S. A.*, 2009, **110**, 258–263.
30. J. K. Leela, A. H. Syeda, K. Anupama and J. Gowrishankar, Rho-dependent transcription termination is essential to prevent excessive genome-wide R-loops in *Escherichia coli*, *Proc. Natl. Acad. Sci. U. S. A.*, 2013, **110**, 258–263.
31. M. N. Ragheb, M. K. Thomason, C. Hsu, P. Nugent, J. Gage, A. N. Samadpour, A. Kariisa, C. N. Merrikh, S. I. Miller, D. R. Sherman and H. Merrikh, Inhibiting the evolution of antibiotic resistance, *Mol. Cell*, 2019, **73**, 157–165.e5.
32. J. S. Fassler and I. Tessman, Relation between UV suppression of polarity in phiX174 and UV sensitivity of rho mutants, *J. Virol.*, 1981, **37**, 955–962.
33. E. M. Witkin, Radiation-induced mutations and their repair, *Science*, 1966, **152**, 1345–1353.
34. H. Wimberly, C. Shee, P. C. Thornton, P. Sivaramakrishnan, S. M. Rosenberg and P. J. Hastings, R-loops and nicks initiate DNA breakage and genome instability in non-growing *Escherichia coli*, *Nat. Commun.*, 2013, **4**, 2115–2125.

35. J. Fan, M. Leroux-Coyau, N. J. Savery and T. R. Strick, Reconstruction of bacterial transcription-coupled repair at single-molecule resolution, *Nature*, 2016, **536**, 234–237.
36. F. Brueckner, U. Hennecke, T. Carell and P. Cramer, CPD damage recognition by transcribing RNA polymerase II, *Science*, 2007, **315**, 859–862.
37. N. M. Haines, Y.-I. T. Kim, A. J. Smith and N. J. Savery, Stalled transcription complexes promote DNA repair at a distance, *Proc. Natl. Acad. Sci. U. S. A.*, 2014, **111**, 4037–4042.
38. J. T. Wade and D. C. Grainger, Pervasive transcription: illuminating the dark matter of bacterial transcriptomes, *Nat. Rev. Microbiol.*, 2014, **12**, 647.
39. O. Porrua, M. Boudvillain and D. Libri, Transcription termination: Variations on common themes, *Trends Genet.*, 2016, **32**, 508–522.
40. C. P. Selby and A. Sancar, Structure and function of transcription-repair coupling factor. I. Structural domains and binding properties, *J. Biol. Chem.*, 1995, **270**, 4882–4889.
41. A. J. Smith and N. J. Savery, RNA polymerase mutants defective in the initiation of transcription-coupled DNA repair, *Nucleic Acids Res.*, 2005, **33**, 755–764.
42. A. M. Deaconescu, A. L. Chambers, A. J. Smith, B. E. Nickels, A. Hochschild, N. J. Savery and S. A. Darst, Structural basis for bacterial transcription-coupled DNA repair, *Cell*, 2006, **124**, 507–520.
43. L. F. Westblade, E. A. Campbell, C. Pukhrambam, J. C. Padovan, B. E. Nickels, V. Lamour and S. A. Darst, Structural basis for the bacterial transcription-repair coupling factor/RNA polymerase interaction, *Nucleic Acids Res.*, 2010, **38**, 8357–8369.
44. D. B. Srivastava and S. A. Darst, Derepression of bacterial transcription-repair coupling factor is associated with a profound conformational change, *J. Mol. Biol.*, 2011, **406**, 275–284.
45. A. M. Deaconescu, A. Sevostyanova, I. Artsimovitch and N. Grigorieff, Nucleotide excision repair (NER) machinery recruitment by the transcription-repair coupling factor involves unmasking of a conserved intramolecular interface, *Proc. Natl. Acad. Sci. U. S. A.*, 2012, **109**, 3353–3358.
46. T. T. Le, Y. Yang, C. Tan, M. M. Suhanovsky, R. M. Fulbright, J. T. Inman, M. Li, J. Lee, S. Perelman, J. W. Roberts, A. M. Deaconescu and M. D. Wang, Mfd dynamically regulates transcription *via* a release and catch-up mechanism, *Cell*, 2018, **172**, 344–357.
47. K. Howan, A. J. Smith, L. F. Westblade, N. Joly, W. Grange, S. Zorman, S. A. Darst, N. J. Savery and T. R. Strick, Initiation of transcription-coupled repair characterized at single-molecule resolution, *Nature*, 2012, **490**, 431–434.
48. H. N. Ho, A. M. van Oijen and H. Ghodke, The transcription-repair coupling factor Mfd associates with RNA polymerase in the absence of exogenous damage, *Nat. Commun.*, 2018, **9**, 1570.
49. A. Schmidt, K. Kochanowski, S. Vedelaar, E. Ahrné, B. Volkmer, L. Callipo, K. Knoops, M. Bauer, R. Aebersold and M. Heinemann, The quantitative and condition-dependent *Escherichia coli* proteome, *Nat. Biotechnol.*, 2016, **34**, 104–110.

50. J. Han, O. Sahin, Y.-W. Barton and Q. Zhang, Key role of Mfd in the development of fluoroquinolone resistance in *Campylobacter jejuni*, *PLoS Pathog.*, 2008, **4**, e1000083.
51. J.-S. Park, M. T. Marr and J. W. Roberts, *E. coli* Transcription repair coupling factor (Mfd protein) rescues arrested complexes by promoting forward translocation, *Cell*, 2002, **109**, 757–767.
52. E. T. Graves, C. Duboc, J. Fan, F. Stransky, M. Leroux-Coyau and T. R. Strick, A dynamic DNA-repair complex observed by correlative single-molecule nanomanipulation and fluorescence, *Nat. Struct. Mol. Biol.*, 2015, **22**, 452–457.
53. E. Llewellyn, J. Chen, P. D. B. Olinares, J. Brewer, B. T. Chait, E. A. Campbell, J.-Y. Kang and S. A. Darst, Structural basis for transcription-complex disruption by the Mfd translocase, *eLife*, 2021, **10**, e62117.
54. D. Pakotiprapha, Y. Inuzuka, B. R. Bowman, G. F. Moolenaar, N. Goosen, D. Jeruzalmi and G. L. Verdine, Crystal structure of *Bacillus stearothermophilus* UvrA provides insight into ATP-modulated dimerization, UvrB interaction, and DNA binding, *Mol. Cell*, 2008, **29**(1), 122–133.
55. H. N. Ho, A. M. van Oijen and H. Ghodke, Single-molecule imaging reveals molecular coupling between transcription and DNA repair machinery in live cells, *Nat. Commun.*, 2020, **11**, 1478.
56. H. Ghodke, H. N. Ho and A. M. van Oijen, Single-molecule live-cell imaging visualizes parallel pathways of prokaryotic nucleotide excision repair, *Nat. Commun.*, 2020, **11**, 1477.
57. M. Stracy, M. Jaciuk, S. Uphoff, A. N. Kapanidis, M. Nowotny, D. J. Sherratt and P. Zawadzki, Single-molecule imaging of UvrA and UvrB recruitment to DNA lesions in living *Escherichia coli*, *Nat. Commun.*, 2016, **7**, 12568.
58. J. R. Portman, G. M. Brouwer, J. Bollins, N. J. Savery and T. R. Strick, Cotranscriptional R-loop formation by Mfd involves topological partitioning of DNA, *Proc. Natl. Acad. Sci. U. S. A.*, 2021, **118**, e2019630118.
59. L. F. Liu and J. C. Wang, Supercoiling of the DNA template during transcription, *Proc. Natl. Acad. Sci. U. S. A.*, 1987, **84**, 7024–7027.
60. T. Strick, V. Croquette and D. Bensimon, Homologous pairing in streched supercoiled DNA, *Proc. Natl. Acad. Sci. U. S. A.*, 1998, **95**, 10579–10583.
61. C. Naughton, N. Avlonitis, S. Corless, J. G. Prendergast, I. K. Mati, P. P. Eijk, S. L. Cockroft, M. Bradley, B. Ylstra and N. Gilbert, Transcription forms and remodels supercoiling domains unfolding large-scale chromatin structures, *Nat. Struct. Mol. Biol.*, 2013, **20**, 387–395.
62. C. Troelstra, A. van Gool, J. de Wit, W. Vermeulen, D. Bootsma and J. H. J. Hoeijmakers, Ercc6, a member of a subfamily of putative helicases, is involved in cockayne's syndrome and preferential repair of active genes, *Cell*, 1992, **71**(6), 939–953.
63. M. Caputo, M. Frontini, R. Velez-Cruz, S. Nicolai, G. Prantera and L. Proietti-De-Santis, The CSB repair factor is overexpressed in cancer cells, increases apoptotic resistance, and promotes tumor growth, *DNA Repair*, 2013, **12**, 293–299.

CHAPTER 8

Transcription–Replication Conflicts, Resolution and Coregulation

BIANCA SCLAVI*

LCQB, UMR 7238 CNRS, Sorbonne Université, 4 Place Jussieu, 75005 Paris, France
*E-mail: bianca.sclavi@sorbonne-universite.fr

8.1 Introduction

DNA replication forks work to copy both the leading and lagging strands of the DNA double helix within a very dynamic multi-protein complex moving at about 750 nucleotides per second.[1] During their journey from the origin of replication to the terminus, however, there is a high probability that they will encounter various kinds of barriers under the form of DNA-binding protein complexes more or less stably bound to this shared substrate, leading to fork stalling and disassembly. This is the case in all living organisms, however, the nature of the nucleoprotein barriers may differ between eukarya, archaea and bacteria, mostly because of the different structures organizing their chromatin and their relative stabilities.[2–4] Moreover, while the conflict that can arise between the DNA replication machinery and transcription complexes is shared among the three domains of life, notably its association with the formations of R-loops,[5,6] the specific factors involved in conflict resolution are not necessarily the same. These have been the subject of several

Chemical Biology No. 19
RNA Polymerases as Molecular Motors: On the Road, 2nd Edition
Edited by Robert Landick, Terence Strick and Jue Wang

Published by the Royal Society of Chemistry, www.rsc.org

reviews.[5–10] This chapter focusses on the more recent results that have been obtained on the mechanisms that play a key role in decreasing the probability that this conflict take place in bacteria.

8.2 The DNA Replication Complex

In bacteria DNA replication begins after the assembly of the replication forks at the origin, oriC. The origin is a region of the genome that contains an AT-rich segment called the DNA unwinding element (DUE) where the two DNA strands are separated following the binding of the initiator protein, DnaA, to its adjacent binding sites. The cooperative oligomerization of the adenosine triphosphate (ATP)-bound form of DnaA determines the timing and synchrony of initiation of the origins present in the cell.[11–13] Additional factors such as the binding of nucleoid proteins [*e.g.* integration host factor (IHF), factor for inversion stimulation (FIS), histone-like nucleoid structuring protein (HNS), DNA-binding proteins from starved cells (Dps)...] and the local level of DNA supercoiling can contribute to the regulation of this event as a function of the growth rate of the cell and its adaptation to different growth phases.[14–17] Melting of the DUE results in recruitment of the DNA replication machinery, starting with the interaction between the DnaC helicase loader and the DnaB helicase. The overall organization of the DNA replication complex is for the most part conserved between Gram-positive and Gram-negative bacteria.[18] In Gram-negative bacteria such as *Escherichia coli* the proteins of the replisome include the helicase, DnaB, separating the two strands in the 5′ to 3′ direction on the lagging strand coupled to ATP hydrolysis, a primase, DnaG, synthesizing a single RNA primer on the leading strand and a succession of RNA primers on the lagging strand to initiate, respectively, continuous and discontinuous DNA synthesis by the Pol III core DNA polymerase. The latter, in turn, is associated with the dimeric processivity clamp, β_2, and the clamp loader complex, DnaX, assuring the high processivity of DNA synthesis. On the lagging strand, the Okazaki fragments' RNA primers are replaced by DNA by the activity of DNA Pol I and a ligase, while SSB, the single-stranded DNA binding protein, protects the lagging single-stranded DNA as it awaits DNA synthesis. This is especially important when leading and lagging strand DNA synthesis are decoupled by the presence of DNA damage and/or protein barriers on the DNA[19–21] (Figure 8.1). Recent results obtained by crystallography and cryo-electron microscopy (cryo-EM) are beginning to create a more detailed picture of the protein–protein interactions and their conformational changes in the presence of the DNA substrate at the different steps of the replication process[18,22–24] while *in vitro* and *in vivo* single-molecule experiments nicely complement this level of description by characterizing the dynamic properties of the DNA replication machinery.[18,20,25–28]

Additional proteins and enzymes are associated with the replication forks, forming what have been called replication factories, or hyperstructures, in which more than one replication fork can assemble into a

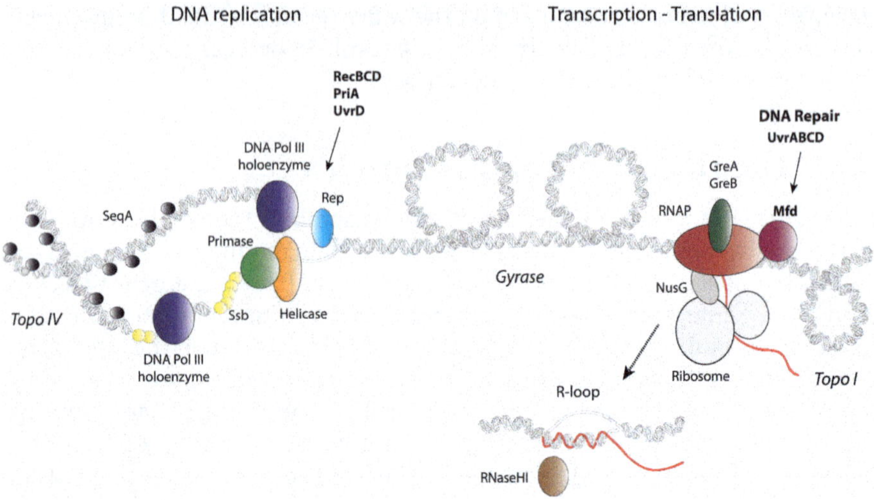

Figure 8.1 **Simplified representation of a DNA replication–transcription conflict.** The head-on encounter of a replication complex and an elongating RNA polymerase (RNAP) can result in the accumulation of positive DNA supercoils between the two machineries, stalling transcription elongation and leading to the formation of a backtracked complex. These positive supercoils can be relaxed by DNA gyrase. A backtracked RNAP complex can be rescued either by elongation factors, such as growth regulator A and B (GreA and GreB) that induce cleavage of the mRNA, or by the ATP-hydrolysis dependent mutation frequency decline (Mfd) motor that can push the RNAP past the backtracked conformation. When the RNAP stalls at a DNA damage site Mfd can recruit the DNA repair complex *via* an interaction with ultraviolet resistance A (UvrA). The negative supercoils behind the transcription complex can favor the formation of an R-loop if the activity of topoisomerase I is impaired. R-loops can constitute a barrier to advancing DNA polymerase within the replication complex. Other factors that can decrease the probability of the formation of R-loops include co-transcriptional translation of the mRNA and binding of the mRNA by the Rho protein (not shown) in the absence of translational coupling *via* N-utilization substance G (NusG). R-loops are recognized by the RNase HI enzyme that cleaves the RNA to facilitate dissociation of the DNA–RNA hybrid by helicases. The Rep helicase, moving ahead of the replication complex on the template strand, can displace DNA-bound proteins, including RNAP. Several factors are involved in the restart of DNA replication forks following a conflict, depending on whether it is followed by the formation of double strand breaks or the induction of the SOS response by the accumulation of single stranded DNA and the activation of RecA by the RecFOR and RecBCD complexes (not shown). The SOS response, in turn, can induce the expression of a set of factors, including helicases, exonucleases and error prone DNA polymerases, that restore the fork at the cost of increased mutation rates. Topoisomerase IV can resolve the catenation of the newly synthesized sister chromatids behind the replication fork. The SeqA protein binds to hemi-methylated GATC sites on newly replicated DNA and may play a role in sister chromatid segregation.

higher-order complex that includes additional enzymes such as ribonucleotide reductase, for the local synthesis of deoxynucleotide triphosphates (dNTPs), and fork reassembly factors such as the replication (Rep) and primosome assembly protein (PriA) helicases.[29–31] Topoisomerases are also part of this higher-order assembly. They are needed to release the topological stress created by fork translocation and synthesis of the new strands, with gyrase removing positive supercoils ahead of the replication fork and topoisomerase I removing negative supercoils behind it. Topoisomerase IV is also required to remove residual topological links between the newly synthesized DNA strands. In addition, Topoisomerase I and gyrase also play a key role in resolving topological stress generated by the transcription machinery[29,30,32–34] (Figure 8.1).

In gamma-proteobacteria, the sequestration A (SeqA) protein is also an integral part of the hyperstructure. DNA methylation can distinguish between newly replicated, sister chromatid DNA and old, mother DNA. The enzyme DNA adenine methyltransferase (Dam) adds a methyl group to adenines when they are found within GATC sequences.[35] In newly replicated DNA, at first the daughter strand is not yet methylated, leading to the temporary presence of hemi-methylated DNA behind the replication forks. These hemi-methylated GATC sites are bound by the SeqA protein, which has been shown to stabilize the replication forks, possibly by organizing the topology of the DNA behind the fork, guiding the process of sister chromosome segregation and maintaining the overall fork structure when fork stalling occurs, resulting in a more efficient reassembly of the hyperstructure and restart process.[36,37]

8.3 A Highly Dynamic Multi-protein Complex

While results of earlier studies had established the strong requirement for fork reassembly factors, and thus had provided indirect evidence for a high probability of fork stalling during the DNA replication cycle in different growth conditions,[38–40] more recent techniques using single-molecule imaging and tracking have been employed to directly measure the processivity of individual replication forks, or fork pairs, *in vivo* and to observe the effect of the transcription process and other barriers, such as DNA damage sites, on replisome stability.[27,41,42] They have indicated that the replisome rarely reaches the end of its replicore without running into a conflict and that the occurrence of fork stalling can be significantly reduced in the presence of rifampicin, an inhibitor of transcription initiation.[42] Furthermore, it was possible to measure the stoichiometry of individual components and to observe their diffusion in and out of the replisome.[27,41] The results of these studies have indicated the DNA replication complexes to be at the same time stable enough to be highly processive as well as very dynamic, where several of the protein factors can rapidly exchange with others in solution.

In vitro biochemistry experiments on the T7 phage replisome had provided evidence for a very stable, highly processive DNA replication machinery, which appears to be in contrast with the dynamic and stochastic process observed *in vivo*[43] (reviewed by *Mueller et al.*[25]). The results indicated that once the replisome was assembled, the complex could be diluted and still maintain processivity, continuing DNA synthesis without the need for DNA Pol rebinding, and thus without losing the enzyme for at least 40 kb. Later, single-molecule experiments were designed to test whether the DNA Pol could exchange with free enzymes in solution at higher concentrations that more closely reproduce the *in vivo* situation (for a recent review on the use of single-molecule techniques for the study of barriers to replication see Sun[44]). Two different approaches were taken. In the first, DNA polymerization was measured by labeling the DNA with a quantum dot and attaching it to the surface of a coverslip, keeping the DNA under flow to pull it towards a stretched configuration.[45] The real time rate of DNA polymerization was measured by the compaction of the DNA due to the accumulation of single stranded DNA of the lagging strand, as single single-stranded DNA tends to collapse onto itself and is thus less easily stretched by the flow than double-stranded DNA. A mixture of wild type and a mutant, slow polymerizing, DNA Pol enzyme was used. By tracking DNA polymerization at the single molecule level, it was possible to observe the change in replication speed on a given DNA molecule as one enzyme left and a new one came in. The kinetics of the exchange were dependent on the concentration of free enzyme in solution, thus indicating an induced dissociation process, also called facilitated dissociation.[46] This is possible when the presence of multiple binding sites within the complex for the same protein allows either fast rebinding, or, in the presence of a competitor, reduced affinity when a competitor is bound to one of the other sites. These results were interesting because they could also indicate the presence of DNA polymerase that was loosely associated with the replisome before exchange took place. This leads to an increased local concentration of enzyme, increasing the speed of exchange as soon as the active DNA Pol falls off. These results were later confirmed in a similar assay with the DNA bound to a surface but using a mixture of enzymes labeled with fluorophores of different colors detected by single molecule fluorescence microscopy.[47] In this case, the authors observed a faster pace of exchange kinetics that correlates with the synthesis of Okazaki fragments on the lagging strand. Rapid exchange of DNA pol was also observed with the *E. coli* enzyme.[48] Here, as in the phage system, the stable interaction of the helicase with the DNA provides a support for the recruitment of polymerases. Other replisome components are known to exchange with the pool of free proteins in the cell, such as the β clamp and SSB.[28,49] These dynamics are indicative of the kind of interactions that define the replication factories, or hyperstructures, described above. Cryo-EM structures have enabled the identification of a set of multiple interactions formed by each component of the replication fork that can mediate facilitated dissociation.[18,22–24]

8.4 The Gene Expression Complex

In addition to certain types of DNA lesions which cannot be processed by canonical replicative polymerase, one of the major causes for replication fork stalling is the presence of a transcription elongation complex on the DNA, particularly when it is itself blocked by the presence of a DNA lesion or by RNA polymerase (RNAP) backtracking. The RNAP holoenzyme ($\alpha_2\beta\beta'\omega + \sigma$ required for promoter recognition) initiates transcription from the promoter sequence(s) found upstream of the gene. The initiation of mRNA synthesis is a multi-step process that is regulated by the presence of other DNA-binding proteins, specific transcription factors and abundant nucleoid proteins, as well as by changes in DNA topology and in the concentration of small metabolites, such as ppGpp, that bind directly to the enzyme.[50] Once the transition from the promoter-bound initiating complex to the transcription elongation complex has taken place, RNAP can synthetize RNA at up to 20–90 nucleotides per second, at least 10 times slower than the replication fork, and depends on a set of protein factors to decrease the probability of pausing and backtracking, increasing transcription processivity.[51–53]

8.5 The Conflict: the Transcription Complex as a Barrier for Replication Forks

When the RNAP elongation complex pauses during mRNA synthesis, for example following ribonucleotide misincorporation, it can lead to backtracking, in which the 3′ end of the mRNA dissociates from the active site of the enzyme and is displaced, leading to the formation of a stable, immobile RNAP complex on the DNA template and creating a strong barrier to elongation by the replication machinery. Furthermore, when this takes place on a highly expressed gene it can lead to a traffic jam on the DNA, the formation of a train of blocked RNAPs. Recently, cryo-EM structures of the paused and backtracked complexes have been obtained, revealing the conformational changes taking place in the polymerase, the nucleic acids and the cofactors that allow for the process to resume.[54,55]

Pausing and backtracking can be prevented or rapidly resolved by the activity of a set of transcription elongation processivity factors. By avoiding conflicts with replication forks, these factors not only contribute to the regulation of gene expression levels but also to increased genome stability.[56] These factors include those that decrease the probability of forming a paused complex, such as NusG and NusA, and those that resolve backtracked complexes by cleaving the RNA to allow resumption of mRNA synthesis, such as GreA and GreB (see Chapter 4). DksA has also been shown to prevent conflicts by sustaining elongation and decreasing nucleotide misincorporation, especially under conditions of amino acid starvation and increases in guanosine tetraphosphate (ppGpp).[57–59] Another RNAP processivity factor is the RNA polymerase-associated protein (RapA),[60] which is found associated with the core enzyme during exponential growth.[61] RapA is related to the switch 2 (Swi2)–sucrose non fermentation 2 (Snf2) family of chromatin remodelers. An allosteric interaction with RNAP coupled to a displacement of RapA's N-terminal domain activates its ATPase activity[53] which induces dissociation of the core RNAP from its template by a backtranslocation mechanism *via* its interaction with the RNAP at the RNA exit channel.[62] This is particularly useful to dislocate stalled RNAP or RNAP that has reached a terminator RNA hairpin, thus contributing to maintaining the free pool of RNAP available for transcription and preventing conflicts with the replication fork.

In addition to its role in transcription coupled repair (TCR), Mfd has recently been shown to play a more general function as a transcription elongation factor as it is also found to be associated with RNAP in the absence of DNA damage, playing a role in prevention of RNAP pausing and backtracking, particularly at sites where the mRNA is highly structured.[63–65] Le *et al.* have shown that upon a transcription–replication conflict the transcription elongation complex can backtrack up to 100 bases and that Mfd can either push RNAP to avoid backtracking or rescue the complex after backtracking has taken place.[64]

In prokaryotes translation of mRNA by the ribosome takes place while the mRNA is still being synthesized by RNAP. This also contributes to reducing transcription–replication conflict by decreasing pausing and backtracking and increasing processivity.[66–73] This can take place by preventing the mRNA from forming secondary structures, such as hairpins, that can pause transcription, but also *via* a physical interaction between the ribosome and the RNAP, in some cases mediated by NusG, that can push the transcription complex forward when it slows down.

8.6 The Formation of R-loops and their Role in Replication Fork Stalling and Restart

Backtracked RNAP can lead to the formation of R-loops, where the newly synthesized RNA invades the DNA double helix upstream of the polymerase and hybridizes to its complementary DNA template strand. The resulting structure is an A-form RNA–DNA double helix, which is more stable than the standard B-form DNA double helix. These DNA–RNA helices are the targets of a dedicated enzyme, RNase HI, that cleaves the RNA so that it can more easily be dissociated by helicases. The non-template strand of DNA within an R-loop is unpaired and thus more accessible to cleavage by nucleases, resulting in nicked DNA once the R-loop has been resolved, these nicks can lead to the formation of a double strand break upon arrival of the next replication fork[74,75] (Figure 8.1).

R-loops not only can create a barrier causing replication fork disassembly and genome instability[76,77] but have also been shown to become substrates for initiation of unprogrammed DNA replication, as they mimic the DNA–RNA hybrid produced by the DnaG primase.[78] In fact, R-loops had been first identified as sites of replication complex assembly and replication initiation in strains where the initiation of DNA replication cannot take place at OriC, either because OriC itself or DnaA have been deleted. The viability of these strains depends on the absence of the RNase HI enzyme and thus on the accumulation of R-loops.[79,80]

This spurious replication will take place co-directionally with the replication fork if the R-loop is on the leading strand, but it will take place in a head-on direction if the gene with the R-loop is on the lagging strand. The latter will result in head-on conflicts between replication forks, requiring DNA recombination and fork reassembly factors for their resolution. Co-directional conflict has also been shown to result in fork restart[81] or, if unsuccessful, the creation of double stranded breaks.[70] In this case, when a replication fork restarts using an R-loop as substrate, a gap in the newly synthetized DNA strand is created from the suspension in replication activity and its resumption downstream, at the 3′ end of the mRNA–DNA hybrid. If the gap is not repaired, the replication fork from the next replication round will synthesize a new strand of DNA up to the gap, generating a double strand break (see Pomerantz and O'Donnel[75] for a review).

Binding of mRNA by the ribosome or by the termination factor Rho keeps the mRNA away from the DNA double helix and prevents the formation of R-loops.[82,83] The role of Rho in preventing R-loop formation is particularly important for untranslated antisense transcripts[83] and has been recently shown to play a role, together with Mfd, in preventing replication–transcription conflict resulting from RNAP stalled at DNA lesions.[84]

8.7 The Role of Topoisomerases in Avoiding Conflict and the Formation of R-loops

While it is possible for a direct physical protein–protein conflict to take place between the transcription and replication machineries, transcription in a head-on orientation relative to the replication fork can also interfere with ongoing DNA replication by the accumulation of positive supercoils ahead of both the replication fork and RNAP, requiring the activity of DNA topoisomerases, such as gyrase, to resolve the resulting positive torque which inhibits polymerase progression[29,30,32–34] (Figure 8.1). In a recent study, the presence of gyrase at the replication fork has been measured by single-molecule *in vivo* fluorescence tracking. The results indicated that there can be up to 12 gyrase molecules per replisome that are more stably bound than those elsewhere on the genome.[34] Furthermore, the accumulation of positive supercoiling between the two complexes will induce the formation of a reversed replication fork[85] as well as stalling the RNAP transcription elongation complex, leading to backtracking and thus to the formation of a more stable, immobile complex.[86,87] As described above, a backtracked complex can then lead to the formation of an R-loop.[88] In a conceptually related fashion, an increase of negative supercoiling upstream of the transcription complex can favor DNA melting and invasion of the mRNA, especially in cases whereby topoisomerase I is compromised in its ability to relax negative supercoiling, and/or the opposing activities of topoisomerase I and gyrase are not properly balanced.[76,86–88]

8.8 Additional Factors Required for Conflict Resolution

In the cell, the high concentration of the different replisome components in solution allows for an exchange which is mediated, or induced, *via* an interaction of the competing enzymes with secondary interaction sites. This is important for the ability of the fork to rapidly disassemble and reassemble in the presence of possible replication blocks and can explain the rapid replacement of DNA Pol III by translesion synthesis polymerases when the replication complex is destabilized by the encounter with a DNA damage site.[25,89] Furthermore, *in vitro* single molecule experiments have revealed that the DNA polymerase can jump over barriers, dissociating from the stalled fork and reassociating downstream, *via* an interaction with the sliding clamp. Once reassociated, the polymerase can restart DNA synthesis from the 3′ end of an R-loop left over from a transcription elongation complex of a leading strand gene, and thus codirectional with the replisome. Alternatively, a primer synthetized by DnaG on the single-stranded DNA exposed by decoupling of the helicase and DNA polymerase can also serve to reinitiate DNA synthesis.[90] This is a nice example of the dynamic nature of the complex, which can be a key property to help bypass a block and quickly restart replication after a collision, however, in most cases additional factors are required *in vivo* for DNA replication to successfully resume its activity after the conflict (reviewed by Michel *et al.*[38]).

A set of factors associated with the replication hyperstructure can help dislodge DNA-bound protein blocks, such as RNAP,[91] or promote fork reversal and reassembly.[40,92,93] In the former case, accessory helicases of the replication complex, Rep and UvrD can displace proteins bound to the DNA, including RNAP elongating codirectionally with the replisome. The helicases accomplish this by unwinding the DNA in the 3′ to 5′ direction on the leading strand (they will be found ahead of the main DnaB helicase which is moving in the 5′ to 3′ direction on the lagging strand). This activity is even robust enough to clear the numerous RNAPs which will be found on highly transcribed genes[91,92,94] (Figure 8.1).

This is one of the reasons why transcription complexes of leading strand genes, elongating co-directionally with the replication fork, pose less of a problem for DNA synthesis than genes transcribed on the lagging strand, and thus in a head-on direction.[75,95] In addition, it is also easier to dislodge elongating RNAP by pushing it from the back than from the front. Pushing from the front, such as by DnaB helicase on the lagging strand, can lead to backtracking and the formation of R-loops, while pushing from the back, such as by Rep and UvrD helicases, dislodges the RNAP leaving only the eight base pairs DNA–RNA hybrid of the transcription bubble which is less stable than an extended RNA–DNA hybrid formed in a backtracked complex.[70]

There can thus be different outcomes when a replication fork encounters a block, such as a transcribing RNAP, depending on whether the RNAP is oriented head-on or codirectionally with the fork, whether the transcription elongation complex itself is blocked on a DNA damage site, whether it is part of a "train" of RNAP enzymes (such as those found on ribosomal RNA operons in rich media), or whether the conflict has resulted in the formation of an R-loop (see below). While it is possible to maintain continuous DNA replication if only a small number of RNAPs need to be displaced, this is no longer the case when a stronger, more long-lived, block is present. This has been studied, for instance, by taking a highly active ribosomal RNA operon, which directs transcription co-directional with the fork, and inverting its orientation so that it initiates transcription into the fork. In such circumstances, fork reversal and the associated activity of homologous recombination factors is required to restart DNA synthesis, however this can also lead to partial DNA degradation and increased mutation rates at the stall site.[38] For example, it has been recently shown that the activity of the RecBCD helicase–exonuclease complex can help resolve severely blocked forks by recognizing the blunt end double stranded DNA that can result from reversed forks, opening it by helicase activity and degrading each strand until it reaches a *chi* sequence, where the degradation from the 3′ end stops thus creating a 3′ overhang where RecA can bind and facilitate homologous recombination and DNA repair so that the new replication forks arriving from the origin will not find a double strand break or a nicked DNA.[96,97] The lack of RecBCD activity can lead to DNA degradation because of the formation of a replication intermediate that can be targeted by nucleases.[97] The recruitment of RecA to stalled replication forks can lead to the induction of the SOS response and a delay in cell division to adjust for the increased time required to complete DNA repair and resume genome replication.[98]

8.9 DNA Damage and Transcription Elongation

DNA damage sites can cause the transcription elongation complex to stop and create a block to advancing DNA replication forks if it is not resolved quickly. When elongating RNAP stalls at a DNA damage site, Mfd will bind to it and undergo a conformational change, revealing a specific RNAP-interacting surface, activating the ATP hydrolysis-dependent motor activity that can displace the transcription complex off the DNA damage site[99] (see Chapter 7). Mfd can then recruit the DNA repair machinery *via* an interaction with UvrA in the context of the process of transcription-coupled DNA repair (TCR) (reviewed by Strick and Portman[100]).

8.10 The Effect of the Conflict on Genome Organization and Evolution

Since resolution of transcription–replication conflicts requires DNA recombination and repair, including DNA degradation and DNA synthesis by error prone DNA polymerases,[9,38] it has been shown to lead to a higher mutation rate in actively transcribed genes in both prokaryotes and eukaryotes.[101] Compared with eukaryotes however, bacterial genomes are very gene dense, with about 85% of the genome made of open reading frames,[102] increasing the probability of a conflict with a transcription complex. While most genes are only expressed once or at maximum a few times per cell cycle,[103] the probability of a conflict is accentuated at the sites of highly expressed genes coding for abundant proteins.

In accordance with head-on conflicts being more dangerous for replication forks, the most highly expressed genes, such as those encoding ribosomal protein, and ribosomal RNA operons, are oriented codirectionally to the replication program.[102] However, expression rate is not the only gene property that can influence strand orientation. In fact, Rocha and Danchin have shown that in *Bacillus subtilis* a higher percentage of essential genes are found on the leading strand (94%) than the percentage of highly expressed ones (78%) with similar results in *E. coli*, where strand bias is weaker.[104] Therefore, there is a higher selection pressure to preserve the sequence of essential genes than to protect the DNA replication forks from conflict. Similarly, increased mutation rate rather than conflict avoidance might be selected for in the case of genes that are required for adaptation to stress, more often found on the lagging strand. The activity of the Mfd protein can contribute to this strand-specific difference in mutation rates in response to stress. Despite its role in recruiting the DNA repair machinery to DNA lesions located on the transcribed strand, its association with the higher mutation rate of head-on oriented genes indicates that its activity in some cases might increase the probability of transcription-replication conflict.[100,105] The reason might be that its RNAP-displacing activity can favor the formation of an R-loop, as evidenced by a decrease in mutation rates when the RNase HI enzyme is overexpressed.[77,100] Many of these head-on genes are involved in stress response and pathogenicity, indicating that their orientation in the genome may have a selective advantage by providing a wider fitness landscape over which to find a route of escape from the stress.[77,106,107] The inherent mutagenicity of lagging strand genes can thus provide a way to achieve increased evolution rates for a specific subset of genes.[108] It is interesting to note that Mfd is named after the mutation frequency decline phenotype, and was isolated by a screen selecting for an increase in mutation rates after UV exposure, indicating that, under these conditions, the main role of Mfd might be to reduce mutations by allowing the cell to quickly respond to UV damage by increasing the efficiency of transcription-coupled DNA repair rather than avoiding replication conflict and the mutations that might be associated with it (reviewed by Strick and Portman[100]).

Finally, gene strand orientation can also be constrained by the need to co-regulate the expression of neighboring genes, either by shared transcription factors or by DNA topology.[109] As described earlier in this chapter, the transcription process results in the accumulation of positive supercoiling in front of elongating RNAP and negative supercoiling behind it.[110] Since the expression of a majority of genes in *E. coli* is sensitive to changes in the levels of supercoiling,[111] the expression of one gene can influence the activity of neighboring promoters by altering the local levels of supercoiling. Relative gene orientation, whether the neighboring gene is divergently or convergently expressed, will determine whether positive or negative supercoiling will increase at the neighboring promoter during transcription elongation (see Sobetzko[112] and references therein). For example, Sobetzko found that in Enterobacteriaceae most of the genes whose expression is increased by negative supercoiling are more frequently found to be divergently expressed relative to their neighbors.[112]

The constraint created by the frequency of replication–transcription conflicts not only affects gene orientation, but also the number of replication origins per genome. The significance of having a single origin of replication in this context is that it will result in the replication forks always replicating a given gene from the same direction and therefore will allow for selection to act on gene orientation.[113–115]

In this chapter, some of the mechanisms that are required to optimize both genome stability and regulation of gene expression by either preventing transcription–replication conflict or by rescuing the two macromolecular machineries have been described. There is a trade-off between the immediate positive effect of the DNA repair process, allowing the forks to proceed through the damage site, and the secondary, potentially costly, effect from the mutations that can arise from the repair process itself. This is exemplified by the Mfd protein, playing a key role in transcription-coupled repair in the response of UV damage by dislodging the RNAP and recruiting the DNA repair machineries, while at the same time also increasing the mutation rate of lagging strand genes. Finally, the mutations and increased recombination rates associated with the conflict can also play a role in varying the rate of horizontal gene transfer and gene evolution with a potentially positive effect on survival in changing or stressful environments.[106]

References

1. A. M. Breier, H.-U. G. Weier and N. R. Cozzarelli, Independence of replisomes in Escherichia coli chromosomal replication, *Proc. Natl. Acad. Sci. U. S. A.*, 2005, **102**, 3942–3947.
2. B. Henneman, C. van Emmerik, H. van Ingen and R. T. Dame, Structure and function of archaeal histones, *PLoS Genet.*, 2018, **14**, e1007582.

3. I. Nikolov and A. Taddei, Linking replication stress with heterochromatin formation, *Chromosoma*, 2016, **125**, 523–533.
4. R. T. Dame, F.-Z. M. Rashid and D. C. Grainger, Chromosome organization in bacteria: mechanistic insights into genome structure and function, *Nat. Rev. Genet.*, 2020, **21**, 227–242.
5. M. P. Crossley, M. Bocek and K. A. Cimprich, R-Loops as Cellular Regulators and Genomic Threats, *Mol. Cell*, 2019, **73**, 398–411.
6. R. Stuckey, N. García-Rodríguez, A. Aguilera and R. E. Wellinger, Role for RNA:DNA hybrids in origin-independent replication priming in a eukaryotic system, *Proc. Natl. Acad. Sci. U. S. A.*, 2015, **112**, 5779.
7. B. Gómez-González and A. Aguilera, Transcription-mediated replication hindrance: a major driver of genome instability, *Genes Dev.*, 2019, **33**, 1008–1026.
8. T. García-Muse and A. Aguilera, Transcription–replication conflicts: how they occur and how they are resolved, *Nat. Rev. Mol. Cell Biol.*, 2016, **17**, 553–563.
9. K. S. Lang and H. Merrikh, The Clash of Macromolecular Titans: Replication-Transcription Conflicts in Bacteria, *Annu. Rev. Microbiol.*, 2018, **72**, 71–88.
10. J. Brochu, É.-V. Breton and M. Drolet, Supercoiling, R-loops, Replication and the Functions of Bacterial Type 1A Topoisomerases, *Genes*, 2020, **11**, 249.
11. K. E. Duderstadt, M. L. Mott, N. J. Crisona, K. Chuang, H. Yang and J. M. Berger, Origin remodeling and opening in bacteria rely on distinct assembly states of the DnaA initiator, *J. Biol. Chem.*, 2010, **285**, 28229–28239.
12. S. Chodavarapu and J. M. Kaguni, Replication Initiation in Bacteria, *Enzymes*, 2016, **39**, 1–30.
13. A. C. Leonard and J. E. Grimwade, The orisome: structure and function, *Front. Microbiol.*, 2015, **6**, 545.
14. K. Skarstad and T. Katayama, Regulating DNA replication in bacteria, *Cold Spring Harbor Perspect. Biol.*, 2013, **5**, a012922.
15. L. Fernández-Coll, M. Maciag-Dorszynska, K. Tailor, S. Vadia, P. A. Levin, A. Szalewska-Palasz and M. Cashel, The Absence of (p)ppGpp Renders Initiation of Escherichia coli Chromosomal DNA Synthesis Independent of Growth Rates, *mBio*, 2020, **11**, e03223-19.
16. J. A. Kraemer, A. G. Sanderlin and M. T. Laub, The Stringent Response Inhibits DNA Replication Initiation in E. coli by Modulating Supercoiling of oriC, *mBio*, 2019, **10**, e01330-19.
17. K. H. Jameson and A. J. Wilkinson, Control of Initiation of DNA Replication in Bacillus subtilis and Escherichia coli, *Genes*, 2017, **8**, 22.
18. Z.-Q. Xu and N. E. Dixon, Bacterial replisomes, *Curr. Opin. Struct. Biol.*, 2018, **53**, 159–168.
19. J. C. Bell, B. Liu and S. C. Kowalczykowski, Imaging and energetics of single SSB-ssDNA molecules reveal intramolecular condensation and insight into RecOR function, *eLife*, 2015, **4**, e08646.

20. P. McInerney and M. O'Donnell, Functional Uncoupling of Twin Polymerases Mechanism of polymerase dissociation from a lagging-strand block, *J. Biol. Chem.*, 2004, **279**, 21543–21551.
21. J. E. Graham, K. J. Marians and S. C. Kowalczykowski, Independent and Stochastic Action of DNA Polymerases in the Replisome, *Cell*, 2017, **169**, 1201–1213.e17.
22. A. J. Oakley, A structural view of bacterial DNA replication, *Protein Sci.*, 2019, **28**, 990–1004.
23. A. W. Kulczyk, A. Moeller, P. Meyer, P. Sliz and C. C. Richardson, Cryo-EM structure of the replisome reveals multiple interactions coordinating DNA synthesis, *Proc. Natl. Acad. Sci. U. S. A.*, 2017, **114**, E1848–E1856.
24. Y. Gao, Y. Cui, T. Fox, S. Lin, H. Wang, N. de Val, Z. H. Zhou and W. Yang, Structures and operating principles of the replisome, *Science*, 2019, **363**, eaav7003.
25. S. H. Mueller, L. M. Spenkelink and A. M. van Oijen, When proteins play tag: the dynamic nature of the replisome, *Biophys. Rev.*, 2019, **11**, 641–651.
26. D. Trojanowski, J. Hołówka and J. Zakrzewska-Czerwińska, Where and When Bacterial Chromosome Replication Starts: A Single Cell Perspective, *Front. Microbiol.*, 2018, **9**, 2819.
27. M. C. Moolman, S. Tiruvadi Krishnan, J. W. J. Kerssemakers, R. de Leeuw, V. Lorent, D. J. Sherratt and N. H. Dekker, The progression of replication forks at natural replication barriers in live bacteria, *Nucleic Acids Res.*, 2016, **44**, 6262–6273.
28. L. M. Spenkelink, J. S. Lewis, S. Jergic, Z.-Q. Xu, A. Robinson, N. E. Dixon and A. M. van Oijen, Recycling of single-stranded DNA-binding protein by the bacterial replisome, *Nucleic Acids Res.*, 2019, **47**, 4111–4123.
29. E. C. Guzmán, J. L. Caballero and A. Jiménez-Sánchez, Ribonucleoside diphosphate reductase is a component of the replication hyperstructure in Escherichia coli, *Mol. Microbiol.*, 2002, **43**, 487–495.
30. F. Molina and K. Skarstad, Replication fork and SeqA focus distributions in Escherichia coli suggest a replication hyperstructure dependent on nucleotide metabolism, *Mol. Microbiol.*, 2004, **52**, 1597–1612.
31. V. Norris, M. Demarty, D. Raine, A. Cabin-Flaman and S. L. Le, Hypothesis: hyperstructures regulate initiation in Escherichia coli and other bacteria, *Biochimie*, 2002, **84**, 341–347.
32. V. Norris, M. Demarty, D. Raine, A. Cabin-Flaman and S. L. Le, Hypothesis: hyperstructures regulate initiation in Escherichia coli and other bacteria, *Biochimie*, 2002, **84**, 341–347.
33. S. M. Mangiameli, B. T. Veit, H. Merrikh and P. A. Wiggins, The Replisomes Remain Spatially Proximal throughout the Cell Cycle in Bacteria, *PLoS Genet.*, 2017, **13**, e1006582.
34. M. Stracy, A. J. M. Wollman, E. Kaja, J. Gapinski, J. E. Lee, V. A. Leek, S. J. McKie, L. A. Mitchenall, A. Maxwell, D. J. Sherratt, M. C. Leake and P. Zawadzki, Single-molecule imaging of DNA gyrase activity in living Escherichia coli, *Nucleic Acids Res.*, 2019, **47**, 210–220.

35. M. G. Marinus and A. Løbner-Olesen, DNA Methylation, *EcoSal Plus*, 2014, **6**, DOI: 10.1128/ecosalplus.ESP-0003-2013.
36. I. B. Pedersen, E. Helgesen, I. Flåtten, S. Fossum-Raunehaug and K. Skarstad, SeqA structures behind Escherichia coli replication forks affect replication elongation and restart mechanisms, *Nucleic Acids Res.*, 2017, **45**, 6471–6485.
37. M. C. Joshi, D. Magnan, T. P. Montminy, M. Lies, N. Stepankiw and D. Bates, Regulation of Sister Chromosome Cohesion by the Replication Fork Tracking Protein SeqA, *PLoS Genet.*, 2013, **9**, e1003673.
38. B. Michel, A. K. Sinha and D. R. F. Leach, Replication Fork Breakage and Restart in Escherichia coli, *Microbiol. Mol. Biol. Rev.*, 2018, **82**, e00013-18.
39. S. J. Sandler, H. S. Samra and A. J. Clark, Differential Suppression of priA2::kan Phenotypes in Escherichia coli K-12 by Mutations in priA, lexA, and dnaC, *Genetics*, 1996, **143**, 5–13.
40. C. B. Gabbai and K. J. Marians, Recruitment to stalled replication forks of the PriA DNA helicase and replisome-loading activities is essential for survival, *DNA Repair*, 2010, **9**, 202–209.
41. R. Reyes-Lamothe, D. J. Sherratt and M. C. Leake, Stoichiometry and architecture of active DNA replication machinery in Escherichia coli, *Science*, 2010, **328**, 498–501.
42. S. M. Mangiameli, C. N. Merrikh, P. A. Wiggins and H. Merrikh, Transcription leads to pervasive replisome instability in bacteria, *eLife*, 2017, **6**, e19848.
43. Z. Debyser, S. Tabor and C. C. Richardson, Coordination of leading and lagging strand DNA synthesis at the replication fork of bacteriophage T7, *Cell*, 1994, **77**, 157–166.
44. B. Sun, Rescuing Replication from Barriers: Mechanistic Insights from Single-Molecule Studies, *Mol. Cell. Biol.*, 2019, **39**, e00576-18.
45. J. J. Loparo, A. W. Kulczyk, C. C. Richardson and A. M. van Oijen, Simultaneous single-molecule measurements of phage T7 replisome composition and function reveal the mechanism of polymerase exchange, *Proc. Natl. Acad. Sci. U. S. A.*, 2011, **108**, 3584–3589.
46. A. Erbaş and J. F. Marko, How do DNA-bound proteins leave their binding sites? The role of facilitated dissociation, *Curr. Opin. Chem. Biol.*, 2019, **53**, 118–124.
47. H. J. Geertsema, A. W. Kulczyk, C. C. Richardson and A. M. van Oijen, Single-molecule studies of polymerase dynamics and stoichiometry at the bacteriophage T7 replication machinery, *Proc. Natl. Acad. Sci. U. S. A.*, 2014, **111**, 4073–4078.
48. T. R. Beattie, N. Kapadia, E. Nicolas, S. Uphoff, A. J. Wollman, M. C. Leake and R. Reyes-Lamothe, Frequent exchange of the DNA polymerase during bacterial chromosome replication, *eLife*, 2017, **6**, e21763.
49. N. A. Tanner, G. Tolun, J. J. Loparo, S. Jergic, J. D. Griffith, N. E. Dixon and A. M. van Oijen, E. coli DNA replication in the absence of free β clamps, *EMBO J.*, 2011, **30**, 1830–1840.

50. D. F. Browning and S. J. Busby, The regulation of bacterial transcription initiation, *Nat. Rev. Microbiol.*, 2004, **2**, 57–65.
51. A. Mustaev, J. Roberts and M. Gottesman, Transcription elongation, *Transcription*, 2017, **8**, 150–161.
52. G. A. Belogurov and I. Artsimovitch, Regulation of Transcript Elongation, *Annu. Rev. Microbiol.*, 2015, **69**, 49–69.
53. S. Kakar, X. Fang, L. Lubkowska, Y. N. Zhou, G. X. Shaw, Y.-X. Wang, D. J. Jin, M. Kashlev and X. Ji, Allosteric Activation of Bacterial Swi2/Snf2 (Switch/Sucrose Non-fermentable) Protein RapA by RNA Polymerase: Biochemical and Structural Studies, *J. Biol. Chem.*, 2015, **290**, 23656–23669.
54. M. Abdelkareem, C. Saint-André, M. Takacs, G. Papai, C. Crucifix, X. Guo, J. Ortiz and A. Weixlbaumer, Structural Basis of Transcription: RNA Polymerase Backtracking and Its Reactivation, *Mol. Cell*, 2019, **75**, 298–309.e4.
55. J. Y. Kang, R. A. Mooney, Y. Nedialkov, J. Saba, T. V. Mishanina, I. Artsimovitch, R. Landick and S. A. Darst, Structural Basis for Transcript Elongation Control by NusG Family Universal Regulators, *Cell*, 2018, **173**, 1650–1662.e14.
56. J. Y. Kang, T. V. Mishanina, R. Landick and S. A. Darst, Mechanisms of Transcriptional Pausing in Bacteria, *J. Mol. Biol.*, 2019, **431**, 4007–4029.
57. M. Roghanian, N. Zenkin and Y. Yuzenkova, Bacterial global regulators DksA/ppGpp increase fidelity of transcription, *Nucleic Acids Res.*, 2015, **43**, 1529–1536.
58. K. K. Myka, K. Küsters, R. Washburn and M. E. Gottesman, DksA-RNA polymerase interactions support new origin formation and DNA repair in Escherichia coli, *Mol. Microbiol.*, 2019, **111**, 1382–1397.
59. A. K. Tehranchi, M. D. Blankschien, Y. Zhang, J. A. Halliday, A. Srivatsan, J. Peng, C. Herman and J. D. Wang, The Transcription Factor DksA Prevents Conflicts between DNA Replication and Transcription Machinery, *Cell*, 2010, **141**, 595–605.
60. D. J. Jin, Y. N. Zhou, G. Shaw and X. Ji, Structure and function of RapA: a bacterial Swi2/Snf2 protein required for RNA polymerase recycling in transcription, *Biochim. Biophys. Acta*, 2011, **1809**, 470–475.
61. I. R. Eason, H. P. Kaur, K. A. Alexander and M. V. Sukhodolets, Growth phase-specific changes in the composition of E. coli transcription complexes, *J. Chromatogr. B: Anal. Technol. Biomed. Life Sci.*, 2019, **1109**, 155–165.
62. B. Liu, Y. Zuo and T. A. Steitz, Structural basis for transcription reactivation by RapA, *Proc. Natl. Acad. Sci. U. S. A.*, 2015, **112**, 2006–2010.
63. H. N. Ho, A. M. van Oijen and H. Ghodke, The transcription-repair coupling factor Mfd associates with RNA polymerase in the absence of exogenous damage, *Nat. Commun.*, 2018, **9**, 1570.
64. T. T. Le, Y. Yang, C. Tan, M. M. Suhanovsky, R. M. Fulbright, J. T. Inman, M. Li, J. Lee, S. Perelman, J. W. Roberts, A. M. Deaconescu and M. D. Wang, Mfd Dynamically Regulates Transcription *via* a Release and Catch-Up Mechanism, *Cell*, 2018, **172**, 344–357.e15.

65. M. N. Ragheb, C. Merrikh, K. Browning and H. Merrikh, Mfd regulates RNA polymerase association with hard-to-transcribe regions *in vivo*, especially those with structured RNAs, *Proc. Natl. Acad. Sci. U. S. A.*, 2021, **118**, e2008498118.
66. J. Gowrishankar and R. Harinarayanan, Why is transcription coupled to translation in bacteria?, *Mol. Microbiol.*, 2004, **54**, 598–603.
67. S. Proshkin, A. R. Rahmouni, A. Mironov and E. Nudler, Cooperation Between Translating Ribosomes and RNA Polymerase in Transcription Elongation, *Science*, 2010, **328**, 504–508.
68. K. McGary and E. Nudler, RNA polymerase and the ribosome: the close relationship, *Curr. Opin. Microbiol.*, 2013, **16**, 112–117.
69. M. Zhu, M. Mori, T. Hwa and X. Dai, Disruption of transcription–translation coordination in Escherichia coli leads to premature transcriptional termination, *Nat. Microbiol.*, 2019, **4**, 2347–2356.
70. D. Dutta, K. Shatalin, V. Epshtein, M. E. Gottesman and E. Nudler, Linking RNA polymerase backtracking to genome instability in E. coli, *Cell*, 2011, **146**, 533–543.
71. R. Kohler, R. A. Mooney, D. J. Mills, R. Landick and P. Cramer, Architecture of a transcribing-translating expressome, *Science*, 2017, **356**, 194–197.
72. S. Saxena, K. K. Myka, R. Washburn, N. Costantino, D. L. Court and M. E. Gottesman, Escherichia coli transcription factor NusG binds to 70S ribosomes, *Mol. Microbiol.*, 2018, **108**, 495–504.
73. H. Fan, A. B. Conn, P. B. Williams, S. Diggs, J. Hahm, H. B. Gamper, Y.-M. Hou, S. E. O'Leary, Y. Wang and G. M. Blaha, Transcription–translation coupling: direct interactions of RNA polymerase with ribosomes and ribosomal subunits, *Nucleic Acids Res.*, 2017, **45**, 11043–11055.
74. M. Thomas, R. L. White and R. W. Davis, Hybridization of RNA to double-stranded DNA: formation of R-loops, *Proc. Natl. Acad. Sci. U. S. A.*, 1976, **73**, 2294–2298.
75. R. T. Pomerantz and M. O'Donnell, What happens when replication and transcription complexes collide?, *Cell Cycle*, 2010, **9**, 2537–2543.
76. M. Drolet and J. Brochu, R-loop-dependent replication and genomic instability in bacteria, *DNA Repair*, 2019, 102693.
77. H. Wimberly, C. Shee, P. C. Thornton, P. Sivaramakrishnan, S. M. Rosenberg and P. J. Hastings, R-loops and nicks initiate DNA breakage and genome instability in non-growing Escherichia coli, *Nat. Commun.*, 2013, **4**, 2115.
78. R. T. Pomerantz and M. O'Donnell, The replisome uses mRNA as a primer after colliding with RNA polymerase, *Nature*, 2008, **456**, 762.
79. T. Kogoma, Stable DNA replication: interplay between DNA replication, homologous recombination, and transcription, *Microbiol. Mol. Biol. Rev.*, 1997, **61**, 212–238.
80. M. K. von, E. Boye, K. Skarstad, L. Koppes and T. Kogoma, Mode of initiation of constitutive stable DNA replication in RNase H-defective mutants of Escherichia coli K-12, *J. Bacteriol.*, 1987, **169**, 2650–2658.

81. H. Merrikh, C. Machón, W. H. Grainger, A. D. Grossman and P. Soultanas, Co-directional replication-transcription conflicts lead to replication restart, *Nature*, 2011, **470**, 554–557.
82. E. Nudler, RNA polymerase backtracking in gene regulation and genome instability, *Cell*, 2012, **149**, 1438–1445.
83. N. Raghunathan, R. M. Kapshikar, J. K. Leela, J. Mallikarjun, P. Bouloc and J. Gowrishankar, Genome-wide relationship between R-loop formation and antisense transcription in Escherichia coli, *Nucleic Acids Res.*, 2018, **46**, 3400–3411.
84. S. Jain, R. Gupta and R. Sen, Rho-dependent transcription termination in bacteria recycles RNA polymerases stalled at DNA lesions, *Nat. Commun.*, 2019, **10**, 1207.
85. L. Postow, C. Ullsperger, R. Keller, C. Bustamante, A. Vologodskii and N. Cozzarelli, Positive torsional strain causes the formation of a four-way junction at replication forks, *J. Biol. Chem.*, 2001, **276**, 2790–2796.
86. M. Drolet, P. Phoenix, R. Menzel, E. Massé, L. F. Liu and R. J. Crouch, Overexpression of RNase H partially complements the growth defect of an Escherichia coli delta topA mutant: R-loop formation is a major problem in the absence of DNA topoisomerase I, *Proc. Natl. Acad. Sci. U. S. A.*, 1995, **92**, 3526–3530.
87. E. Massé and M. Drolet, Escherichia coli DNA topoisomerase I inhibits R-loop formation by relaxing transcription-induced negative supercoiling, *J. Biol. Chem.*, 1999, **274**, 16659–16664.
88. M. Drolet, S. Broccoli, F. Rallu, C. Hraiky, C. Fortin, E. Massé and I. Baaklini, The problem of hypernegative supercoiling and R-loop formation in transcription, *Front. Biosci.*, 2003, **8**, d210–221.
89. J. E. Kath, S. Chang, M. K. Scotland, J. H. Wilbertz, S. Jergic, N. E. Dixon, M. D. Sutton and J. J. Loparo, Exchange between Escherichia coli polymerases II and III on a processivity clamp, *Nucleic Acids Res.*, 2016, **44**, 1681–1690.
90. R. E. Georgescu, N. Y. Yao and M. O'Donnell, Single-molecule analysis of the Escherichia coli replisome and use of clamps to bypass replication barriers, *FEBS Lett.*, 2010, **584**, 2596–2605.
91. M. Hawkins, J. U. Dimude, J. A. L. Howard, A. J. Smith, M. S. Dillingham, N. J. Savery, C. J. Rudolph and P. McGlynn, Direct removal of RNA polymerase barriers to replication by accessory replicative helicases, *Nucleic Acids Res.*, 2019, **47**, 5100–5113.
92. A. L. D. Septenville, S. Duigou, H. Boubakri and B. Michel, Replication Fork Reversal after Replication–Transcription Collision, *PLoS Genet.*, 2012, **8**, e1002622.
93. G. M. Weaver, K. A. Mettrick, T.-A. Corocher, A. Graham and I. Grainge, Replication fork collapse at a protein-DNA roadblock leads to fork reversal, promoted by the RecQ helicase, *Mol. Microbiol.*, 2019, **111**, 455–472.
94. H. Boubakri, A. L. de Septenville, E. Viguera and B. Michel, The helicases DinG, Rep and UvrD cooperate to promote replication across transcription units in vivo, *EMBO J.*, 2010, **29**, 145–157.

95. R. T. Pomerantz and M. O'Donnell, Direct restart of a replication fork stalled by a head-on RNA polymerase, *Science*, 2010, **327**, 590–592.
96. A. H. Syeda, J. Atkinson, R. G. Lloyd and P. McGlynn, The Balance between Recombination Enzymes and Accessory Replicative Helicases in Facilitating Genome Duplication, *Genes*, 2016, **7**(8), 42.
97. J. U. Dimude, S. L. Midgley-Smith and C. J. Rudolph, Replication-transcription conflicts trigger extensive DNA degradation in Escherichia coli cells lacking RecBCD, *DNA Repair*, 2018, **70**, 37–48.
98. O. Huisman and R. D'Ari, Effect of suppressors of SOS-mediated filamentation on sfiA operon expression in Escherichia coli, *J. Bacteriol.*, 1983, **153**, 169–175.
99. A. J. Smith, M. D. Szczelkun and N. J. Savery, Controlling the motor activity of a transcription-repair coupling factor: autoinhibition and the role of RNA polymerase, *Nucleic Acids Res.*, 2007, **35**, 1802–1811.
100. T. R. Strick and J. R. Portman, Transcription-Coupled Repair: From Cells to Single Molecules and Back Again, *J. Mol. Biol.*, 2019, **431**, 4093–4102.
101. H. Gaillard and A. Aguilera, Transcription as a Threat to Genome Integrity, *Annu. Rev. Biochem.*, 2016, **85**, 291–317.
102. M. Touchon and E. P. C. Rocha, Coevolution of the Organization and Structure of Prokaryotic Genomes, *Cold Spring Harbor Perspect. Biol.*, 2016, **8**, a018168.
103. M. Bon, S. J. McGowan and P. R. Cook, Many expressed genes in bacteria and yeast are transcribed only once per cell cycle, *FASEB J.*, 2006, **20**, 1721–1723.
104. E. P. C. Rocha and A. Danchin, Essentiality, not expressiveness, drives gene-strand bias in bacteria, *Nat. Genet.*, 2003, **34**, 377–378.
105. M. Ragheb and H. Merrikh, The enigmatic role of Mfd in replication-transcription conflicts in bacteria, *DNA Repair*, 2019, **81**, 102659.
106. M. N. Ragheb, M. K. Thomason, C. Hsu, P. Nugent, J. Gage, A. N. Samadpour, A. Kariisa, C. N. Merrikh, S. I. Miller, D. R. Sherman and H. Merrikh, Inhibiting the Evolution of Antibiotic Resistance, *Mol. Cell*, 2019, **73**, 157–165.e5.
107. S. Paul, S. Million-Weaver, S. Chattopadhyay, E. Sokurenko and H. Merrikh, Accelerated gene evolution through replication-transcription conflicts, *Nature*, 2013, **495**, 512–515.
108. H. Merrikh, Spatial and temporal control of evolution through replication-transcription conflicts, *Trends Microbiol.*, 2017, **25**, 515–521.
109. J. O. Korbel, L. J. Jensen, C. von Mering and P. Bork, Analysis of genomic context: prediction of functional associations from conserved bidirectionally transcribed gene pairs, *Nat. Biotechnol.*, 2004, **22**, 911–917.
110. L. F. Liu and J. C. Wang, Supercoiling of the DNA template during transcription, *Proc. Natl. Acad. Sci. U. S. A.*, 1987, **84**, 7024–7027.
111. N. Blot, R. Mavathur, M. Geertz, A. Travers and G. Muskhelishvili, Homeostatic regulation of supercoiling sensitivity coordinates transcription of the bacterial genome, *EMBO Rep.*, 2006, **7**, 710–715.

112. P. Sobetzko, Transcription-coupled DNA supercoiling dictates the chromosomal arrangement of bacterial genes, *Nucleic Acids Res.*, 2016, **44**, 1514–1524.
113. J. U. Dimude, A. Stockum, S. L. Midgley-Smith, A. L. Upton, H. A. Foster, A. Khan, N. J. Saunders, R. Retkute and C. J. Rudolph, The Consequences of Replicating in the Wrong Orientation: Bacterial Chromosome Duplication without an Active Replication Origin, *mBio*, 2015, **6**, e01294-15.
114. D. Ivanova, T. Taylor, S. L. Smith, J. U. Dimude, A. L. Upton, M. M. Mehrjouy, O. Skovgaard, D. J. Sherratt, R. Retkute and C. J. Rudolph, Shaping the landscape of the Escherichia coli chromosome: replication-transcription encounters in cells with an ectopic replication origin, *Nucleic Acids Res.*, 2015, **43**, 7865–7877.
115. A. H. Syeda, J. U. Dimude, O. Skovgaard and C. J. Rudolph, Too Much of a Good Thing: How Ectopic DNA Replication Affects Bacterial Replication Dynamics, *Front. Microbiol.*, 2020, **11**, 534.

CHAPTER 9

Using Single-cell RNA Measurements to Decipher the Stochastic Kinetics of Transcription

MENGYU WANG[†a,b], JING ZHANG[†a,b] AND IDO GOLDING[*a,b]

[a]Department of Physics, Department of Microbiology and Center for the Physics of Living Cells, University of Illinois at Urbana-Champaign, Urbana, IL 61801, USA; [b]Verna & Marrs McLean Department of Biochemistry and Molecular Biology, Baylor College of Medicine, Houston, TX 77030, USA
*E-mail: igolding@illinois.edu

9.1 Premise: "Understanding" Transcription

The aim of the studies described in this chapter is to better understand prokaryotic transcription as it occurs *in vivo*. By "understand", what we typically mean is, form a satisfying narrative for an observed phenomenon. As reductionist scientists, this narrative typically consists of describing (or more strictly, predicting) complex observables in terms of simpler components and interactions. As physicists, we add the requirement that the narrative is formulated mathematically rather than verbally, and is able to describe and predict quantitative, rather than qualitative, observables.

[†]Equal contribution authors

Chemical Biology No. 19
RNA Polymerases as Molecular Motors: On the Road, 2nd Edition
Edited by Robert Landick, Terence Strick and Jue Wang

Published by the Royal Society of Chemistry, www.rsc.org

Even within the parameters defined above, what constitutes a satisfying narrative depends entirely on a subjective, personal choice of the desired scale, resolution and precision of the sought-after description. One may set as the goal the ability to predict the identity (identified, *e.g.* by their 5′ and 3′ ends) and copy-number of all the different RNA molecules in a given cell, at any moment in time, given the *a-priori* knowledge of the genome sequence and the cellular concentration of all the relevant molecules—RNA polymerase (RNAP), transcription factors, nucleotides, *etc.* Once the infeasibility of this level of detail becomes obvious (or does it?), the narrative's requirements may be relaxed in different ways. For example, one may aim to predict only the population-averaged, rather than the single-cell, values; or the relative change in levels between two different conditions (*e.g.* temperature) rather than the absolute numbers under a given condition.

It must be said that, using these definitions, we cannot claim to "understand transcription" in all but the coarsest resolution and precision. Not at all. At the same time, as in other areas of scientific investigation, the identification of scales at which we can and cannot provide a successful narrative helps us define the current limits to our knowledge, our rate of progress, and how far we still have to go. In this chapter, we attempt to do that. We report on some recent progress towards the formation of a quantitative narrative of transcription, and point to open questions, addressing which will lead to further refinements of this narrative.

9.2 Single-cell Measurements of RNA Copy-number Can be Used to Learn About the Stochastic Kinetics of Transcription

The approaches for using fluorescence-based, single-cell RNA measurements to study transcription kinetics build on earlier studies that used analogous methods to analyze single-cell *protein* data in bacteria.[1–3] The analysis comprises the following steps.[4–6] First, the copy number of RNA from a gene of interest is measured in many individual cells, under well-defined growth conditions. In chemically fixed cells, this can be reliably done using single-molecule fluorescence *in situ* hybridization (smFISH), in which a set of fluorescently-labeled oligonucleotides target the RNA of interest[7,8] (Figure 9.1A). Under the microscope, the labeled RNAs will appear as fluorescent foci ("spots"). The total signal intensity of all spots in a given cell can then be measured and, properly calibrated, provides an estimate of the copy-number of the RNA of interest. The measured values from a few hundreds or thousands of cells then yield the distribution of RNA copy-number within the population (Figure 9.1A).

The key to the investigative approach is the statistical concept of ergodicity. Ergodicity posits an equivalence between the above "snapshot" of RNA numbers from a population of cells at a given instant, and the stochastic trajectory of an individual cell over time—namely the kinetics of RNA copy

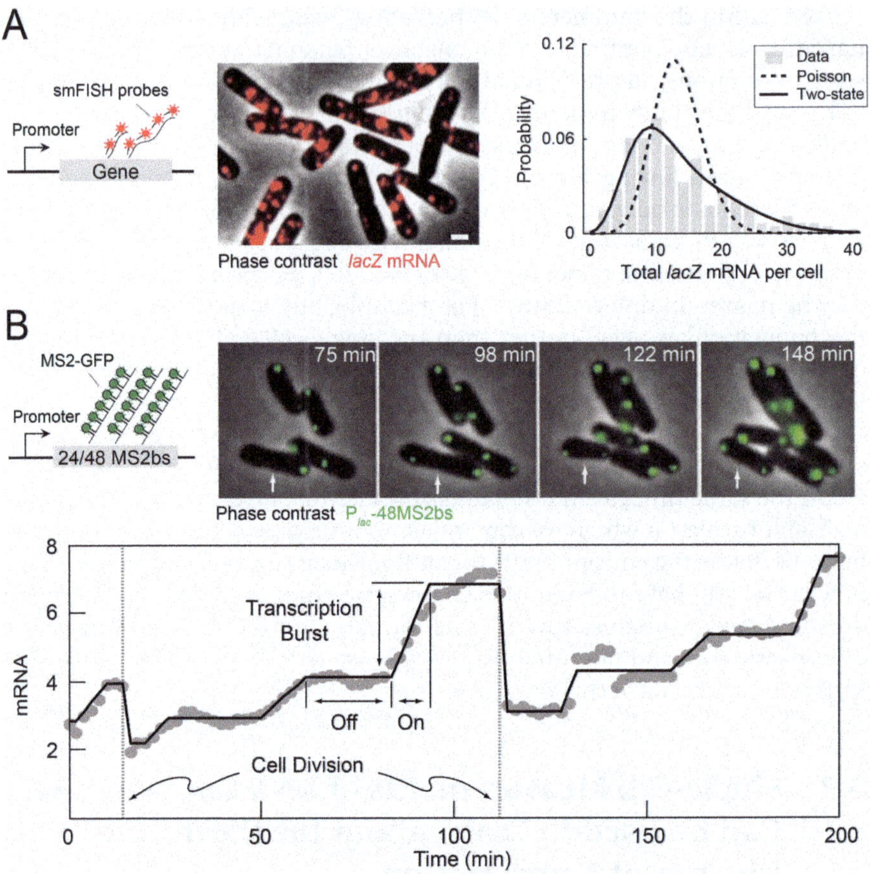

Figure 9.1 Probing the stochastic kinetics of transcription using single-cell RNA measurements. (A) The copy number of endogenous mRNA is measured in individual *E. coli* cells using single-molecule fluorescence *in situ* hybridization (smFISH). The resulting distribution is used to discriminate between alternative models for mRNA kinetics. The Poisson model, corresponding to the observed population average of ≈ 13 *lacZ* mRNA per cell, does not account properly for the observed variation in copy-number between individual cells. In contrast, the two-state model, which includes state-switching in the transcriptional state of the gene, successfully captures the observed variability. (B) In live cells, transcription is followed in real time *via* MS2–GFP labeling. The resulting time series shown below exhibits bursts of transcription, consistent with the broader-than-Poisson mRNA distribution found in (A). Panel B: Adapted from ref. 86 with permission from Taylor & Francis, Copyright 2012.

numbers, which we are seeking. The deduction of these single-cell properties is performed by writing down a theoretical model for the proposed stochastic kinetics of RNA. This model is then solved to yield the predicted distribution of RNA copy number in a population of cells. Comparing this prediction to the experimentally measured distribution allows testing the model's assumptions and estimating its kinetic parameters.

In fact, the assumptions regarding RNA kinetics can also be tested more directly, by measuring these kinetics in individual live cells (Figure 9.1B). RNA detection and quantification in living cells is achieved using an RNA binding protein (most commonly, the phage MS2 coat protein), fused to a fluorescent protein. The RNA of interest is transcriptionally fused to multiple tandem copies of the coat protein's binding site. When the RNA is expressed, it serves as a target for the binding of multiple copies of a fluorescent form of the coat protein [*e.g.* MS2–green fluorescent protein (GFP)].[9,10] The end result, as in the case of smFISH above, is the appearance of fluorescent "spots", whose signal intensity can be quantified and converted to RNA copy-number.[4] Performing this measurement in a single cell over time provides the stochastic time series of RNA levels, and the characteristics of this time series can then be compared with the expectations from the proposed model.[4]

An early example of how the approach above can be used to illuminate transcription kinetics was the demonstration of transcription bursts in *Escherichia coli.*[4] The prevailing model prior to that study was that, for a fully induced ("constitutive") promoter, RNA production can be approximated as a Poisson process, namely, molecules are produced one at a time, with a constant probability over time.[2] Consequently, the RNA copy-number distribution in a population of cells at a steady state level (*i.e.* when RNA production and elimination are balanced) should follow the Poisson distribution. However, when the distribution was measured experimentally, it was found to be much broader than expected according to the model (Figure 9.1A). Other predictions of the Poisson assumption were also contradicted by the experimental data, for example, the fraction of cells having no RNA during the transition from promoter inactivity to full induction.[4]

Confronted with the failure of the Poisson hypothesis, Golding *et al.*[4] then resorted to an alternative kinetic scheme known as the two-state (or "telegraph") model. In this model, the promoter switches stochastically between two states. RNA production takes place, stochastically, only in one of these states (the "active state"), whereas the other promoter state is inactive. Promoter state transitions, RNA production in the active state, and RNA degradation, all have constant probabilities, *i.e.* are Poissonian. The two-state model, while being more elaborate, can still be solved to predict RNA copy-number statistics. Unlike the Poisson distribution, the predicted copy-number distribution from the two-state model is consistent with the experimentally measured histograms (Figure 9.1A). The two-state model is also able to reliably predict how the fraction of zero-RNA cells will diminish during gene induction.[4]

In light of its success in reproducing the population statistics, the authors next attempted to test whether the two-state model is consistent with the RNA kinetics observed at the single-cell level. A key prediction of the model is that, when the promoter spends most of its time in the inactive state, RNA production will be "bursty", characterized by short periods of activity where multiple RNAs are produced, interspersed by longer periods of inactivity. This prediction was borne out by following RNA production in individual cells over time (Figure 9.1B). Moreover, the two-state model predicts that, in the bursty regime, the durations of active (RNA production) and inactive (no RNA production) periods will be exponentially distributed, reflecting the stochastic Poissonian switching between the active and inactive states. The experimentally measured durations were consistent with this prediction.[4]

In addition to providing strong evidence for the two-state hypothesis, the comparison of model predictions and experimental data yields an estimation of the relevant physiological parameters. In the example above, the transcriptional burst size (*i.e.* the average number of RNA molecules produced during each active period) can be evaluated, not only from the single-cell kinetics, but also from the RNA copy-number distribution. Specifically, it can be shown that the burst size is approximately equal to the ratio of population variance in RNA number per cell to the mean of this number, a ratio known as the "Fano factor" (note that measuring this ratio requires the ability to count the *absolute* number of molecules, rather than merely quantifying relative levels of expression[4]). In fact, measuring the RNA Fano factor has become the go-to method for estimating the transcription burst size. It was used, for example, to evaluate how burstiness varies when gene expression level changes, and to compare the burstiness of different promoters.[6,11] The effects of gene activity regulators (*e.g.* transcription factors[12] or DNA supercoiling[13]) on transcription kinetics could also be deciphered by examining the changes in RNA copy-number distributions. Similar analyses have been successfully applied for the analysis of bursty transcription in eukaryotes as well.[14–17]

9.3 Caveats

As with any method of scientific investigation, it is crucial to note the limitations of the approach described above. These reflect what can be thought of (in analogy to quantum physics) as the "uncertainty principle of single-cell biology", namely, the recognition that each measurement is also a *perturbation* to the system under study. This results in inherent, unavoidable limitations to the fidelity of the measurements performed.

In that context, we note that the MS2 method for live-cell RNA detection involves a significant modification of the endogenous system: The presence of an array of binding sites downstream of the promoter of interest, and the subsequent binding of MS2–GFP to the nascent RNA, are likely to perturb multiple aspects of the gene expression process, including transcription elongation, RNA translation, degradation, and spatial organization. Some of these expected perturbative effects have already been demonstrated, for example, the fact that RNA degradation in *E. coli*[4] and yeast[18] is inhibited by MS2–GFP binding. Furthermore, there are considerable limitations to the sensitivity, accuracy, and dynamic range of the MS2 method.[4] Thus, while live-cell RNA measurements provide a vivid impression of "the thing itself" by allowing real-time characterization of RNA kinetics, they rarely reflect the dynamics of the endogenous system in a precise manner.

As for measurements on fixed samples, these offer the advantage of being applied to the endogenous (genetically unmodified) system. In addition, smFISH typically provides better RNA counting performance than MS2.[6,7] This is due to the lower variability in the label-to-target stoichiometry, and the lower levels of background cellular fluorescence. Of course, it is not possible to track the temporal trajectory of individual cells which have been chemically fixed. Instead, as described above, kinetic information is deduced by examining "snapshots" of copy-number statistics under different conditions and comparing those snapshots to the prediction of a theoretical model for RNA kinetics. Our deduction in this case is thus only as good as the model we conceived of. Critically, the kinetic models used are rarely if ever unique, *i.e.*, more than one model can fit the data equally well.[19] In choosing among models, one typically applies Occam's razor and selects the simplest one, but it is doubtful whether this criterion, so powerful in physics, is foolproof when it comes to living systems.[20] When modeling cell-to-cell variability, we also commonly attribute to stochasticity (randomness) features that may actually reflect *deterministic* differences between individual cells, *e.g.*, cell cycle phase or gene copy number, resulting in mistaken conclusions regarding RNA kinetics.[21] We will return to this point later.

It is also important to remember that the models used for analyzing single-cell data are typically phenomenological. In other words, they are constructed, "top-down", to reproduce the experimental data, rather than "bottom-up", beginning from our knowledge of microscopic interactions. Consequently, the level of detail in the model is not determined by the actual complexity of the biological system, or by how well we know the particulars of the system. Rather, model complexity reflects the quality of available experimental data. Thus, for example, data describing single-molecule resolution counting of RNA in individual cells (using MS2 or smFISH) has been used to test and refine models that describe the discrete events of RNA production,[4–6] improving on earlier models that relied on relative measurements of protein fluorescence level.[1,2] However, the above single-cell RNA measurements were still limited to a resolution of one whole RNA. The corresponding models, accordingly, were unable to describe the kinetics of RNA elongation and degradation, which were approximated as instantaneous (occurring at infinite speed). Only further improvements in the resolution of RNA measurements, down to the sub-molecule level (partial transcripts), can allow for testable models that describe elongation and degradation kinetics. Such measurements, and the corresponding models, are now emerging.[22–26]

9.4 From Individual Cells to Individual Gene Copies

In addition to the general caveats discussed above, measurements that yield the total number of RNA copies in a single cell suffer specific limitations in terms of their ability to inform us about RNA kinetics. This is because whole-cell RNA measurements are blind to multiple features of RNA life history. One such feature is the identity of the individual gene copy from which the

RNA is transcribed. In a population of exponentially growing bacteria, individual cells will sample different cell-cycle phases, and thus exhibit a twofold range of gene copy numbers (between 1 and 2 during slow growth, but up to 8 or even 16 in fast growth medium[27]). Since whole-cell measurements yield no information on individual gene copies, the analysis of the resulting data relies on assumptions regarding the number of copies and the degree to which individual copies act independently or are, instead, correlated with each other.[6,12,28]

Whole-cell RNA measurements are also blind to the age of individual RNA molecules present in the cell. Note that a typical mRNA in *E. coli* spends a significant fraction of its minutes-long life still tethered to the template gene as an actively transcribed (nascent) message, before being released as a complete (mature) mRNA and degraded.[22] Nascent and mature mRNAs are subject to different kinetics and spatiotemporal dynamics, but those differences are all hidden from whole-cell measurements that cannot distinguish the two mRNA species.

In light of these limitations of the current single-cell measurements, it should not surprise us that fundamental questions regarding the spatiotemporal dynamics of mRNA in *E. coli* are still unresolved. First, temporally, there is still no consensus as to what determines the stochastic kinetics of mRNA production from the promoter (*i.e.* the transcriptional time series), and, in particular, what is the mechanism leading to transcription bursts. The leading hypotheses are that bursts reflect transcription-factor binding and unbinding at the promoter,[28] the coupling of transcription to DNA supercoiling,[13] or an as-yet unknown genome-wide process leading to non-gene-specific (universal) relation between gene expression level and the observed burst size.[6,29] Each of these theories can claim some support from whole-cell RNA data. Therefore, improved measurement capabilities may be required to resolve them.

As for the spatial dimension, there is likewise still a debate as to where in the cell mRNA spends the duration of its lifetime. The textbook picture is that, upon completion, each transcript is immediately released from the template gene to diffuse freely in the cytoplasm, where it soon gets degraded.[30] However, single-cell experiments have provided conflicting evidence regarding the veracity of this picture, with results of one study indicating that mRNA never leaves the vicinity of the gene,[31] while results of later studies revealed mRNA distributed across the cytoplasm.[6,32] Messenger RNAs encoding inner-membrane proteins exhibit a tendency to localize to the membrane, but there is no consensus regarding the mechanism by which this localization is achieved.[32–34] Here, too, the current limitations on probing the life history of mRNA in a single cell stand in our way of resolving these disagreements.

9.5 Detecting Active Transcription from a Single Gene

To probe the spatiotemporal life history of mRNA in *E. coli* and address the open questions above, one would need to go beyond merely labeling and counting all mRNA from a given gene in the cell. Instead, it would be necessary to measure the transcriptional activity of each individual copy of the gene, in order to examine how this activity depends on factors such as the presence of other copies of the same gene and the progression of the cell cycle. One would likewise need to discriminate nascent from mature mRNA, in order to characterize their different kinetics and spatial preferences.

In eukaryotic cells, active gene copies have been successfully identified on the basis of the presence of a strong RNA signal in the nucleus, corresponding to multiple nascent RNAs at the gene.[15,35–38] The presence of intronic RNA was likewise used to distinguish actively transcribed (pre-spliced) RNA from the mature species, where only exonic sequences can be detected.[39–41] However, the absence of a nucleus or introns prevents these methods from being directly applicable to bacteria.

As an alternative approach, inspired again by work in eukaryotes[42] as well as by a previous study in bacteria,[31] we decided to use the physical position of the gene locus as a fiducial marker for where active transcription takes place. We proposed the hypothesis that identifying and quantifying gene-proximal RNA would allow us to measure active transcription, whereas RNA further away would correspond to mature transcripts, which we would also measure in the same cell.[43]

We labeled the gene locus using the fluorescent repressor operator system (FROS) method,[44] in which an array of binding sites (here, 140 copies of *tetO*), engineered next to the gene of interest, is detected through the binding of the cognate transcription factor, fused to a fluorescent protein [here, tetracycline repressor–yellow fluorescent protein (TetR–YFP)] (Figure 9.2A). Other pairs of binding sites and proteins (*lacO*–LacI, *parS*–ParB) have been successfully used for the same purpose.[45,46] Consistent with previous reports, we find that the labeling scheme allows us to reliably measure the copy-number of the gene of interest in individual cells. By sorting the cells according to their length, we can track the gene copy number through the progression of the cell cycle (Figure 9.2A). Furthermore, by measuring the signal intensity of individual fluorescent foci (spots), we can detect unseparated sister copies (Figure 9.2A), which appear soon after gene replication.[44] This serves as a useful means of identifying the timing of gene replication, in both time-lapse movies of live cells (Figure 9.2B) and length-sorted fixed cells (see later).[43]

We next combined FROS gene tagging with the RNA labeling methods described above, namely MS2 in live cells and smFISH in fixed samples[43] (Figure 9.3). In the following paragraphs, we focus on our finding for the lactose promoter, P_{lac}. We have obtained similar results for P_{lac} and phage lambda P_R and P_{RM}. Applying the dual labeling approach to any promoter of interest is straightforward. In the analysis below, the live and fixed-cell methods for gene-plus-RNA detection are used to complement each other, as established previously for enumeration of RNA, to wit, precise measurements of an endogenous gene are made using smFISH, and interpreted using a theoretical model to yield kinetic schemes and parameter estimation. The live approach, less quantitative but more independent of model assumptions, is then used to validate the kinetic features inferred from fixed-cell data.

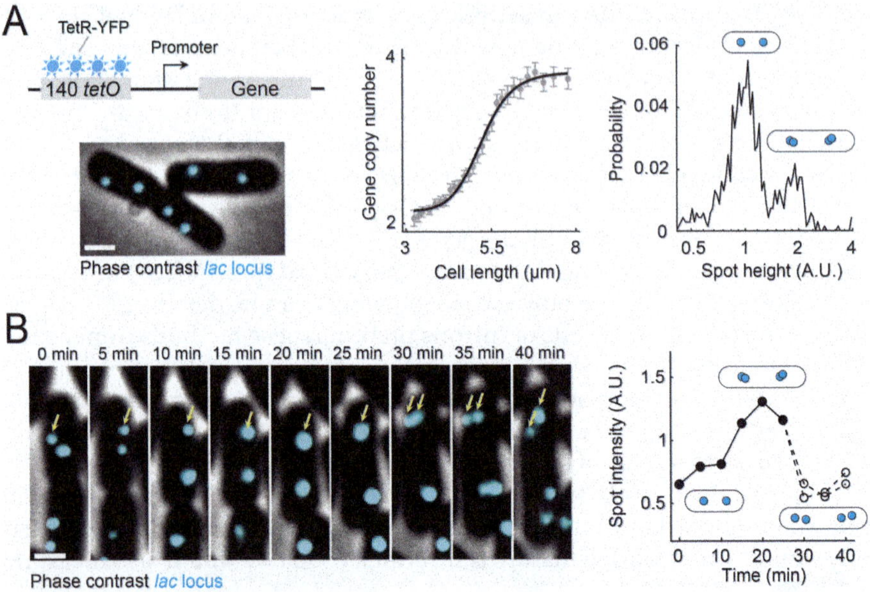

Figure 9.2 Detecting an individual DNA locus and the event of gene replication. (A) The locus of interest is labeled using the fluorescent repressor operator system (FROS). Sorting cells by length yields the locus copy number during the phase of the cell cycle. Measuring the fluorescent intensity of individual FROS spots allows the identification of newly replicated loci. (B) In time-lapse movies, gene replication is identified on the basis of the doubling of spot intensity followed by the spatial separation of the two sister copies.[43]

Our first test of the gene-plus-RNA labeling system is to ask whether it allows us to identify actively transcribed mRNA and distinguish it from the mature population. An indication that this is indeed the case is provided by examining the distance between each *lacZ* mRNA spot and the gene that encodes this mRNA (Figure 9.3A). The distribution of these distances revealed two distinct mRNA populations, one close to the gene and the other one further away. It is then plausible to propose that the gene-proximal population corresponds to nascent RNA, and this picture is supported by the following findings[43] (Figure 9.3A): (1) Under conditions of high expression, the gene-proximal signal is stronger than that from distal mRNA spots, consistent with the presence of multiple nascent mRNA molecules at the gene.[47] These nascent molecules are expected to appear as a single, diffraction-limited fluorescent spot, due to the limits of optical resolution. Previous smFISH-only experiments had already identified these putative transcription sites on the basis of their signal intensity.[7] The proximity of these strong RNA signals to the encoding gene, as seen here, solidifies this interpretation. (2) Two-color smFISH reveals that the gene-proximal mRNA is enriched for the 5′ region of the gene, as would be expected from the presence of partially transcribed

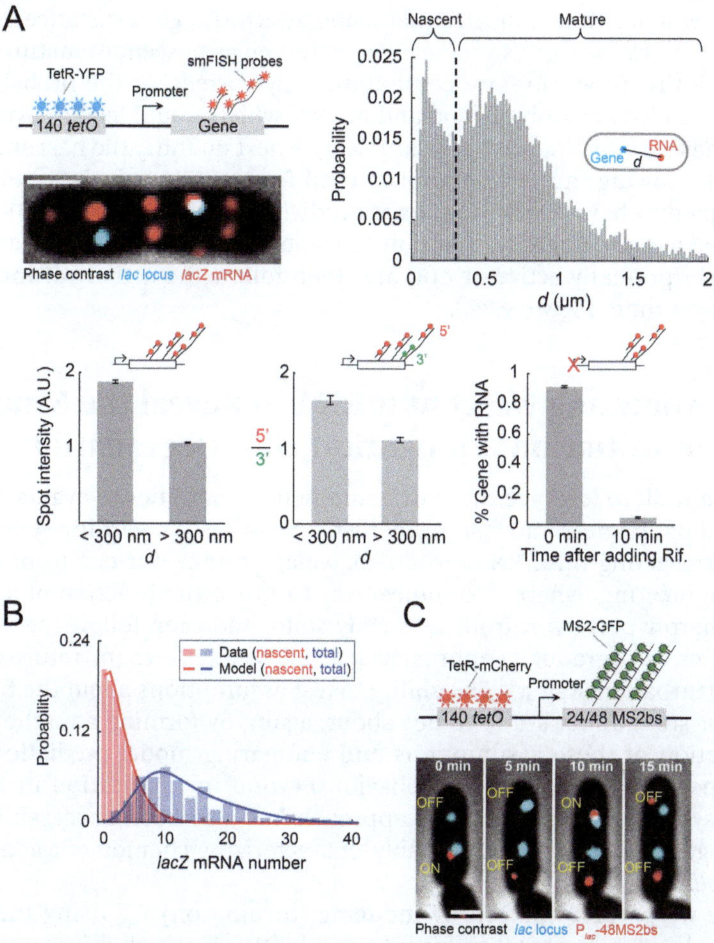

Figure 9.3 Detecting active transcription at a single gene copy. (A) Simultaneous labeling of endogenous mRNA and the gene that encodes it in fixed cells. The distribution of RNA-to-gene distances indicates the existence of gene-proximal and gene-distal mRNA populations. Bottom: Evidence supporting the identification of gene-proximal mRNA as actively transcribed (nascent) molecules: the gene-proximal RNA signal is stronger, is enriched for the 5′ region of the gene, and is eliminated following rifampicin treatment. (B) Nascent (per gene copy) and total (per cell) mRNA are measured to yield the copy-number distributions, which are then compared with the prediction by a theoretical model for mRNA kinetics. (C) In live cells, the transcriptional state (ON/OFF) of each gene copy can be followed over time.[43]

(incomplete) mRNA.[47] (3) On the other hand, when transcription initiation is inhibited using rifampicin, the gene-proximal mRNA signal disappears within a few minutes, consistent with the completion and release of the transcripts already initiated.[22] (4) The gene-proximal population is depleted altogether when the labeled locus and mRNA correspond to different genes.[43]

The evidence thus indicates that using an RNA-to-gene distance criterion allows us to classify mRNA molecules in the cell as nascent or mature. When applying this procedure, we computationally correct for the probability of random co-localization of gene and mRNA, which would lead to overcounting of nascent mRNA.[43] In fixed samples, we next quantify the nascent mRNA signal (following the same approach used for smFISH-only data[7]) to obtain the copy-number distribution across individual cells (Figure 9.3B). In live cells, we use the distance criterion to establish whether a given gene copy is transcriptionally active or not, and then follow the activity of individual genes over time (Figure 9.3C).

9.6 Analyzing Nascent mRNA to Reveal the Kinetics of Initiation, Elongation, and Degradation

We next wish to leverage these measurement capabilities to evaluate physiological parameters and test mechanistic hypotheses regarding some of the processes acting on mRNA. To do so, we again take our cue from physics and engineering, where a canonical way to probe the function of a system is to sharply perturb it from its steady state, and then follow the system's dynamics as it gradually approaches a new steady state (or returns to the pre-perturbation state).[48] The ability to test assumptions about the function of the system under study comes about, again, by forming a mathematical abstraction of these assumptions and comparing model predictions with the experimentally measured behavior. Beyond its application in human-made systems, the perturbative approach has also been successful in the interrogation of living ones, notably, characterizing chemotactic adaptation in *E. coli.*[49,50]

Here we perturb the cell by inducing (turning on) P_{lac} using the gratuitous inducer isopropyl-β-D-thiogalactoside (IPTG), which drives unbinding of the Lac repressor.[51] This procedure is commonly used to probe transcription kinetics[4,52] (the inverse scenario, turning the gene off, can also be used to obtain kinetic insights[51]). P_{lac} induction using IPTG provides a generous dynamic range of more than 100-fold (from a mean expression level of less than 0.1 to more than 10 *lacZ* mRNA per cell) provided that cyclic adenosine monophosphate (cAMP) metabolism is genetically abrogated and cAMP receptor protein (CRP) is instead activated by providing cAMP exogenously.[6,53]

Figure 9.4A depicts the results of a typical induction experiment. At different times after adding IPTG, samples were taken and the cells fixed, labeled, and imaged. The analysis described earlier in this chapter was used to measure the amounts of nascent, mature, and total *lacZ* mRNA per cell, which were then population-averaged. Even before performing detailed analysis, a few features can be observed in the data. The amount of total *lacZ* mRNA per cell shows the expected continuous increase, eventually saturating (after a few mRNA lifetimes) at a steady-state level that reflects the balance of mRNA production

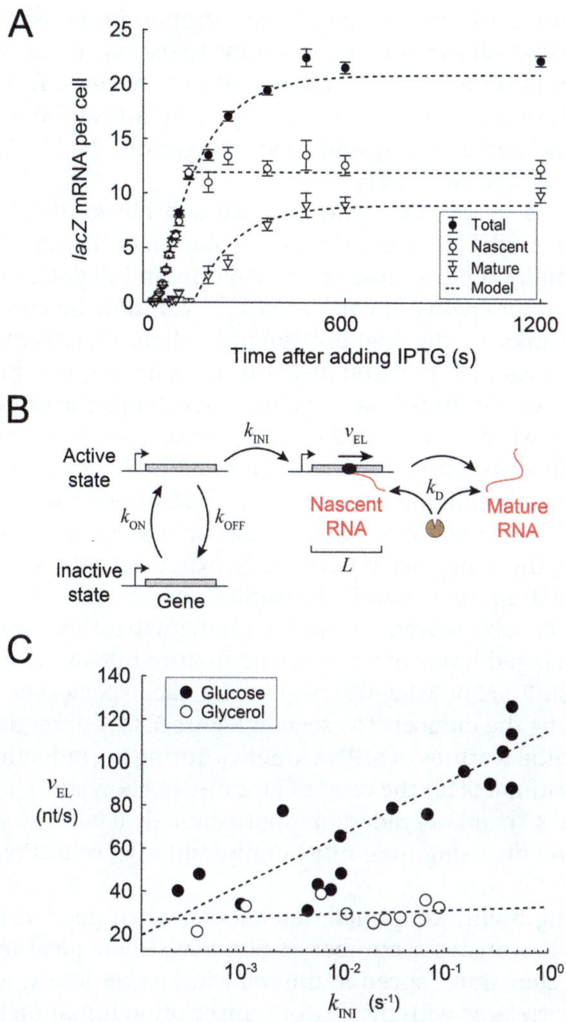

Figure 9.4 Analyzing nascent mRNA reveals the stochastic kinetics of transcript initiation, elongation, and degradation. (A) The levels of total, nascent, and mature *lacZ* mRNA following induction using IPTG, in cells grown in glucose. (B) A stochastic model for mRNA kinetics. The model is able to reproduce the experimental data in panel A. (C) The estimated mRNA elongation speed as a function of the transcription initiation rate (corresponding to different *lacZ* expression levels), for cells grown in glucose and glycerol.[43]

and degradation.[4] Mature mRNA, in contrast, only appears after a delay, corresponding to the time it takes to complete and release the first *lacZYA* transcript. At that same moment, nascent *lacZ* mRNA reaches its steady-state level, reflecting the balance of transcription initiation and release.[25]

To render these interpretations more rigorous and to estimate the underlying kinetic parameters, we formulated a mathematical model for the transcription process[43] (Figure 9.4B). In the model, transcription initiation (at rate k_{INI}) is followed by mRNA elongation at a constant speed v_{EL}. Upon completion of the full transcript (of length L), nascent mRNA is released from the gene to become a mature one. Both nascent and mature mRNA are degraded, with degradation initiated at rate k_D and proceeding, from the 5′ end of the gene, at a speed that is also equal to v_{EL}.[22]

Here again, as in previous efforts discussed above, the level of detail included in the model is dictated by the resolution of the experimental data. Thus, for example, the processes of elongation and degradation, which we modeled as instantaneous in earlier work,[4–6] can now be considered more realistically thanks to the higher-resolution data. Our formalism, which explicitly describes each position (0–L) on the gene, follows the model used by Chen *et al.*[22] to interpret RNA-Seq data, but extends the original model to differentiate between nascent and mature species,[25] a distinction which the authors' experiments (and therefore, their model) did not allow. At the same time, the strong simplifying assumptions made in our model, for example, that the speeds of elongation and degradation are equal to each other and constant along the gene, also reflect the fact that our data is not sufficiently resolved to constrain more detailed assumptions.

Our model can be solved to obtain mathematical expressions for the population-averaged levels of nascent and mature mRNA (more specifically, the expected smFISH or MS2–GFP signals for each species) as a function of time after adding the inducer.[43] As seen in Figure 9.4A, our model indeed captures the essential features of mRNA kinetics during P_{lac} induction. Moreover, the resulting estimates for the rates of *lacZ* mRNA elongation (approximately 42 nucleotides s^{-1}) and degradation (approximately 0.008 s^{-1}) are consistent with measurements using total mRNA only, and with values reported in the literature.[6,22,32,54]

Despite being highly simplified, our model, when applied to the experimental data, illuminates a number of non-trivial biological features. First, by evaluating elongation speed at different induction levels, we found that this speed is correlated with the rate of transcription initiation (Figure 9.4C). This confirms a previous report, which attributed this correlation to RNAP–ribosome interactions.[54,55] Second, the assumptions made in the model (*i.e.* the existence of co-transcriptional degradation[22] and the equality between rate of elongation and degradation[22]) are strengthened by the failure to fit our experimental data using alternative models that do not include these assumptions.[43]

9.7 Evidence for Partial Retention of Mature mRNA at the Gene

We next performed similar measurements for cells grown in glycerol, a slow growth medium (Figure 9.5A). Surprisingly, we found that our model failed to predict the level of gene-proximal (presumably, nascent) *lacZ* mRNA, both

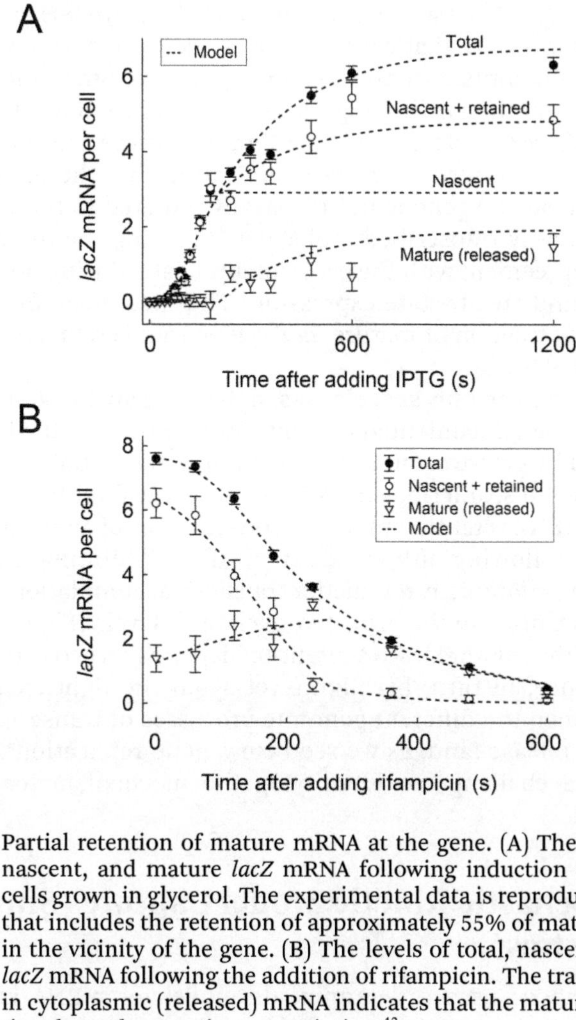

Figure 9.5 Partial retention of mature mRNA at the gene. (A) The levels of total, nascent, and mature *lacZ* mRNA following induction using IPTG, in cells grown in glycerol. The experimental data is reproduced by a model that includes the retention of approximately 55% of mature *lacZ* mRNA in the vicinity of the gene. (B) The levels of total, nascent, and mature *lacZ* mRNA following the addition of rifampicin. The transient increase in cytoplasmic (released) mRNA indicates that the mature mRNA retention depends on active transcription.[43]

during P_{lac} induction and at the steady-state level of expression. Specifically, the fraction of cellular mRNA that was localized to the encoding gene was higher than expected on the basis of the time it takes to complete one transcript. This observation was reminiscent of an earlier report of strong DNA–RNA colocalization at the *lac* locus under similar growth conditions.[31] In contrast to these results however, where the authors did not detect any non-gene-proximal mRNA, our data revealed both proximal and cytoplasmic populations, albeit not at the theoretically expected ratio.

To interpret these results, we added a feature to our theoretical model, where upon completion of transcript elongation a fraction of mature mRNAs remained in the vicinity of the gene rather than being released to diffuse in the cytoplasm. The hypothesis of mature mRNA retention is supported by a few pieces of data.[43] First, the gene-proximal *lacZ* smFISH signal showed a lower 5′:3′ ratio than that expected for nascent mRNA, consistent with the presence of complete transcripts among the proximal mRNA. Second, we found that rapidly centrifuging the cells (4500 *g* for 5 min) restored the gene–proximal *lacZ* levels, and the 5′:3′ ratios, to the expected values for nascent (actively transcribed) mRNA, indicating that the mature population retained near the gene is not physically tethered to the DNA, in contrast to the actively transcribed molecules.[43] Solving the modified model yielded good agreement with the experimental data, during both induction (Figure 9.5A) and steady-state expression. The theoretical fit provided an estimate for the fraction of mature *lacZ* mRNA that is retained at the gene, approximately 55%.

At this stage, we can only speculate as to the mechanism of mature mRNA retention. It is well documented that multiple features of the bacterial cell are modulated by growth conditions, including the spatial organization of the nucleoid, RNAPs, and ribosomes,[56] as well as the fluidity of the bacterial cytoplasm.[57] mRNA retention may be a consequence of any of those effects. Interestingly, following mRNA kinetics after inhibiting transcription initiation using rifampicin revealed a transient accumulation of cytoplasmic *lacZ* mRNA prior to the expected decline in levels (Figure 9.5B). This suggests to us that mature mRNA retention depends on active transcription. This dependence, in turn, highlights yet again the tight spatiotemporal coupling between the different genomic processes of transcription, translation, mRNA release (and, as we see below, gene replication), a coupling which poses a challenge to identifying the mechanism for any single phenomenon.

9.8 A Stochastic Kinetic Model Captures Single-cell Statistics

The analysis above used a deterministic model for mRNA kinetics, and was applied to the population-averaged experimental values. All stochastic effects, and the resulting differences between individual cells, were ignored. As in the case of whole-cell RNA measurements discussed earlier, we can next formulate a stochastic version of the mRNA model (Figure 9.4B), and use it to capture the measured copy-number statistics of nascent mRNA. Specifically, in this new model, we describe the events of transcription initiation, mRNA release, and degradation, as stochastic. In addition, promoter state now follows the two-state kinetics introduced earlier.[4] Inclusion of this latter feature is further supported by the observed statistics of gene "on" and "off" durations when using the live MS2 FROS reporter.[43] We solved the stochastic model of mRNA kinetics and fitted it to the measured distributions of nascent and total *lacZ* mRNA, during induction[43] and at steady-state levels of expression (Figure 9.3B above). This fitting procedure yielded estimates for the model parameters. In particular, it revealed that the probability of switching from inactive to active promoter state, k_{ON}, is the main feature modulated as P_{lac} expression is increased. This result is supported by the direct observation of promoter kinetics in live cells,[43] and reminiscent of similar reports in other organisms.[16,36,41,58]

9.9 Sister Gene Copies Can be Highly Correlated in Their Transcriptional Activity

Two or more sister copies of a gene are often present in the same cell. Under slow-growth conditions, this will only happen during the latter part of the cell cycle, after the gene has replicated. At shorter doubling times, the simultaneous presence of multiple replication forks results in the presence of two or more copies of each gene even in newborn cells. Consequently, a locus may be present at anywhere from 1 to 16 copies per cell, depending on the genome position, bacterial growth rate, and cell-cycle phase.[27]

Whereas each copy of the gene is regulated locally, through the binding of transcription factors and RNAP, early studies of stochastic gene expression evaluated the hypothesis that fluctuations in the cytoplasmic concentration of these actors would result in correlations between the activities of individual gene copies.[59] These correlations were conceptualized using the term "extrinsic noise" (to contrast from "intrinsic noise", which describes the uncorrelated fluctuations of individual gene copies),[59] and were probed using a two-color reporter, comprising two spectrally-distinct fluorescent proteins, driven by identical promoters, and placed symmetrically on the left and right sides of the chromosome from the origin of replication.[1] Measuring the relative signals of the two proteins provided a window into possible copy–copy correlations. However, despite the ingenuity of the engineered system, the two reporter genes, unlike true sister copies, do not share the same genomic context. It is thus not obvious that the measurements reflect the behavior of endogenous sister copies in the replicating chromosome.

Our novel ability to examine transcription at a single gene copy now allows us to directly measure the correlation between sister copies. When doing so, we observed a dramatic difference in the behavior of P_{lac} between cells grown in glucose and glycerol. Focusing on induction conditions where the fraction of active gene copies is approximately ½, and examining only those cells having two P_{lac} copies, we found that, in glucose, the number of active copies per cell closely followed a binomial distribution, as would be expected if gene copies are independent in their activity (Figure 9.6). However, this was not the case in glycerol. There, most two-copy cells had both P_{lac} copies in the same transcriptional state (either active or inactive), with only a small minority of cells showing mixed activity of the two copies (Figure 9.6). Consistent with these observations, the measured correlation in activity was low in glucose and high in glycerol. A similar trend was seen when comparing the measured correlation in nascent *lacZ* levels, as well as when measuring the temporal cross-correlation between P_{lac} copies in live cells.[43]

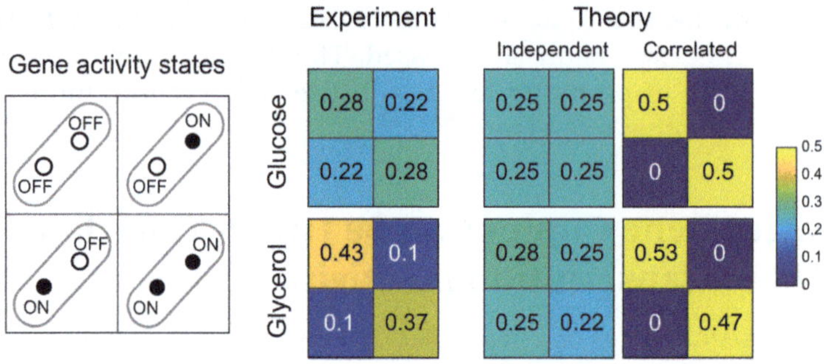

Figure 9.6 The correlation in transcriptional activity of two sister gene copies. The activity state (ON/OFF) of each P_{lac} copy was measured in cells having two copies of the locus. In glucose (top row), the distribution of number of active copies is consistent with each copy acting independently. In glycerol (bottom row), the two sister copies are highly correlated.[43]

Expanding our measurements beyond these two samples revealed a complex dependence on the experimental parameters.[43] Some of these dependencies indicated that attributing sister-copy correlations to fluctuations in the cytoplasmic concentration of an upstream regulator may be misplaced, or at least incomplete. For one, we found that a different promoter, phage lambda P_R, exhibited a similar trend of higher correlation in a slower growth medium. The behavior of P_R, which in the absence of the lambda repressor is considered "constitutive",[60] indicates that a gene-specific transcription factor is not the regulator whose fluctuations create correlations. This still leaves open the possibility of fluctuations in a genome-wide regulator, such as RNAP[61] or guanosine tetraphosphate (ppGpp).[62] However, we also found that sister-copy correlation strongly depended on the genomic locus of the promoter.[43] This appears to go against any cytoplasmic regulator as the single driver of correlations, and points, perhaps, to a possible role for temporal changes in the local chromosomal environment,[63] which would create a common regulatory effect on sister copies in two chromosomes.

9.10 Transcription from a Repressed Promoter is Coupled to the Event of Gene Replication

The way in which promoter activity in *E. coli* depends on the cell cycle remains an open question. *A priori*, cell-cycle progression involves multiple changes that could plausibly affect transcription. Some of these changes are, to a good approximation, continuous, *e.g.* cell volume, genome size, the concentrations of RNAP and other regulators, whereas others are discrete, *e.g.* cell division and the replication of the gene of interest and the genes encoding its upstream regulators. The possible consequences of these cell-cycle events for bacterial gene expression have received recent theoretical attention.[64–66] but experimental progress has lagged.

Of particular interest, the event of gene replication has long been speculated to affect that gene's transcription.[67] Early studies, using populations of synchronized bacteria, detected an increase in expression (beyond the obvious doubling of dosage) at a time corresponding to gene replication,[68,69] although no similar effect has been reported in the years since using newer experimental approaches. Multiple mechanisms that could couple gene replication and its transcription have been put forward.[67] These include changes in supercoiling of DNA surrounding the replication fork; movement of the replicated gene to the surface of the nucleoid, increasing its accessibility to RNAP; and transient de-repression of the new gene copy (or of both old and new copies). Regarding the latter, the idea that a tightly-bound transcription factor would only unbind DNA during passage of the replication fork is supported by single-cell measurements of P_{lac} repression by LacI[70] and of the binding kinetics of catalytically inactive Cas9 (dCas9).[71]

To probe the possible coupling between promoter activity and gene replication, we needed to identify the replication event within the cell cycle. In chemically-fixed cells, we first used length sorting as a proxy for cell-cycle progression.[72] Gene replication was then detected through the increase in FROS signal intensity, corresponding to the appearance of replicated, but not yet separated, sister copies (see Figure 9.2; note that the increase in the number of spots, indicating separation of the cohesive sister copies, is a lagging, and thus inefficient, indicator of replication[44]). Using this approach, we detected gene replication events at the expected approximately twofold length intervals,[72] and in a manner consistent with the expected dependence on distance from the origin of replication.[43]

We next proceeded to examine how the degree of active transcription, as indicated by the amount of nascent mRNA, changes during the cell cycle. The null hypothesis would be that the activity of each gene copy is constant throughout the cell cycle, and, therefore, that the level of nascent mRNA per cell closely follows gene dosage. Measuring the activity of a P_R reporter (in cells lacking the transcription factors to repress it), revealed just that: A doubling in the amount of nascent RNA is observed at the cell-cycle phase corresponding to gene replication (Figure 9.7A).

Very different behavior was revealed in P_{lac}, under conditions where the promoter is strongly repressed. Nascent mRNA levels increased transiently around the estimated time of gene replication (Figure 9.7A). This was the case for both the endogenous promoter and for reporters placed at other genomic loci.[43] A similar pulse of activity could be detected near the event of gene replication in live cells (Figure 9.7B). In fact, even in wild-type, genetically unmodified cells (strain MG1655), a higher propensity for transcription was observed in cells whose length corresponded to the cell-cycle phase of gene replication.[43]

Thus, our data provides evidence for an instance in which transcription from a repressed promoter is coupled to the replication of the gene. The data, however, does not offer an obvious means to distinguish between the

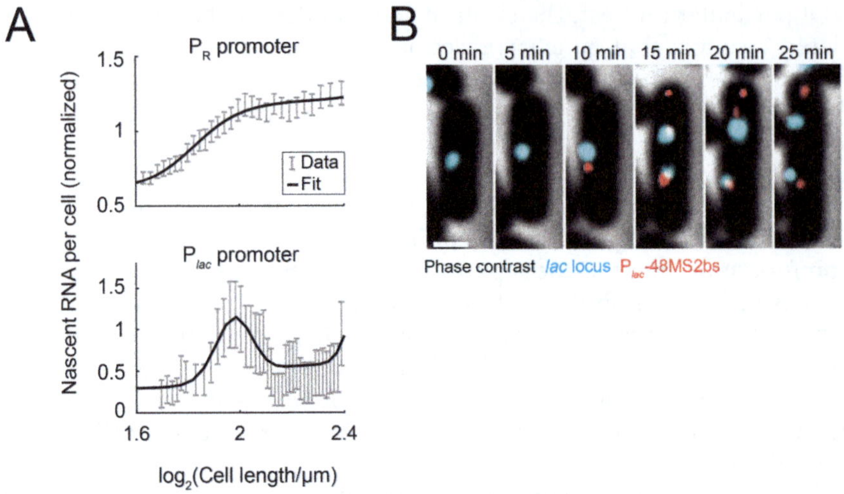

Figure 9.7 Promoter activity is coupled to gene replication. (A) In length-sorted fixed cells, nascent RNA from the P_R promoter (top) exhibits the expected doubling at the cell-cycle phase corresponding to gene replication. In contrast, a repressed copy of P_{lac} exhibits a pulse of transcription around gene replication. (B) In time-lapse movies, transcription from a repressed P_{lac} copy is observed close to the time of gene replication.[43]

proposed mechanisms for this coupling. Our current resolution does not allow us to determine the relative timing between replication and the increase in transcription, an observable that could yield useful information regarding the mechanistic questions posed. In terms of the magnitude of the effect, however, the phenomenology is quite clear, with the relative amplitude of the replication-induced pulse monotonically decreasing as gene expression level increases, until the effect becomes undetectable, as in highly expressed P_{lac} and P_R.[43]

9.11 Summary

In this chapter, we described our recent progress in using single-cell RNA measurements, interpreted using stochastic models of the underlying kinetics, to elucidate the spatiotemporal life history of mRNA in *E. coli*. As is evident from the discussion above, fundamental questions regarding those dynamics remain open.

In the temporal dimension, it is still unclear what drives the stochastic kinetics of mRNA production from the promoter. A large body of data is consistent with the picture of two-state kinetics, but the molecular underpinning of the active and inactive states is still debated.[13,28,29] Improvements in the temporal resolution of live-cell RNA measurements could help illuminate this question, and may reveal more elaborate kinetic schemes.[73] Importantly, it is also unclear how much of the observed population heterogeneity in mRNA levels actually reflects the stochastic aspects of transcription, as opposed to other drivers of cell-to-cell differences. As described earlier, we demonstrated two such drivers, namely the coupling of transcription to gene replication and the correlation between sister copies. In the absence of proper analysis, the cellular heterogeneity in mRNA numbers that stems from these effects may be erroneously attributed to stochastic transcription.[21]

Spatially, the life history of mRNA is just as mysterious, part of our greater ignorance of how the flow of genetic information in *E. coli* is organized within the cell.[74] For example, beyond the partitioning of the cell between nucleoid and cytoplasm, is the cytoplasm itself well mixed, or do key gene-expression players localize to improve function? RNAP "transcription factories", localized to ribosomal RNA operons, have been well-documented,[75,76] but their functional significance is still unclear.[77] The *lac* repressor, too, has been reported to exhibit a heterogeneous spatial distribution in the cell.[78] Considering the exciting recent observations of RNAP and transcription-factor "hubs" in eukaryotes,[79–81] this promises to be a fruitful direction for future research. Subsequent to transcription, mRNA translation has been shown to be modulated by the different spatial preferences of DNA, RNAP, and ribosomes in the cell,[56,82,83] and would plausibly be affected by the partial retention of mature mRNA discussed above. Finally, mRNA degradation has been suggested to take place mostly at the cell membrane,[32] but direct evidence for this is still lacking. The spatial dimension of the Central Dogma thus still needs to be experimentally elucidated. The theoretical modeling of these aspects, too, is only at an early stage.[84,85]

Acknowledgements

Work in the Golding lab is supported by grants from the National Institutes of Health (R35 GM140709), the National Science Foundation (PHY 1147498, PHY 1430124 and PHY 1427654), the Welch Foundation (Q-1759) and the John S. Dunn Foundation (Collaborative Research Award). We gratefully acknowledge the computing resources provided by the CIBR Center of Baylor College of Medicine.

References

1. M. B. Elowitz, A. J. Levine, E. D. Siggia and P. S. Swain, *Science*, 2002, **297**, 1183–1186.
2. E. M. Ozbudak, M. Thattai, I. Kurtser, A. D. Grossman and A. van Oudenaarden, *Nat. Genet.*, 2002, **31**, 69–73.
3. N. Friedman, L. Cai and X. S. Xie, *Phys. Rev. Lett.*, 2006, **97**, 168302.
4. I. Golding, J. Paulsson, S. M. Zawilski and E. C. Cox, *Cell*, 2005, **123**, 1025–1036.
5. C. Zong, L. H. So, L. A. Sepúlveda, S. O. Skinner and I. Golding, *Mol. Syst. Biol.*, 2010, **6**, 440.

6. L. H. So, A. Ghosh, C. Zong, L. A. Sepulveda, R. Segev and I. Golding, *Nat. Genet.*, 2011, **43**, 554–560.
7. S. O. Skinner, L. A. Sepúlveda, H. Xu and I. Golding, *Nat. Protoc.*, 2013, **8**, 1100–1113.
8. A. Raj, P. van den Bogaard, S. A. Rifkin, A. van Oudenaarden and S. Tyagi, *Nat. Methods*, 2008, **5**, 877–879.
9. I. Golding and E. C. Cox, *Proc. Natl. Acad. Sci. U. S. A.*, 2004, **101**, 11310–11315.
10. E. Bertrand, P. Chartrand, M. Schaefer, S. M. Shenoy, R. H. Singer and R. M. Long, *Mol. Cell*, 1998, **2**, 437–445.
11. Y. Taniguchi, P. J. Choi, G. W. Li, H. Chen, M. Babu, J. Hearn, A. Emili and X. S. Xie, *Science*, 2010, **329**, 533–538.
12. L. A. Sepúlveda, H. Xu, J. Zhang, M. Wang and I. Golding, *Science*, 2016, **351**, 1218–1222.
13. S. Chong, C. Chen, H. Ge and X. S. Xie, *Cell*, 2014, **158**, 314–326.
14. A. Raj, C. S. Peskin, D. Tranchina, D. Y. Vargas and S. Tyagi, *PLoS Biol.*, 2006, **4**, e309.
15. D. Zenklusen, D. R. Larson and R. H. Singer, *Nat. Struct. Mol. Biol.*, 2008, **15**, 1263–1271.
16. A. Senecal, B. Munsky, F. Proux, N. Ly, F. E. Braye, C. Zimmer, F. Mueller and X. Darzacq, *Cell Rep.*, 2014, **8**, 75–83.
17. K. Bahar Halpern, S. Tanami, S. Landen, M. Chapal, L. Szlak, A. Hutzler, A. Nizhberg and S. Itzkovitz, *Mol. Cell*, 2015, **58**, 147–156.
18. E. Tutucci, M. Vera, J. Biswas, J. Garcia, R. Parker and R. H. Singer, *Nat. Methods*, 2018, **15**, 81–89.
19. G. Neuert, B. Munsky, R. Z. Tan, L. Teytelman, M. Khammash and A. van Oudenaarden, *Science*, 2013, **339**, 584–587.
20. I. Golding, *Annu. Rev. Virol.*, 2016, **3**, 1–20.
21. I. Golding, *Curr. Opin. Microbiol.*, 2017, **43**, 9–13.
22. H. Chen, K. Shiroguchi, H. Ge and X. S. Xie, *Mol. Syst. Biol.*, 2015, **11**, 808.
23. S. Choubey, J. Kondev and A. Sanchez, *PLoS Comput. Biol.*, 2015, **11**, e1004345.
24. S. Iyer, B. R. Park and M. Kim, *Nucleic Acids Res.*, 2016, **44**, e142.
25. H. Xu, S. O. Skinner, A. M. Sokac and I. Golding, *Phys. Rev. Lett.*, 2016, **117**, 128101.
26. S. Kim and C. Jacobs-Wagner, *Biophys. J.*, 2018, **114**, 1718–1729.
27. H. J. Nielsen, B. Youngren, F. G. Hansen and S. Austin, *J. Bacteriol.*, 2007, **189**, 8660–8666.
28. D. L. Jones, R. C. Brewster and R. Phillips, *Science*, 2014, **346**, 1533–1536.
29. A. Sanchez and I. Golding, *Science*, 2013, **342**, 1188–1193.
30. B. Alberts, *Molecular Biology of the Cell*, Garland Science, New York, 4th edn, 2002.
31. P. M. Llopis, A. F. Jackson, O. Sliusarenko, I. Surovtsev, J. Heinritz, T. Emonet and C. Jacobs-Wagner, *Nature*, 2010, **466**, 77–81.
32. J. R. Moffitt, S. Pandey, A. N. Boettiger, S. Wang and X. Zhuang, *Elife*, 2016, **5**, e13065.

33. K. Nevo-Dinur, A. Nussbaum-Shochat, S. Ben-Yehuda and O. Amster-Choder, *Science*, 2011, **331**, 1081–1084.
34. S. Kannaiah, J. Livny and O. Amster-Choder, *Mol. Cell*, 2019, **76**, 574–589.
35. D. R. Larson, D. Zenklusen, B. Wu, J. A. Chao and R. H. Singer, *Science*, 2011, **332**, 475–478.
36. H. Xu, L. A. Sepúlveda, L. Figard, A. M. Sokac and I. Golding, *Nat. Methods*, 2015, **12**, 739–742.
37. S. C. Little, M. Tikhonov and T. Gregor, *Cell*, 2013, **154**, 789–800.
38. K. Tantale, F. Mueller, A. Kozulic-Pirher, A. Lesne, J. M. Victor, M. C. Robert, S. Capozi, R. Chouaib, V. Backer, J. Mateos-Langerak, X. Darzacq, C. Zimmer, E. Basyuk and E. Bertrand, *Nat. Commun.*, 2016, **7**, 12248.
39. D. Y. Vargas, K. Shah, M. Batish, M. Levandoski, S. Sinha, S. A. Marras, P. Schedl and S. Tyagi, *Cell*, 2011, **147**, 1054–1065.
40. C. H. Hansen and A. van Oudenaarden, *Nat. Methods*, 2013, **10**, 869–871.
41. S. O. Skinner, H. Xu, S. Nagarkar-Jaiswal, P. R. Freire, T. P. Zwaka and I. Golding, *eLife*, 2016, **5**, e12175.
42. X. Darzacq, Y. Shav-Tal, V. de Turris, Y. Brody, S. M. Shenoy, R. D. Phair and R. H. Singer, *Nat. Struct. Mol. Biol.*, 2007, **14**, 796–806.
43. M. Wang, J. Zhang, H. Xu and I. Golding, *Nat. Microbiol.*, 2019, **4**, 2118–2127.
44. M. C. Joshi, A. Bourniquel, J. Fisher, B. T. Ho, D. Magnan, N. Kleckner and D. Bates, *Proc. Natl. Acad. Sci. U. S. A.*, 2011, **108**, 2765–2770.
45. Y. Li, K. Sergueev and S. Austin, *Mol. Microbiol.*, 2002, **46**, 985–996.
46. I. F. Lau, S. R. Filipe, B. Soballe, O. A. Okstad, F. X. Barre and D. J. Sherratt, *Mol. Microbiol.*, 2003, **49**, 731–743.
47. S. L. French and O. L. Miller Jr, *J. Bacteriol.*, 1989, **171**, 4207–4216.
48. S. H. Strogatz, *Nonlinear Dynamics and Chaos: with Applications to Physics, Biology, Chemistry, and Engineering*, Addison-Wesley Pub., Reading, Massachusetts, 1994.
49. J. E. Segall, S. M. Block and H. C. Berg, *Proc. Natl. Acad. Sci. U. S. A.*, 1986, **83**, 8987–8991.
50. T. L. Min, P. J. Mears, I. Golding and Y. R. Chemla, *Proc. Natl. Acad. Sci. U. S. A.*, 2012, **109**, 9869–9874.
51. J. Elf, G. W. Li and X. S. Xie, *Science*, 2007, **316**, 1191–1194.
52. S. Kim, B. Beltran, I. Irnov and C. Jacobs-Wagner, *Cell*, 2019, **179**, 106–119 e116.
53. T. Kuhlman, Z. Zhang, M. H. Saier Jr and T. Hwa, *Proc. Natl. Acad. Sci. U. S. A.*, 2007, **104**, 6043–6048.
54. S. Proshkin, A. R. Rahmouni, A. Mironov and E. Nudler, *Science*, 2010, **328**, 504–508.
55. V. Epshtein and E. Nudler, *Science*, 2003, **300**, 801–805.
56. S. Mohapatra and J. C. Weisshaar, *Mol. Microbiol.*, 2018, **110**, 262–282.
57. B. R. Parry, I. V. Surovtsev, M. T. Cabeen, C. S. O'Hern, E. R. Dufresne and C. Jacobs-Wagner, *Cell*, 2014, **156**, 183–194.
58. O. Padovan-Merhar, G. P. Nair, A. G. Biaesch, A. Mayer, S. Scarfone, S. W. Foley, A. R. Wu, L. S. Churchman, A. Singh and A. Raj, *Mol. Cell*, 2015, **58**, 339–352.

59. P. S. Swain, M. B. Elowitz and E. D. Siggia, *Proc. Natl. Acad. Sci. U. S. A.*, 2002, **99**, 12795–12800.
60. N. Rosenfeld, J. W. Young, U. Alon, P. S. Swain and M. B. Elowitz, *Science*, 2005, **307**, 1962–1965.
61. S. Yang, S. Kim, Y. Rim Lim, C. Kim, H. J. An, J. H. Kim, J. Sung and N. K. Lee, *Nat. Commun.*, 2014, **5**, 4761.
62. V. Hauryliuk, G. C. Atkinson, K. S. Murakami, T. Tenson and K. Gerdes, *Nat. Rev. Microbiol.*, 2015, **13**, 298–309.
63. J. K. Fisher, A. Bourniquel, G. Witz, B. Weiner, M. Prentiss and N. Kleckner, *Cell*, 2013, **153**, 882–895.
64. J. R. Peterson, J. A. Cole, J. Fei, T. Ha and Z. A. Luthey-Schulten, *Proc. Natl. Acad. Sci. U. S. A.*, 2015, **112**, 15886–15891.
65. J. Lin and A. Amir, *Nat. Commun.*, 2018, **9**, 4496.
66. C. A. Vargas-Garcia, K. R. Ghusinga and A. Singh, *Curr. Opin. Syst. Biol.*, 2018, **8**, 109–116.
67. P. Guptasarma, *Bioessays*, 1995, **17**, 987–997.
68. P. Hanawalt and R. Wax, *Science*, 1964, **145**, 1061–1063.
69. J. M. Mitchison, *Science*, 1969, **165**, 657–663.
70. P. Hammar, M. Wallden, D. Fange, F. Persson, O. Baltekin, G. Ullman, P. Leroy and J. Elf, *Nat. Genet.*, 2014, **46**, 405–408.
71. D. L. Jones, P. Leroy, C. Unoson, D. Fange, V. Curic, M. J. Lawson and J. Elf, *Science*, 2017, **357**, 1420–1424.
72. M. Wallden, D. Fange, E. G. Lundius, O. Baltekin and J. Elf, *Cell*, 2016, **166**, 729–739.
73. C. Scholes, A. H. DePace and A. Sanchez, *Cell Syst.*, 2017, **4**, 97–108 e109.
74. M. Campos and C. Jacobs-Wagner, *Curr. Opin. Microbiol.*, 2013, **16**, 171–176.
75. T. Gaal, B. P. Bratton, P. Sanchez-Vazquez, A. Sliwicki, K. Sliwicki, A. Vegel, R. Pannu and R. L. Gourse, *Genes Dev.*, 2016, **30**, 2272–2285.
76. X. Weng, C. H. Bohrer, K. Bettridge, A. C. Lagda, C. Cagliero, D. J. Jin and J. Xiao, *bioRxiv*, 2018, 320481.
77. X. Weng, C. H. Bohrer, K. Bettridge, A. C. Lagda, C. Cagliero, D. J. Jin and J. Xiao, *Proc. Natl. Acad. Sci. U. S. A.*, 2019, **116**, 20115–20123.
78. T. E. Kuhlman and E. C. Cox, *Mol. Syst. Biol.*, 2012, **8**, 610.
79. M. Mir, A. Reimer, J. E. Haines, X.-Y. Li, M. Stadler, H. Garcia, M. B. Eisen and X. Darzacq, *Genes Dev.*, 2017, **31**, 1784–1794.
80. S. Chong, C. Dugast-Darzacq, Z. Liu, P. Dong, G. M. Dailey, C. Cattoglio, A. Heckert, S. Banala, L. Lavis and X. Darzacq, *Science*, 2018, eaar2555.
81. W.-K. Cho, J.-H. Spille, M. Hecht, C. Lee, C. Li, V. Grube and I. I. Cisse, *Science*, 2018, eaar4199.
82. A. Sanamrad, F. Persson, E. G. Lundius, D. Fange, A. H. Gynnå and J. Elf, *Proc. Natl. Acad. Sci. U. S. A.*, 2014, **111**, 11413–11418.
83. M. Stracy, C. Lesterlin, F. Garza de Leon, S. Uphoff, P. Zawadzki and A. N. Kapanidis, *Proc. Natl. Acad. Sci. U. S. A.*, 2015, **112**, E4390–E4399.

84. M. Castellana, S. Hsin-Jung Li and N. S. Wingreen, *Proc. Natl. Acad. Sci. U. S. A.*, 2016, **113**, 9286–9291.
85. T. M. Earnest, J. A. Cole, J. R. Peterson, M. J. Hallock, T. E. Kuhlman and Z. Luthey-Schulten, *Biopolymers*, 2016, **105**, 735–751.
86. R. Phillips, H. Garcia, J. Theriot and J. Kondev, *Physical Biology of the Cell*, Taylor & Francis Group, 2nd edn, 2012.

CHAPTER 10

Molecular Dynamics Simulations and Kinetic Network Models Elucidate RNA Polymerase Transcription Elongation Mechanisms

ILONA CHRISTY UNARTA[a,b], KIRILL A. KONOVALOV[b,c], XIAOWEI WANG[b,c], GIANMARCO DOMENICO SUAREZ[a,b] AND XUHUI HUANG*[a,b,c]

[a]Bioengineering Graduate Program, Clear Water Bay, Kowloon, Hong Kong; [b]State Key Laboratory of Molecular Neuroscience, Clear Water Bay, Kowloon, Hong Kong; [c]Department of Chemistry, The Hong Kong University of Science and Technology, Clear Water Bay, Kowloon, Hong Kong
*E-mail: xuhuihuang@ust.hk

10.1 Introduction

Structural features of RNA polymerase (RNAP) are crucial to its catalytic function. RNAPs of bacteria, archaea, and eukaryotes (except mitochondrial RNAP of eukaryotes) are homologous multi-subunit complexes, sharing a high degree of similarity in the core enzyme. The structures of multi-subunit RNAP has been likened to a crab's claw, which is able to clamp

Chemical Biology No. 19
RNA Polymerases as Molecular Motors: On the Road, 2nd Edition
Edited by Robert Landick, Terence Strick and Jue Wang

Published by the Royal Society of Chemistry, www.rsc.org

and stabilize the transcription bubble. Furthermore, distinct structural features of these polymerases (trigger loop and bridge helix, Figure 10.1A right) play important roles in orchestrating the transcription elongation cycle (TEC). In contrast to multi-subunit RNAPs, the RNAP of T7 bacteriophage (T7 RNAP) is a single subunit that is structurally similar to

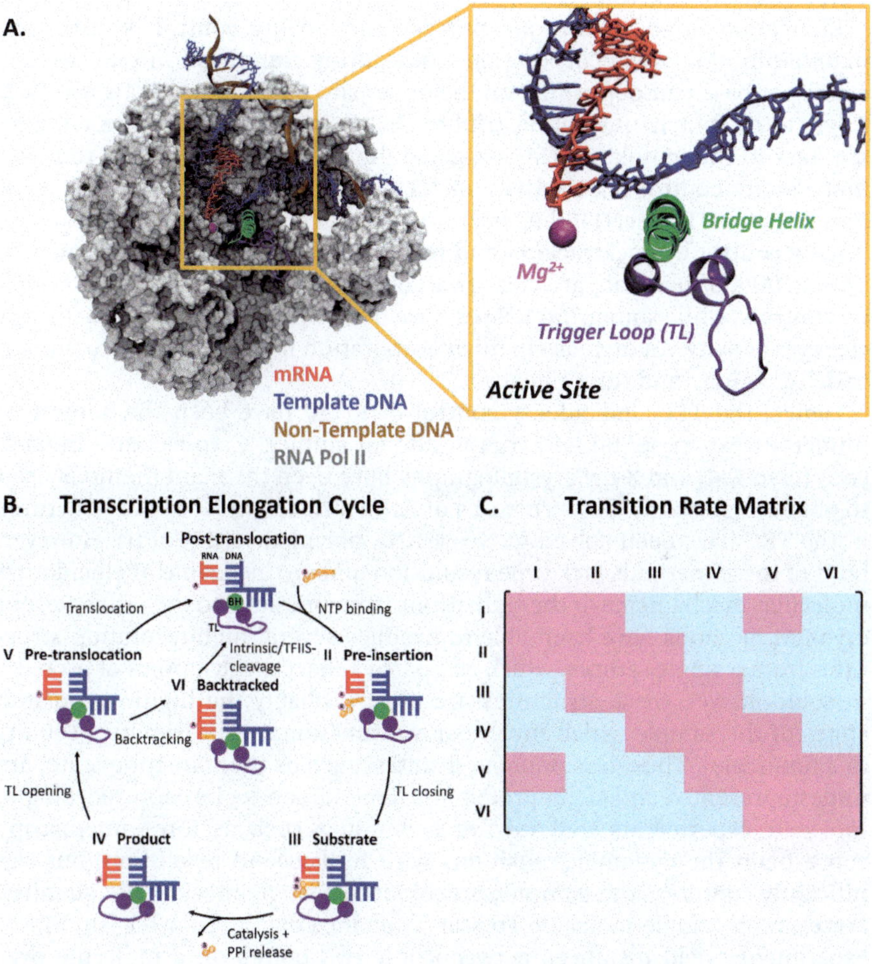

Figure 10.1 (A) The elongation complex of RNA polymerase II (Pol II) (Left). Pol II (Gray) is in surface representation, template DNA (Blue) and nascent RNA (Red) are in stick representation, and non-template DNA (Brown) is in cartoon representation. The active site is enlarged on the right. The Mg^{2+} ion (Magenta) is shown as a sphere, and bridge helix (Green) and trigger loop (Purple) are in cartoon representation. (B) The transcription elongation cycle (TEC) of Pol II represented by six states involving structural changes in the active site of Pol II. (C) The transition rate matrix which describes the transition rates between each step of TEC. Reproduced from ref. 15 with permission from Elsevier, Copyright 2018.

DNA polymerase I from bacteria and mitochondrial RNAP from eukaryotes. Despite the differences in sequence and structure, the TEC of T7 RNA polymerase is remarkably similar to that of multi-subunit RNA polymerases, where it proceeds with the same discrete steps to add nucleotides to the elongating RNA strand. These structural features play a significant role in efficient nucleotide triphosphate (NTP) addition as well as fidelity control.

Elongation is an error-prone process and fidelity control is crucial to maintaining low misincorporation rates during transcription (10^{-4} to 10^{-6} errors per base pair depending on the organism). Fidelity control is two-fold. Cognate or match nucleotide triphosphates (cNTPs), which are complementary to the template DNA, are generally preferred in the TEC, thus the first fidelity control is selectivity of NTPs. However, when a mismatch or non-cognate NTP (ncNTP) has been incorporated, RNAP employs a second fidelity control by backtracking and removing the misincorporated nucleotide. RNAPs of bacteria, archaea, and eukaryotes are capable of both fidelity controls, while some viral RNAPs have not demonstrated backtracking or cleavage activity. An increase in misincorporation has been linked to proteotoxicity,[1] aging,[2] and tumorigenesis.[3,4]

Hence, the TEC and fidelity control of RNAP have been the subject of numerous experimental and computational studies. Cryo-electron microscopy (cryo-EM) and X-ray crystallography have been the main methods used to obtain high resolution structures of RNAP. In fact, most of the structures in the TEC have been solved for yeast RNA polymerase II (Pol II). However, both of these methods lack time resolution and are incapable of elucidating molecular mechanisms of the transitions from one state to the other. Recent cryo-EM methods have been able to produce several high-resolution structures from a single sample, which can partly reflect the dynamics of proteins in solution. Yet, these structures are most probably the highly populated states of the sample which involve slow transitions (approximately 100 ms to s timescale). Thus, less populated states, such as the transition state, are underrepresented and faster protein movements cannot be captured. Single-molecule experiments that involve techniques such as force microscopy can explain the dynamic transitions with millisecond resolution, but significantly compress the complex process into a single measurable quantity. For example, single-molecule Forster resonance energy transfer (sm-FRET) experiments yield distances between only two points on a molecule over time. As in all experimental approaches, there is a trade-off between detailed spatial resolution and time resolution.

Multiscale-modelling approaches, which incorporate quantum and classical mechanics, can fill in the gaps in knowledge. Quantum mechanics (QM) theory, which considers electronic degrees of freedom, can be used to investigate chemical catalysis in the TEC of RNAPs (*i.e.* NTP addition catalysis and intrinsic cleavage). However, QM theory is too computationally expensive to directly simulate the entire RNAP complex, which contains approximately 50 000 atoms. Therefore, hybrid QM and molecular mechanics (MM)[5] approaches are popular, since a small part of the system (*e.g.* the active site) can be simulated with QM, while the remaining part of the system can be modelled with classical molecular mechanics methods. QM–MM simulations can describe the atomistic reaction mechanism and calculate the energy barrier of these chemical reactions.

For studying conformational changes without chemical reactions, classical molecular dynamics (MD) simulations are a powerful tool to model dynamic transitions between static structures, without compromising either temporal or spatial resolution. A typical MD simulation system includes a biomolecule of interest in a solvated simulation box. This system is evolved according to Newton's equations of motion, which produces a trajectory of the system in time. The trajectories of all-atom MD simulations (wherein all atoms in the system, including hydrogen, are simulated) are usually produced with femtosecond resolution. This implies that in order to observe slow dynamics in the TEC (μs to s timescale), the number of steps must be very large, taking an unfeasibly long time to obtain results. However, through advances in both hardware and mathematical theory, researchers can now simulate the dynamics of elongation complexes at all-atom resolution and gain useful insights into the mechanisms of transcription in less time. In recent years, Markov state models (MSM), a mathematical kinetic network model (KNM), have become a popular approach to bridge the above mentioned timescale gap by modelling long timescale dynamics based on many short MD simulations.[6–13] In MSMs, dynamic processes are modelled as a series of Markovian transitions among metastable conformational states (or free energy minima) at discrete time intervals. This allows estimation of transitions at timescales unreachable by individual MD simulations. MSM can be used to obtain the thermodynamic and kinetic properties of TEC of RNAP,[14,15] detailing stationary populations of metastable states and transition rate constants between the metastable states.

The rates obtained by MSMs can be combined with experimental observables to build KNMs of the whole TEC (Figure 10.1B). Using the transition rate matrix of the KNM, which describes the transition between each state (Figure 10.1C), one can derive the elongation rate equations and fit them with experimentally measured overall elongation rates. This would help in predicting unknown rates which are inaccessible by experiment or computation. Importantly, by incorporating sequence-dependent effects into the complete KNM, it can be further used to predict the error-prone sequences in the genome.

In this chapter, we review recent insights into transcription elongation gained through atomistic simulations. We cover the intermediate states of the TEC and the pathways that connect them, considering how the structural elements of the elongation complex contribute to each step. We review mechanistic studies of nucleotide addition and pyrophosphate release.

We examine how proofreading and fidelity are achieved by RNAP backtracking and nucleotide selectivity, as well as some effects of DNA damage. We also briefly discuss the construction of KNMs of the entire TEC.

10.2 Atomistic Mechanism of the Transcription Elongation Cycle

10.2.1 Structural Transitions During the Transcription Elongation Cycle

During the TEC, multi-subunit RNAP undergoes structural changes at the active site and surrounding regions, such as the bridge helix (BH) and trigger loop (TL) (Figure 10.1A right). While the exact mechanism of the TEC is still debatable, there are five main steps for Pol II TEC: (1) NTP binding to the active site of RNAP, (2) TL folding–closing, (3) the chemical addition of NTP to the RNA chain accompanied by pyrophosphate release, (4) TL unfolding–opening, and (5) RNAP translocation along the DNA strand to empty the active site and complete the cycle.

First, in the post-translocated state, the template DNA at the active site is unpaired (State I, Figure 10.1B). NTP then needs to travel through RNAP to reach the active site. During the pre-insertion stage (State II, Figure 10.1B), NTP binds at the active site and pairs with the template DNA. The triphosphate of NTP is stabilized by two Mg^{2+} ions, with the base occupying either the E or A site, which are near the active site. When bound to the E-site, the base is outside of the active site and cannot bind to the template DNA. When bound to the A-site, it can pair with the next unpaired template DNA nucleotide. After an NTP binds to the A-site, it interacts with the TL, an unstructured loop flanked by two helical segments. The TL and the C-terminus of the BH form a BH–TL bundle. The TL loop segment can fold and form numerous contacts with the NTP bound in the A-site. This isolates the active site and brings together residues which are critical for catalysis (State III, Figure 10.1B). Then, RNAP catalyses nucleotide addition to the 3′ hydroxyl group of the RNA strand, releasing a pyrophosphate as by-product (State IV, Figure 10.1B). In the next step, the TL swings open and the RNAP transitions to the pre-translocated state (State V, Figure 10.1B). Finally, RNAP translocates along the DNA to start a new cycle of elongation. The downstream DNA nucleotides are separated from the active site by the BH – an approximately 30 residue long alpha-helix. A complex sequence of BH uncoiling, bending, and oscillation supports translocation. In the Brownian ratchet model, the BH acts as a pawl that inserts in between the DNA bases and controls their progression through the active site during translocation. BH is also found to help in identifying misincorporated nucleotides in the RNA chain during backtracking.

RNAP undergoes transient backtracks to maintain transcription fidelity. A misincorporated nucleotide at the active site can stall the elongation complex and promotes backtracking. When backtracked (State VI, Figure 10.1B), the base of the 3′ end RNA is frayed into the secondary channel. Cleavage of the P–O bond then occurs at the active site removing two nucleotides from the 3′ end of the RNA. The RNA polymerase itself can catalyse cleavage of the frayed end of the RNA, but specialized proteins, such as growth regulator B (GreB) or transcription factor IIS (TFIIS) in bacteria or eukaryotes, respectively, can also enhance cleavage of the frayed RNA. Although the structure of T7 RNAP is largely different from that of multi-subunit RNAP, T7 RNAP has structural features that perform similar functions as in the multi-subunit RNAP. The architecture of the T7 RNAP resembles a hand shape with three distinct domains: fingers, thumb, and palm. These domains have striking functional similarities to those of the multi-subunit RNAPs. The palm domain contains the active site which holds two magnesium ions required for catalysis. The fingers domain includes a conserved helix–loop–helix structure (the helices are called O and Y-helices). Akin to the trigger loop of the bacterial and eukaryotic RNAP, the O and Y helices can shift away from the active site. This is the open conformation, which allows NTPs to bind but prevents the template DNA base from reaching the active site. The NTP first binds to the pre-insertion site on the O-helix. Then, the O and Y helices can move towards the active site, positioning the NTP in the active site and allowing the template to enter the active site. This is the closed conformation, which is catalytically active. In the closed form, the Y helix acts as a separator between the downstream DNA and the template in the active site, resembling the bridge helix of the bacterial and eukaryotic polymerases. Both single and multiple subunit RNA polymerases move in single nucleotide steps along the template DNA, extending the 3′ end of the RNA strand by adding nucleotides and releasing pyrophosphate. They also have a dynamic structure (TL–O-helix) that can open or close the active site and a helix (BH–Y-helix) that inserts between the bases of the template strand and regulates translocation.

10.2.2 Translocation

At the translocation step, the RNAP moves one base pair towards the template strand's 5′ end, to expose unpaired template DNA and vacate the active site for the incoming NTP. Results of previous MD simulation studies on the Pol II elongation complex have indicated that forward translocation can occur only when the TL is open.[16] Using MSM, it was revealed that there are two metastable states (S2 and S3) connecting the pre-translocation state and the post-translocation state.[17] In the pre-translocation state (S1), the elongating end of the DNA–RNA hybrid is at the active site. Oscillation in BH bending can drive forward translocation of the DNA–RNA hybrid. Specifically, when the BH is bent, residue T831 of the BH is in contact with the template DNA. This is corroborated by correlation of movement in BH residues 831–836 with the template DNA. The transition from S1 to the first intermediate state (S2) involves the translocation of the DNA–RNA hybrid (Figure 10.2).

The translocating template DNA (TN, orange spheres in Figure 10.2) has not crossed over the BH yet in S2. Instead, it base stacks with residue Y836 of the BH. The subsequent steps involve the TN crossing over the bridge helix to reach the active site. The fluctuation of the BH was also found to be important in S2 and S3. There is strong correlation between the motion in residues 831–836 of the BH and that of the TN as it crosses over the BH. In the next transition to S3, the TN loses the interaction with Y836 and interacts with T831, moving the TN base closer to the active site. Finally, the last transition is the movement of the TN base into the active site, completing the forward translocation step. From the free energy landscape

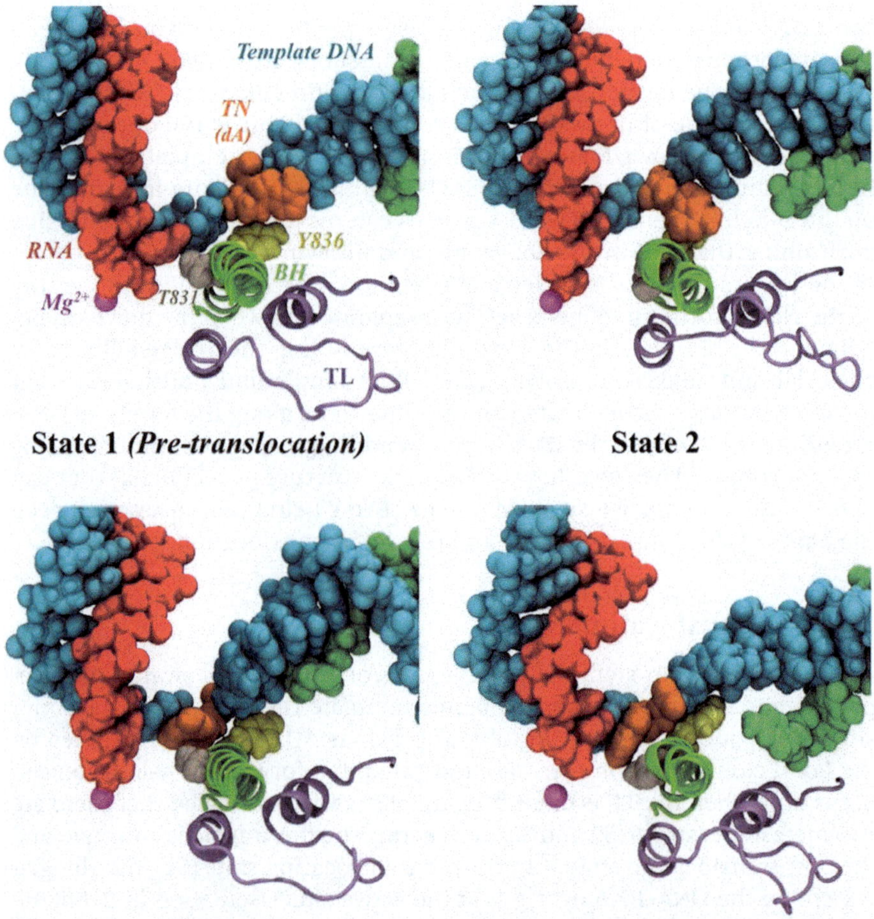

Figure 10.2 Four states in the translocation process as described by a Markov State Model (MSM). Reproduced from ref. 17 with permission from PNAS, Copyright 2014.

and the estimated mean first passage time (MFPT), the rate-limiting step (approximately 20 μs) is from S1 to S2, which involves numerous interactions between Pol II and the DNA–RNA hybrid breaking and reforming. The forward and backward rate of the rate-limiting step (S1 ⇔ S2) are almost equal, which implies that forward and backward translocation is equiprobable. This result agrees with the Brownian ratchet model of elongation, as the translocation of Pol II is reversible until the incoming NTP binds at the active site and stabilizes the post-translocated state.

The translocation of T7 RNAP is largely different from that of multi-subunit RNAPs. First, unlike with Pol II, the DNA–RNA hybrid translocates asynchronously, with RNA translocating first, followed by template DNA. The rate limiting step is the translocation of the RNA chain (approximately 20 μs). In the pre-translocated state, the TN base stacks with F644 of the Y-helix. This interaction must be broken to reach the intermediate state. Results of further analysis indicate that opening and closing of the O-helix in the pre-translocated state brings residue Y639 closer to the DNA–RNA hybrid and helps in the hybrid's initial translocation. Opening of the Y-helix is correlated with the motion of Y639, further encouraging the translocation of only the RNA chain. In the intermediate state, due to the open conformation of the Y-helix, F644 of the Y-helix can base stack with downstream template DNA, which leads to lagging of template DNA translocation. Thus, this stacking must be broken, to permit the translocation of template DNA. From the intermediate state to the post-translocation state, opening of the O-helix allows residue Y639 to transition from the pre-translocation to the post-translocation state (from far away to close to the DNA–RNA hybrid). Once residue Y639 base stacks with the DNA–RNA hybrid, the base stacking stabilizes the complex and helps the TN cross over to the active site. Finally, the template DNA chain relaxes, and TN occupies the active site. The transition of the Y-helix from closed to open conformations is reminiscent to the bending of the bridge helix in Pol II, which promotes the translocation process. In addition, residue F644 of the Y-helix may have a similar function to Y836 in the bridge helix of Pol II, which stabilizes the TN and the downstream template DNA. Similar to Pol II, the forward and backward rates of the rate-limiting step also occur at similar rates, indicating that T7 RNAP may also have a Brownian mechanism.

10.2.3 NTP Entry

To continue the elongation process, NTP must enter Pol II and bind the active site, which is deeply buried inside RNAP. According to the results of examination of crystal structures of elongating Pol II, there are two possible pathways for NTP to reach the active site: the secondary channel and the main NTP-entry channel (Figure 10.3A and B). The NTP can go through the secondary channel to bind at the entry site (E-site). Alternatively, NTP can first bind to the pre-catalytic downstream template DNA and enter the active site *via* the main NTP-entry channel. Molecular modelling and MD simulations can

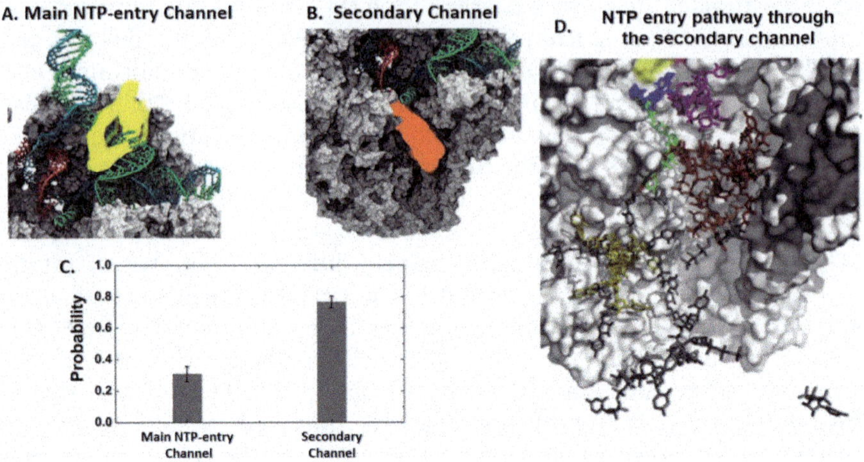

Figure 10.3 (A) The main channel of NTP entry, in which NTPs pass through the non-template DNA, shown in yellow. (B) The secondary channel of NTP entry which is under the bridge helix (green in cartoon representation), shown in orange. (C) Probability of finding MD structures with enough space in the main NTP-entry channel and secondary NTP-entry channel. Reproduced from ref. 18, https://doi.org/10.1371/journal.pcbi.1004354, under the terms of the CC BY 4.0 license, https://creativecommons.org/licenses/by/4.0/. (D) NTP entry pathway through the secondary channel of Pol II, as shown by NTPs in sticks representation. Each color represents the different NTP binding sites in the secondary channel pathway, E, blue; G, green; T, purple; F, brown; H, tan; bulk, gray. Reproduced from ref. 19 with permission from Elsevier, Copyright 2017. Figure D is adapted with permission from ref. 19. Copyright (2017) Elsevier.

help with further understanding of how these channels facilitate NTP entry. By cavity analysis based on structures from MD simulations, 77% of MD structures were found to have enough space for NTPs to diffuse through the secondary NTP-entry channel, whereas only 31% of the structures provided enough space in the main NTP-entry channel. Furthermore, calculation of electrostatic and van der Waals interactions between NTP and the RNAP environment indicate that NTPs in the main channel are energetically unfavourable, specifically due to repulsion between NTPs and the non-template DNA.[18] These calculations, however, were done without consideration of solvation energy, which may screen most of the repulsion between NTPs and the non-template DNA strand in the main NTP-entry channel.

The mechanism of NTP entry through the secondary channel of Pol II has been studied by MD simulations and MSM.[19] Between the bulk solvent and the active site, there are five NTP binding sites identified in the secondary NTP-entry channel. Besides the well-known E site near the active site, four more binding sites were identified, listed in the order of nearest to farthest from the active site: G (gate), T (trigger loop), F (funnel), and H (hind funnel)

(Figure 10.3D). At all binding sites, the NTP triphosphate tail interacts with positively charged residues, while the base remains flexible. In the E site, NTP interacts with R766, R1020, and K752. At the G site, it forms salt bridges with R619 and K752 and a partial salt bridge with K518. The interactions at the T site involve R731 and protonated H1085. All residues involved in the binding site are conserved, except K518. The F and H sites, however, are composed of non-conserved positively-charged residues. By incorporating these binding sites into kinetic network models, theoretical elongation rates can be obtained which are consistent with experimental ones.

10.2.4 Trigger Loop Closing

Once NTP binds to the active site in multi-subunit RNAPs, the TL must fold and close to contain the NTP inside the active site and bring the residues that stabilize the active site closer.[16] The closed-TL state is energetically favoured over the open-TL state by approximately 2 kcal mol^{-1}, on the basis of MD simulation structures of yeast Pol II.[20] This indicates that, at least with cognate NTP in the active site, TL closing should be spontaneous. An intermediate state (IS) exists between the open-TL state (OS) and the closed-TL state (CS). The IS is slightly higher in energy than both the OS and CS. Two transition states (TS), TS1 and TS2, exist on the pathway from OS to CS and are associated with significant kinetic barriers of 5 kcal mol^{-1} and 4.5 kcal mol^{-1}, respectively, relative to the OS. The transition from an entropy-dominated flexible open state to an ordered closed state is likely to be an enthalpy-driven process, which is fostered by the interactions between TL residues with their surroundings. As it transitions from OS to CS, the TL residues break and re-form salt-bridge interactions with several residues of Rpb1 (a subunit of Pol II).

The OS is mainly stabilized by salt bridges between TL residues and surrounding residues in Rpb1. During the OS–IS transition, interactions between TL residue K1093 and Rpb1 residues are broken, along with the severance of interactions between D1359 and T1083 or H1085, partially releasing the N-terminal region of TL. This allows the TL to expand and enables F1084 or F1086 to form extensive contacts with K726, I756, N757, or Q760, stabilizing the IS. Furthermore, L1081 and N1082 also begin engaging in distal contacts with the active-site NTP. During the IS–CS transition, the K1093–D1359 salt bridge that is formed first in TS2 is largely replaced by the K1092–D1359 salt bridge in the CS. Meanwhile, H1085 flips towards the active site and disrupts F1084–F1086 hydrophobic packing, and contacts between F1086 and surrounding residues in TS2. Finally, in the CS, the contacts between F1086 and the surrounding residues, such as K752, S754, and N757, are formed again. Residues 1093–1096 adopt a helical conformation, whereas S1091 forms a hydrogen bond with the backbone of V1094, essentially locking the conformation of the C-terminal region of TL. Residues 1084–1090 adopt a hairpin conformation, which brings the protonated imidazole group of H1085 closer to the active site to interact with the γ-phosphate oxygen of the NTP, stabilizing the NTP for catalysis.

10.2.5 NTP Addition Catalysis

The addition of NTP to the nascent RNA by RNAP follows the two-metal ion catalysis mechanism, which usually involves Mg^{2+} ions and is used by both DNA and RNA polymerases. In general, the addition of NTP follows a two-step mechanism. First, the 3′ hydroxyl of the elongating end of the RNA chain is deprotonated (nucleophilic formation or activation). Then the resulting 3′ oxide attacks the phosphorus atom of the incoming NTP's α-phosphate (P_α) (nucleophilic attack), yielding a pyrophosphate ion (PPi). In this reaction, Mg^{2+} ions may facilitate the formation of nucleophiles (3′ oxides), stabilize the transition state(s), and facilitate the removal of the by-product (PPi).[21]

To capture the crystal structure of RNAP before the catalytic step, the 3′ hydroxyl at the terminal RNA is often removed. Consequently, it is not known whether the 3′ hydroxyl is coordinated with $Mg_A{}^{2+}$, the magnesium ion which is coordinated with the α and β phosphates of the incoming NTP. To test the stability of the initial active site conformations before reaction, MD simulations of the elongation complex of Pol II and QM calculations based on density functional theory (DFT) were performed.[22] In these MD simulations, the positions of Mg^{2+} ions are the same as in the crystal structure,[23] with the $Mg_A{}^{2+}$ coordinated with the oxygen atoms of P_α and P_β instead of the 3′ hydroxyl group. In addition, $Mg_B{}^{2+}$ is coordinated with the oxygen atoms connected to P_β and P_γ and three aspartic acid, Asp, residues (D483, D481, and D837). This slightly differs from the initial structure of the conventional two-metal mechanism. This is likely to be due to the additional residue in Pol II (four Asp residues instead of the three Asp residues in DNA polymerase near the active site). Thus, without coordination with $Mg_B{}^{2+}$ ions, there will be a charge imbalance or repulsion between the negatively-charged P_γ and Asp residues. The positions of the four Asp residues are conserved for *Escherichia coli* RNAP and *Thermus thermophilus* RNAP, indicating similar active site structures in the NTP addition reaction. Furthermore, DFT calculation attempts with initial structures where the 3′ hydroxyl is coordinated with $Mg_A{}^{2+}$ result in 3′ hydroxyl decoupling from $Mg_A{}^{2+}$, accompanied by a decrease in energy. According to this result, it was concluded that 3′ hydroxyl is only weakly coordinated with $Mg_A{}^{2+}$. Then, using DFT calculations with different initial and intermediate state conditions, it was found that deprotonation of 3′ hydroxyl by hydroxide in the solution results in the lowest transition state energy. This step is then followed by nucleophilic attacks by 3′ oxide towards P_α in order to add the nucleotide, with simultaneous formation of a PPi ion. The PPi ion is then protonated by a general acid. Through DFT calculations, protonated H1085 of TL in Pol II was found to act as the general acid. By MD simulations, this H1085 was shown to be protonated to stabilize the position of NTP in the active site.[24] However, according to results from a mutagenesis experiment, the corresponding histidine residue in *E. coli* RNAP (H936 residue of TL) is not involved in acid–base catalysis.[25] In fact, TL was shown to assist only in stabilizing the position of incoming NTP and not proton transfer.[25] Moreover, results from an MD simulation study indicated that the 3′ hydroxyl can form stable coordination with Mg_A^{2+}.[26] Thus, further analysis of the coordination of Mg_A^{2+} and the general acid that protonates PPi will be needed.

10.2.6 Pyrophosphate Release

Upon successful NTP incorporation, a pyrophosphate (PPi) molecule is produced as a by-product and needs to be released from the active site to provide space for the incoming NTP. Pyrophosphate is thought to be released through the secondary channel, which is composed of a pore and a funnel region in Pol II. Comparing the TL fluctuations before and after the NTP incorporation in MD simulations, it was found that the TL is more flexible after the NTP incorporation while PPi is still in the active site.[27] This fluctuation of TL breaks its interactions with the 3′ end of the RNA chain, while PPi maintains a salt bridge with positively charged Pol II residues (Rpb1 K752, Rpb2 R766, and R1020). An MSM was built to aid understanding of the pathway of Mg_B^{2+} and PPi, $(Mg\text{-}PPi)^{2-}$, release through the secondary channel (Figure 10.4A). This process was found to occur on the microsecond timescale and two intermediate states were identified (Figure 10.4D). In the active site during S1, $(Mg\text{-}PPi)^{2-}$ interacts with protonated H1085 of TL and Rpb1 K752, which facilitates the ion's movement out of the active site (Figure 10.4D). This interaction is crucial for this first step, as $(Mg\text{-}PPi)^{2-}$ becomes much more stable in the active site in both K752A mutant and H1085F mutant simulations. From the first intermediate state (S2) to the next one (S3), $(Mg\text{-}PPi)^{2-}$ breaks its interactions with K752 and H1085 and forms new interactions with two positively charged residues, Rpb1 K619 and K518 in the pore region (Figure 10.4D).

The transition from S2 to S3 is also accompanied with an opening motion of the tip region of TL, where H1085 moves away from the closed TL state and disrupts its interaction with PPi. Finally, the $(Mg\text{-}PPi)^{2-}$ ion then moves to the last position before the funnel region, forming interactions with the K620 and K880 residues. The $(Mg\text{-}PPi)^{2-}$ ion moves through the pore region *via* hopping through several positively charged residues (Figure 10.4D). The $(Mg\text{-}PPi)^{2-}$ ion release seems to correlate with the TL movement, as the increased flexibility of the TL tip facilitates the movement of the $(Mg\text{-}PPi)^{2-}$ ion out of the active site. As the $(Mg\text{-}PPi)^{2-}$ ion leaves the active site, its interaction with H1085 may drag the tip of the TL and open it. Yet, TL opening was not observed in the MD simulations, which implies that PPi release is not coupled with translocation, as translocation requires opening of the TL. In addition, PPi release is unlikely to be the rate-limiting step of NTP addition, as it occurs much faster than other steps in the NTP addition cycle.

A similar MSM study of PPi release was performed for bacterial *T. thermophilus* RNA polymerase (Tth RNAP).[28] Unlike Pol II, the secondary channel of Tth RNAP is shorter due to the lack of a funnel region, leading to threefold faster PPi release (Figure 10.4B). Furthermore, PPi release in Tth RNAP follows

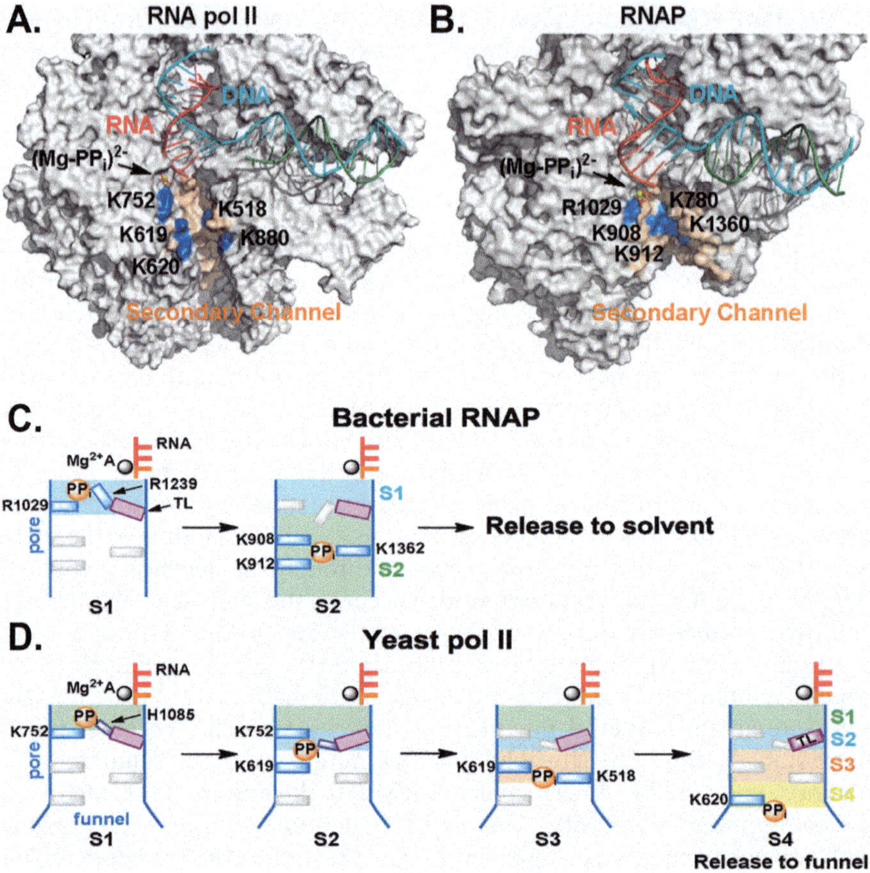

Figure 10.4 (A) Secondary channel of T. thermophilus (Tth) RNAP. (B) Secondary channel of yeast Pol II. (C) PPi release pathway of Tth RNAP. (D) PPi release pathway of yeast Pol II. Reproduced from ref. 28, https://doi.org/10.1371/journal.pcbi.1003020, under the terms of the CC BY 4.0 license, https://creativecommons.org/licenses/by/4.0/.

a simple two-state model, in contrast with the PPi release in Pol II, which goes through a hopping mechanism between several positively charged sites. In the active site, PPi interacts with H1242 and R1239 of TL, and R1029 (Figure 10.4C). Residue R1029 may play a similar role to K752 in Pol II as it is located near the exit of the active site and the secondary channel. Conserved residue H1242 of TL is located at a slightly different position compared with H1085, thus it interacts with (Mg-PPi)$^{2-}$ differently. Instead, residue R1239 of TL, which is located at the same position as H1085, may function similarly to H1085 of TL in Pol II (Figure 10.4C and D). From the first state to the second metastable state, PPi moves from the active site to an elongated channel, where several positively charged residues (K780, K908, K912 and K1362) can interact with the PPi (Figure 10.4C). Similar to H1085F and K752A mutant S1 simulations in Pol II, R1239A and R1029A mutants in Tth RNAP prevent

PPi from leaving the active site. On the other hand, MD simulations with a H1242A mutation indicate that PPi release is encouraged, implying that H1242 may prevent release of PPi from the active site. In MD simulations starting from S2, R1239A drives the PPi back to the active site (S1), indicating that R1239 is important to keep PPi in S2 and for the subsequent release to the solvent. Since R1239 is longer than H1085 in Pol II, it can assist in PPi release all the way to the solvent. Lastly, TL does not unfold during the simulation of PPi release. However, the rotation of R1239 as PPi leaves the active site leads to an increase in TL tip fluctuations. The mean first passage time of PPi release in Tth RNAP is approximately 0.5 μs which is much faster than the expected timescale of TL opening.

In a single-subunit RNAP, such as T7 virus RNA polymerase (T7 RNAP), the PPi release mechanism is even simpler, since the active site is only approximately 15 Å away from bulk water. *Via* MD simulation, it was found that after NTP incorporation, $(Mg\text{-}PPi)^{2-}$ sits inside a cavity and interacts with three positively charged residues (K631, R627, and K472)[29] and Mg^{2+} of $(Mg\text{-}PPi)^{2-}$ interacts with two negatively charged residues (D537 and D812). Unlike PPi release in multi-subunit RNAP, PPi release in T7 RNAP is an activation process where the end state is less energetically favourable. By MSM, the rate-limiting step was found to be PPi release from the cavity, which includes cessation of interactions with D537, D812, and K631, and occurs in approximately 0.2 μs. The end state is only stabilized by the interactions of PPi with R627 and K472. The distance between PPi and K472 decreases at the end state due to dramatic swinging motions of K472, indicating that K472 may act as a swing that pulls PPi out of the cavity. Residue K472 in T7 RNAP plays a similar role to K752 in Pol II and R1029 in Tth RNAP, indicating a common module across RNA polymerases in PPi release. Similarly to multi-subunit RNAP simulations, the O-helix opening is not observed, since PPi release occurs on a much faster timescale. Thus, this simulation indicates that PPi release cannot support the O-helix opening and translocation, favouring the Brownian ratchet mechano-chemical coupling scenario.

10.3 Fidelity Control

10.3.1 Nucleotide Selectivity

To maintain the high accuracy of NTP incorporation, RNAP needs to selectively choose cognate NTPs (cNTPs) over non-cognate NTPs (ncNTPs). Theoretically, active site binding, TL closing, and the catalysis steps of TEC are involved in nucleotide selectivity. The NTP may enter the RNAP active site through the secondary or main NTP-entry channel in Pol II. If NTP enters through the secondary NTP-entry channel, ncNTP and cNTP cannot be distinguished up until the A-site. This is because Pol II interacts with NTP only *via* the triphosphate group along the secondary channel, meaning that bases cannot be discerned.[19] At the A site, ncNTP has less favourable base-pairing energy compared with cognate NTP, providing some level of selectivity. However, relying on base-pairing energy alone is not enough to maintain the low transcription error rate (10^{-4} to 10^{-6} errors per base pair). This is due to some stable mismatch base-pairings, such as wobble base pairing. Thus, high selectivity of the incoming NTP must be achieved by other additional factors. On the basis of MD simulations which place either cNTP, ncNTP, or cognate dNTP (cdNTP) in the active site of RNAP with different TL conformations, there are several structural differences which may explain nucleotide selectivity.[26]

Given an open TL, base-pairing between ncNTPs or cdNTPs with template DNA is not stable. ncNTPs were found to form interactions with RNAP residues resulting in half-rotated NTPs in the exit pathway. In one of the Pol II MD simulations, an ncNTP spontaneously moved from the active site to the E-site over a sub-microsecond timeframe, indicating that the secondary NTP-entry channel is an exit pathway for ncNTPs. Another trajectory reveals that ncNTP rotates toward the entrance of the main NTP-entry channel (see subsection 10.2.3 NTP entry), offering alternative exit pathways for mismatched NTPs. For wobble base pair and cdNTP, the conformational changes of the NTP are not very significant, limited to rotation of the base. Although wobble base-pairing can form stable base pairing, it is only favoured by 0.7 kcal mol^{-1} in the A site over the E site, which is significantly lower than the 3 kcal mol^{-1} stabilization of the cNTP in the A site. Thus, the first fidelity check is indeed the base-pair instability of ncNTP with the template DNA, which eventually may lead to ncNTP exiting the active site.

In the MD simulations of Pol II and *Tth* RNAP with closed TL, residue L1081 of Rpb1 in Pol II (M1238 of Tth β′ subunit) is known to position the NTP at the active site *via* hydrophobic interaction. If cNTP is at the active site, the distance between L1081 or M1238 and cNTP has a narrow distribution centred around 5 Å. However, when ncNTP is at the active site, residue L1081 or M1238 can move closer to or further from the active site, depending on the NTP. If L1081 or M1238 is farther, ncNTP is not stabilized at the optimum position. If L1081 or M1238 is closer, it may introduce unfavourable interactions and lead to nucleotide expulsion or TL reopening. The unstable closed TL conformation when ncNTP occupies the active site indicates that TL closing may be hindered, which may act as the second fidelity checkpoint. At this point, neither wobble base-pair nor cdNTP can form stable interactions with the TL.

Furthermore, on the basis of MD simulations with a closed TL, ncNTP in the active site can disrupt the catalytic geometry of the active site. The addition of NTP to the nascent RNA chain is a reaction between P_α of the NTP and 3′ hydroxyl of the elongating end of the RNA, thus the distance between P_α and the 3′ hydroxyl must be kept close, possibly by Mg_A^{2+}. In the MD simulations of all ncNTPs an cdNTP, Mg_A^{2+} shifted in the active site, separating Mg_A^{2+} from 3′ hydroxyl or P_α depending on NTP, creating an unstable active site and hindering catalysis of NTP. This distortion of the active site geometry serves as the third fidelity checkpoint.

In the Brownian ratchet model, the presence of NTP in the active site stabilizes the post-translocation state. However, when ncNTP or cdNTP is in the active site, the template DNA tends to translocate backward (from post-translocation to pre-translocation state, Figure 10.1B). The presence of ncNTP or cdNTP in the active site also disrupts the terminal DNA–RNA hybrid base-pairing which may hinder catalysis and encourage backward translocation. The fourth and fifth fidelity checkpoints would be backward translocation and disruption of the upstream DNA–RNA hybrid.

The only known fidelity control mechanism of T7 RNAP is the selectivity of cNTP *versus* ncNTP during NTP entry. Similar to the A-site of Pol II, a pre-insertion site has been identified in T7 RNAP in high resolution crystal structures.[30] There are four checkpoints of nucleotide selectivity: (1) NTP dissociation from the pre-insertion site, (2) NTP insertion from the pre-insertion site into the active site, (3) NTP stability in the active site, (4) and NTP addition catalysis. At the post-translocation state, residue Y639 base stacks with the DNA–RNA hybrid, blocking the active site. Residue Y639 can also distinguishes cNTP from ncNTP, by competing with the template DNA to interact with the NTP. Interestingly, such a residue is not identified in multi-subunit RNAP, indicating a different nucleotide selection mechanism in T7 RNAP. As cNTP moves from pre-insertion to active site and base-pairs with the template DNA, cNTP can form stable base-pairing with template DNA, thus replacing Y639 from the active site. When ncNTP binds at the pre-insertion site, the interactions between Y639 with either DNA–RNA hybrid and/or NTP strengthens significantly compared with cNTP at the pre-insertion site. This strong association of Y639 with ncNTP keeps ncNTP away from the template DNA and from subsequently moving into the active site (off-path configuration).[31]

Another free energy calculation of the ncNTP insertion from the off-path configuration results in an off-path conformation, and dissociation of ncNTP from this conformation was found to be important in keeping the misincorporation rate limited to 10^{-4}.[32] Using free energy calculation based on MD structures, cNTP needs to overcome a much higher energy barrier to dissociate from the pre-insertion state compared with ncNTP dissociation from the off-path configuration. This tendency of ncNTP to go to an off-pathway configuration and hence dissociate from this configuration is the major fidelity control for T7 RNAP, lowering the error rate by 10^{-2} errors per base pair.[33] Furthermore, lowering of the error rate to 10^{-3} errors per base pair is achieved by the enhanced barrier of insertion of ncNTP to the active site.[33] The selectivity in the catalysis process would then be the last fidelity check that maintains the low error rate of T7 RNAP (10^{-4} errors per base pair).

10.3.2 Backtracking and Intrinsic Cleavage

When an ncNTP is incorporated into the elongating RNA chain, RNAP must employ a proofreading mechanism, *i.e.* backtracking and cleavage of the misincorporated nucleotide. Therefore, RNAP must detect the misincorporated nucleotide in the nascent RNA chain. Using an MSM, it was revealed that backtracking of Pol II follows a stepwise mechanism, wherein the misincorporated RNA is first frayed, and then the DNA–RNA hybrid is backtracked.[34] In the pre-translocation state, due to the oscillation of BH bending, residue T831 of the BH can act as the sensing probe to examine the stability of the terminal DNA–RNA base pair in the active site (Figure 10.5). When an ncNTP is incorporated, the pre-translocation state is destabilized, lowering the energy barrier to the frayed state, S2. In S2, the RNA 3′-end frays and stacks with residue Y769 in Rpb2 subunit and the BH residue T827. S2 is consistent with the crystal structure of the frayed state.[35] Next, in S3, the misincorporated nucleotide is stacked with Rpb2 residue Y769 and the template nucleotide (TN, Cyan in Figure 10.5) rotates to stack with the BH residue Y836, stabilizing the disrupted base pairs. From S3 to S4, the upstream DNA–RNA hybrid translocates. This is the rate-limiting step (approximately 100 μs), with many interactions between Pol II and the DNA–RNA hybrid breaking and re-forming. Notably, this rate limiting step is ten times longer than forward translocation, in which the movement of DNA–RNA is also the rate-limiting step. This could be due to the loss of stacking between Y769 and the RNA 3′-end in backtracking. In a TFIIS-aided cleavage assay of the T831A mutant, there is a 2.5-fold reduction in the Pol II TFIIS-aided cleavage rate as compared with the wild type, which is consistent with computational prediction.

After backtracking, the misincorporated NTP can be removed either by intrinsic cleavage or TFIIS-aided cleavage. The mechanism of intrinsic cleavage of Pol II has been characterized through QM/MM-MD simulations and biochemical studies.[36] Pol II removes the last two nucleotides from the RNA chain (State VI → I, Figure 10.1B); the intrinsic cleavage mechanism follows a two-step mechanism which includes endonucleolytic cleavage and proton transfer. In the first step, an oxygen atom in the phosphate group (Rp-O) of

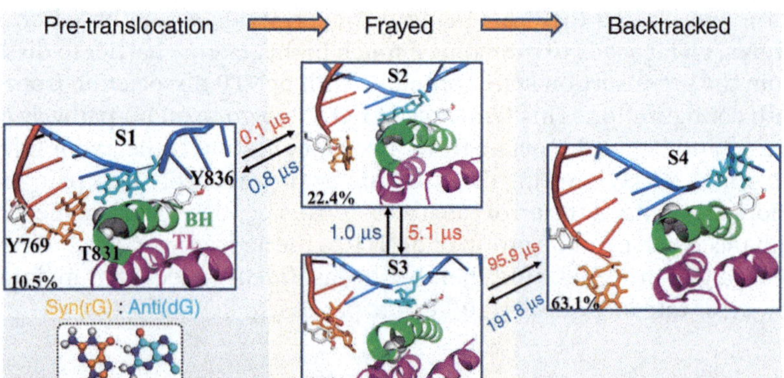

Figure 10.5 Atomistic backtracking mechanism of Pol II as described by a MSM. Reproduced from ref. 34, https://doi.org/10.1038/ncomms11244, under the terms of the CC BY 4.0 license, http://creativecommons.org/licenses/by/4.0/.

the terminal misincorporated nucleotide (position 0, rightmost nucleotide in all panels of Figure 10.6) acts as the general base that attacks the hydrogen atom of an Mg_B^{2+} coordinated water molecule (RS, Figure 10.6). The deprotonated water molecule then attacks the phosphorus atom of the phosphate group between the position −1 and −2 nucleotides (middle and leftmost nucleotide in all panels of Figure 10.6). This results in a metastable intermediate state where the 3′ hydroxyl end of the −2 nucleotide is deprotonated and the Rp-O atom of the position 0 nucleotide is protonated (IS, Figure 10.6). In the proton transfer step, the 3′ oxide 3′ hydroxyl of the −2 nucleotide is protonated by an Mg_A^{2+}-coordinated water molecule. This deprotonated water molecule then attacks another water molecule coordinating Mg_B^{2+}, leading to deprotonation of the position 0 nucleotide's phosphate group. This final step recovers the neutral water molecules in the active site (PS, Figure 10.6).

The role of Rp-O of the position 0 nucleotide as a general base is validated by the results of two sets of biochemical experiments. The first one compares reaction rates of 3′–5′ and 2′–5′ phosphodiester linkages between position −1 and 0 nucleotides. MD simulations of a 2′–5′ linked scaffold indicate that the Rp-O atom of the 0 nucleotide will be farther from the hydrogen atom of the attacking water at 5.47 ± 0.29 Å. This is in comparison to 3.44 ± 0.29 Å for the 3′–5′ linked scaffold. The reaction rate of the 2′–5′ linked scaffold is 14 times slower than that of the 3′–5′ linked scaffold. The next experiment involves the replacement of the Rp-O and Sp-O of the 0 nucleotide with a sulphur atom. This thio-substitution of Rp-O results in a significant

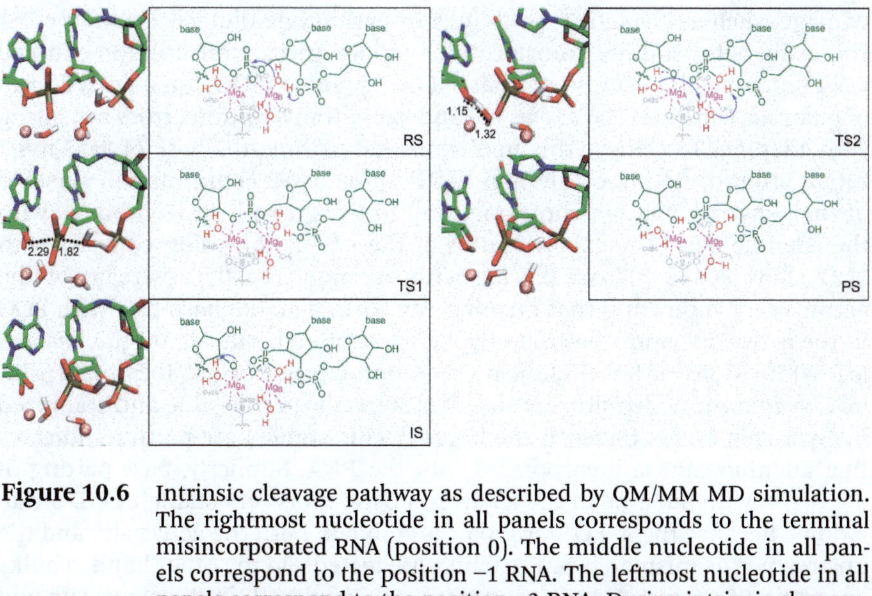

Figure 10.6 Intrinsic cleavage pathway as described by QM/MM MD simulation. The rightmost nucleotide in all panels corresponds to the terminal misincorporated RNA (position 0). The middle nucleotide in all panels correspond to the position −1 RNA. The leftmost nucleotide in all panels correspond to the position −2 RNA. During intrinsic cleavage, RNAP cleaves between nucleotides at position −1 and −2. Reproduced from ref. 36 with permission from Springer Nature, Copyright 2019.

reduction in cleavage rate. The results of these two experiments indicate that the terminal nucleotide Rp-O acts as the general base that first deprotonates the attacking water molecule. In addition, at the proton transfer step, Pol II residues, such as E485, are unlikely to act as the general acid. E485 is the closest residue to the 3′ hydroxyl of the −2 nucleotide, which, if protonated, may act as general acid. However, MD simulations of protonated E485 do not result in a stable active site. Furthermore, the nucleobase of the position 0 nucleotide has been suggested to act as general acid or base. To investigate its role in intrinsic cleavage, the intrinsic cleavage rate using a modified guanine (7-deaza guanine) as the misincorporated base was measured. 7-deaza guanine cannot act as general acid or base, as the N7 of guanine is replaced with carbon atom. The rates for intrinsic cleavage of 7-deaza guanine are similar to those observed with guanine, which indicates that the base of the misincorporated nucleotide has no more than a small effect on the intrinsic cleavage mechanism. Indeed, the rate of intrinsic cleavage predicted by QM/MM calculation is 400 times slower than the TFIIS-assisted cleavage rate, due to the high free energy barrier of the cleavage reaction (approximately 29 kcal mol^{-1}). In the presence of TFIIS, this barrier is lowered, permitting a more efficient cleavage reaction.

10.3.3 Damaged DNA-Induced Misincorporation

So far, the fidelity controls discussed here are focused on misincorporation caused by a mismatched NTP. However, transcription errors can also be caused by damaged template DNA. Structural studies have revealed that bulky DNA damage, such as cisplatin[37] crosslinks or pyrimidine dimers,[38] cannot enter the active site, stalling translocation, inducing the transcription-coupled DNA repair mechanism. In contrast, 8-oxo-guanine (8OG), an oxidized form of guanine, can enter the active site and cause transcription errors by pairing with adenine. To achieve this non-canonical pairing, the base of 8OG must rotate around the glycosidic bond to assume a *syn*-conformation (instead of the conventional *anti* conformation), forming two hydrogen bonds with the adenine base. Crystal structures of the Pol II elongation complex with 8OG show 8OG(*syn*) base-pairing with adenine directly upstream of the active site,[39] indicating that adenine can form a stable base-pair with 8OG in the active site and subsequently be incorporated. Indeed, on the basis of MD simulations of Pol II elongation complex with 8OG, ATP can pair with 8OG(*syn*) in the active site, because it is sterically permissible and stabilized by hydrogen bonds between the bases.[40] This stable base-pairing indicates that adenine can be incorporated into the RNA. Similarly, base-pairing of 8OG (*syn*) with adenine in the adjacent upstream is very stable in MD simulations. Because the A–8OG(*syn*) pair is stable in both the active site and the upstream, this mismatch can remain undetected. On the other hand, a bulky base-pair, 8OG(*anti*)–A(*anti*), cannot be stabilized both in the active site and the adjacent upstream positions due to steric clashing. In fact, 8OG (*anti*)–A(*anti*) is unstable in the adjacent upstream position, disrupting the base pair in the active site. This indicates another layer of proofreading, in which an unstable base pair in the upstream region will prevent Pol II from incorporating the next NTP and probably induce backward translocation.

10.4 Construction of a Kinetic Network Model of the Entire Transcription Elongation Cycle

By combining the rates obtained by MSM and from single-molecule experiments, such as optical tweezers or magnetic tweezers, one can build the KNM of the whole TEC. The objective of building the KNM is to derive the equation for the elongation rate. Then, using the equation, one can tune the transition rates in each step of the TEC and determine how it affects the theoretical elongation rate. Using experimentally measured elongation rates as the boundary condition, different sets of transition rates can be obtained depending on the assumption considered. The number of steps in the TEC can vary from 3 to 5, with more steps increasing complexity.[41]

Building the KNM begins with a master equation, $\frac{\mathrm{d}\Pi}{\mathrm{d}t} = M\Pi$, where Π is the population of each state [given by $(P_I P_{II} P_{III} P_{IV} P_V)^T$ for a five-state KNM], and M is the transition rate matrix (a 5×5 matrix in the case of a five-state KNM).

The transition rate matrix contains the transition rates between each state. At steady state, the master equation is set to 0: $\frac{\mathrm{d}\Pi}{\mathrm{d}t} = M\Pi = 0$. With this equation, the elongation rate, ν (probability flux) can be derived. Then, the theoretically derived ν can be fitted to the experimentally measured elongation rate, to obtain the unknown rate constants in the KNM. Varying the assumptions in the kinetic model may result in different expressions for the elongation rate, and consequently different sets of rate constants. For example: for a five-state KNM of Pol II (Figure 10.7),[41] the rate limiting step after NTP binding can be either TL closing or isomerization or NTP addition catalysis. If using the first assumption, to maintain the elongation rate, NTP must have a high affinity to Pol II since TL closing happens right after NTP binds at the Pol II active site. On the other hand, NTP affinity does not matter if using the second assumption. Therefore, a measurement of NTP affinity to Pol II binding site will help to identify the rate limiting step after NTP binding. If $K_D > 10\ \mu$M, the first model cannot be true ($K_D = \frac{k_{II \to I}}{k^0_{I \to II}}$). Yet as previously mentioned in the NTP entry subsection, NTP association may involve multiple kinetic steps, which may include off-path metastable states that complicate the determination of NTP binding affinity.

KNM can also be built based on an NTP association-centric model.[19] This model considers the multi-step binding of NTP association into the kinetic model. It was shown that the multiple binding sites are crucial to maintaining an elongation rate that is consistent with the results of

$$M\Pi = \begin{pmatrix} -k_t^+ - k_{TLo}^- & k_t^- & 0 & 0 & k_{TLo}^+ \\ k_t^+ & -k_t^- - k_b^+ & k_b^- & 0 & 0 \\ 0 & k_b^+ & -k_b^- - k_{TLc}^+ & k_{TLc}^- & 0 \\ 0 & 0 & k_{TLc}^+ & -k_{TLc}^- - k_c^+ & k_c^- \\ k_{TLo}^- & 0 & 0 & k_c^+ & -k_c^- - k_{TLo}^+ \end{pmatrix} \begin{pmatrix} P_V \\ P_I \\ P_{II} \\ P_{III} \\ P_{IV} \end{pmatrix}$$

Figure 10.7 Master equation for a five-state KNM as shown in Figure 10.1B without backtracking. M is the transition rate matrix describing the transition rate between one state and another k_t^+/k_t^- corresponds to forward/backward translocation rate (State V ⇔ State I); k_b^+/k_b^- corresponds to the association/dissociation of NTP to the binding site (State I ⇔ State II); k_{TLc}^+/k_{TLc}^- corresponds to the closing/opening in the presence of NTP in the active site (State II ⇔ State III); k_c^+/k_c^- corresponds to the forward/reverse catalysis rate (State III ⇔ State IV); k_{TLo}^+/k_{TLo}^- corresponds to the opening/closing in the absence of NTP in the active site (State IV ⇔ V). Π describes the population of each state written as population vector. P_x corresponds to the population of state x, (x = I, II, III, IV, V, numbering of the state is consistent with that shown in Figure 10.1B).

single-molecule experiments. Having metastable binding sites of NTP may also indicate that NTP binding could occur before translocation. Thus, exploring different assumptions and models while considering the experimental proposals will be useful for further understanding the transcription elongation cycle.

10.5 Future Perspectives

MSM and KNM studies of TEC have neglected many other elongation factors involved in transcription. For example: in the genome, DNA is stored as nucleosome complexes. Thus, DNA must unwrap from the histone in order to be transcribed; this may cause transcription pausing and affect elongation rates. In addition, the highly conserved elongation factor suppressor of Ty 4–5 (Spt4–5) of Archaea and Eukaryotes [N-utilization substance G (NusG) of Bacteria] has been shown to regulate transcription pausing.[42,43] Spt4–5 in conjunction with elongation factor 1 (Elf1) has been shown to accelerate transcription in the presence of nucleosomes by adjusting downstream nucleosomes.[44] It will be useful to use MSMs and KNM to assess the atomistic mechanisms of the TEC in the presence of nucleosomes and elongation factors.

With the rapid advancement of MD simulation algorithms and computer hardware, there are more opportunities to study other steps of transcription, such as initiation. Transcription initiation requires large scale conformational changes of RNAP, such as clamp domain motion and DNA melting.[45,46] As mentioned in the introduction, while current cryo-EM is able to reveal large conformational changes and slow transition complexes from a single sample, it is still difficult to obtain atomistic dynamics for fast motions or infrequently populated structures. Therefore, MSMs and KNM will continue to be particularly useful in obtaining atomistic transition states in the initiation process and transcription processes in general. The KNMs obtained can be used to understand how each step in the TEC affects the overall elongation rates obtained using single-molecule experiments.

Lastly, in the long term, KNMs can also potentially be applied to predict genome-wide transcription errors by incorporating sequence-dependency in KNMs. In particular, the stability of the transcription complex can be calculated based on base-pairing energies of DNA–DNA and DNA–RNA complexes, which will determine its tendency to translocate forward or backtrack.[47] For example, if the downstream double-stranded DNA (dsDNA) has more stable energies than the upstream dsDNA. RNAP will tend to backtrack and correct the misincorporated nucleotides. On the other hand, if the downstream dsDNA is less stable than the upstream dsDNA, RNAP will tend to forward translocate with the misincorporated nucleotide. By examining all possible transcription bubbles in the genome, one can rank their tendency to incorporate matched NTPs over mismatched NTPs (Accuracy = $A = P^{c}/P^{nc}$) to find the least accurate transcription bubbles (transcription error hotspots). This study has been done for bacterial RNAP and can be easily applied to other organisms (*i.e.* yeast Pol II). Notably, as previously mentioned in the nucleotide selectivity subsection, RNAP is involved in nucleotide selectivity and backtracking, which is neglected in the transcription accuracy model by Mellenius and Ehrenberg.[47] Thus, the effect of RNAP must be considered in future accuracy calculation models, probably with MSMs using different DNA sequences. At the experimental level genome-wide transcription errors could be mapped using emerging RNA sequencing techniques such as circular sequencing[48] and repeated sequencing[49] to validate the predicted error hotspots.

Acknowledgements

This work was supported by Hong Kong Research Grant Council (16303919, 16307718, AoE/P-705/16, and T13-605/18-W).

References

1. M. Vermulst, *et al.*, Transcription errors induce proteotoxic stress and shorten cellular lifespan, *Nat. Commun.*, 2015, **6**, 8065.
2. A. Paoloni-Giacobino, C. Rossier, M. P. Papasavvas and S. E. Antonarakis, Frequency of replication/transcription errors in (A)/(T) runs of human genes, *Hum. Genet.*, 2001, **109**, 40–47.
3. K. Hubbard, J. Catalano, R. K. Puri and A. Gnatt, Knockdown of TFIIS by RNA silencing inhibits cancer cell proliferation and induces apoptosis, *BMC Cancer*, 2008, **8**, 133.
4. S. Rodin, A. S. Rodin, A. Juhasz and G. P. Holmquist, Cancerous hypermutagenesis in p53 genes is possibly associated with transcriptional bypass of DNA lesions, *Mutat. Res.*, 2002, **510**, 153–168.

5. Y. Zhou, S. Wang, Y. Li and Y. Zhang, Born–Oppenheimer *Ab Initio* QM/MM Molecular Dynamics Simulations of Enzyme Reactions, *Methods Enzymol.*, 2016, **577**, 105–118.
6. J. H. Prinz, *et al.*, Markov models of molecular kinetics: Generation and validation,*J. Chem. Phys.*, 2011, **134**, 174105.
7. J. D. Chodera, N. Singhal, V. S. Pande, K. A. Dill and W. C. Swope, Automatic discovery of metastable states for the construction of Markov models of macromolecular conformational dynamics,*J. Chem. Phys.*, 2007, **126**, 1–17.
8. A. C. Pan and B. Roux, Building Markov state models along pathways to determine free energies and rates of transitions, *J. Chem. Phys.*, 2008, **129**, 1–8.
9. B. W. Zhang, *et al.*, Simulating Replica Exchange: Markov State Models, Proposal Schemes, and the Infinite Swapping Limit, *J. Phys. Chem. B*, 2016, **120**, 8289–8301.
10. F. Morcos, *et al.*, Modeling conformational ensembles of slow functional motions in pin1-WW, *PLoS Comput. Biol.*, 2010, **6**, 1–13.
11. X. Huang, G. R. Bowman, S. Bacallado and V. S. Pande, Rapid equilibrium sampling initiated from nonequilibrium data, *Proc. Natl. Acad. Sci. U. S. A.*, 2009, **106**, 19765–19769.
12. R. D. Malmstrom, C. T. Lee, A. T. Van Wart and R. E. Amaro, Application of molecular-dynamics based markov state models to functional proteins,*J. Chem. Theory Comput.*, 2014, **10**, 2648–2657.
13. N. V. Buchete and G. Hummer, Coarse master equations for peptide folding dynamics,*J. Phys. Chem. B*, 2008, **112**, 6057–6069.
14. L. Zhang, *et al.*, Elucidation of the Dynamics of Transcription Elongation by RNA Polymerase II using Kinetic Network Models, *Acc. Chem. Res.*, 2016, **49**, 687–694.
15. I. C. Unarta, *et al.*, Molecular mechanisms of RNA polymerase II transcription elongation elucidated by kinetic network models, *Curr. Opin. Struct. Biol.*, 2018, **49**, 54–62.
16. M. Feig and Z. F. Burton, RNA polymerase II with open and closed trigger loops: Active site dynamics and nucleic acid translocation, *Biophys. J.*, 2010, **99**, 2577–2586.
17. D.-A. Silva, *et al.*, Millisecond dynamics of RNA polymerase II translocation at atomic resolution, *Proc. Natl. Acad. Sci. U. S. A.*, 2014, **111**, 7665–7670.
18. L. Zhang, D. A. Silva, F. Pardo-Avila, D. Wang and X. Huang, Structural Model of RNA Polymerase II Elongation Complex with Complete Transcription Bubble Reveals NTP Entry Routes, *PLoS Comput. Biol.*, 2015, **11**, 1–21.
19. B. Wang, R. E. Sexton and M. Feig, Kinetics of nucleotide entry into RNA polymerase active site provides mechanism for efficiency and fidelity, *Biochim. Biophys. Acta, Gene Regul. Mech.*, 2017, **1860**, 482–490.
20. B. Wang, A. V. Predeus, Z. F. Burton and M. Feig, Energetic and structural details of the trigger-loop closing transition in RNA polymerase II, *Biophys. J.*, 2013, **105**, 767–775.

21. T. A. Steitz and J. A. Steitz, A general two-metal-ion mechanism for catalytic RNA, *Proc. Natl. Acad. Sci. U. S. A.*, 2006, **90**, 6498–6502.
22. A. T. P. Carvalho, P. A. Fernandes and M. J. Ramos, The catalytic mechanism of RNA polymerase II, *J. Chem. Theory Comput.*, 2011, **7**, 1177–1188.
23. D. Wang, D. A. Bushnell, K. D. Westover, C. D. Kaplan and R. D. Kornberg, Structural Basis of Transcription: Role of the Trigger Loop in Substrate Specificity and Catalysis, *Cell*, 2006, **127**, 941–954.
24. M. Levitt, *et al.*, RNA polymerase II trigger loop residues stabilize and position the incoming nucleotide triphosphate in transcription, *Proc. Natl. Acad. Sci. U. S. A.*, 2010, **107**, 15745–15750.
25. T. V. Mishanina, M. Z. Palo, D. Nayak, R. A. Mooney and R. Landick, Trigger loop of RNA polymerase is a positional, not acid–base, catalyst for both transcription and proofreading, *Proc. Natl. Acad. Sci. U. S. A.*, 2017, 201702383.
26. B. Wang, K. Opron, Z. F. Burton, R. I. Cukier and M. Feig, Five checkpoints maintaining the fidelity of transcription by RNA polymerases in structural and energetic details, *Nucleic Acids Res.*, 2015, **43**, 1133–1146.
27. L. T. Da, D. Wang and X. Huang, Dynamics of pyrophosphate ion release and its coupled trigger loop motion from closed to open state in RNA polymerase II, *J. Am. Chem. Soc.*, 2012, **134**, 2399–2406.
28. L. T. Da, F. Pardo Avila, D. Wang and X. Huang, A Two-State Model for the Dynamics of the Pyrophosphate Ion Release in Bacterial RNA Polymerase, *PLoS Comput. Biol.*, 2013, **9**, e1003020.
29. L. T. Da, *et al.*, A Jump-from-Cavity Pyrophosphate Ion Release Assisted by a Key Lysine Residue in T7 RNA Polymerase Transcription Elongation, *PLoS Comput. Biol.*, 2015, **11**, 1–22.
30. D. Temiakov, V. Patlan, M. Anikin, W. T. Mcallister and S. Yokoyama, Structural Basis for Substrate Selection by T7 RNA Polymerase out all of the steps in the transcription process in a nearly identical manner to that observed for multisubunit RNAPs (Jeruzalmi and Steitz), *Cell*, 2004, **116**, 381–391.
31. B. Duan, S. Wu, L. T. Da and J. Yu, A critical residue selectively recruits nucleotides for T7 RNA polymerase transcription fidelity control, *Biophys. J.*, 2014, **107**, 2130–2140.
32. E. Chao, B. Duan and J. Yu, Nucleotide Selectivity at a Preinsertion Checkpoint of T7 RNA Polymerase Transcription Elongation, *J. Phys. Chem. B*, 2017, **121**, 3777–3786.
33. C. Long, C. E, L. T. Da and J. Yu, Determining selection free energetics from nucleotide pre-insertion to insertion in viral T7 RNA polymerase transcription fidelity control, *Nucleic Acids Res.*, 2019, **47**, 4721–4735.
34. L.-T. Da, *et al.*, Bridge helix bending promotes RNA polymerase II backtracking through a critical and conserved threonine residue, *Nat. Commun.*, 2016, 7, 11244.

35. J. F. Sydow, *et al.*, Structural Basis of Transcription: Mismatch-Specific Fidelity Mechanisms and Paused RNA Polymerase II with Frayed RNA, *Mol. Cell*, 2009, **34**, 710–721.
36. C. K. M. Tse, *et al.*, Intrinsic cleavage of RNA polymerase II adopts a nucleobase-independent mechanism assisted by transcript phosphate, *Nat. Catal.*, 2019, **2**, 228–235.
37. G. E. Damsma, A. Alt, F. Brueckner, T. Carell and P. Cramer, Mechanism of transcriptional stalling at cisplatin-damaged DNA, *Nat. Struct. Mol. Biol.*, 2007, **14**, 1127–1133.
38. F. Brueckner, U. Hennecke, T. Carell and P. Cramer, CPD Damage Recognition by Transcribing RNA Polymerase II, *Science*, 2007, **315**, 859–862.
39. G. E. Damsma and P. Cramer, Molecular basis of transcriptional mutagenesis at 8-oxoguanine, *J. Biol. Chem.*, 2009, **284**, 31658–31663.
40. K. A. Konovalov, *et al.*, 8-Oxo-guanine DNA damage induces transcription errors by escaping two distinct fidelity control checkpoints of RNA polymerase II, *J. Biol. Chem.*, 2019, **294**, 4924–4933.
41. J. Yu, L.-T. Da and X. Huang, Constructing kinetic models to elucidate structural dynamics of a complete RNA polymerase II elongation cycle, *Phys. Biol.*, 2014, **12**, 016004.
42. A. V. Yakhnin, K. S. Murakami and P. Babitzke, NusG is a sequence-specific RNA polymerase pause factor that binds to the non-template DNA within the paused transcription bubble, *J. Biol. Chem.*, 2016, **291**, 5299–5308.
43. J. B. Crickard, J. Fu and J. C. Reese, Biochemical analysis of yeast suppressor of Ty 4/5 (Spt4/5) reveals the importance of nucleic acid interactions in the prevention of RNA polymerase II arrest, *J. Biol. Chem.*, 2016, **291**, 9853–9870.
44. H. Ehara, *et al.*, Structural insight into nucleosome transcription by RNA polymerase II with elongation factors, *Science*, 2019, **363**, 744–747.
45. E. Nogales, R. K. Louder and Y. He, Structural Insights into the Eukaryotic Transcription Initiation Machinery, *Annu. Rev. Biophys.*, 2017, **46**, 59–83.
46. E. A. Hubin, M. Lilic, S. A. Darst and E. A. Campbell, Structural insights into the mycobacteria transcription initiation complex from analysis of X-ray crystal structures, *Nat. Commun.*, 2017, **8**, 16072.
47. H. Mellenius and M. Ehrenberg, DNA template dependent accuracy variation of nucleotide selection in transcription, *PLoS One*, 2015, **10**, 1–21.
48. A. Acevedo and R. Andino, Library preparation for highly accurate population sequencing of RNA viruses, *Nat. Protoc.*, 2014, **9**, 1760–1769.
49. J.-F. Gout, W. K. Thomas, Z. Smith, K. Okamoto and M. Lynch, Large-scale detection of *in vivo* transcription errors, *Proc. Natl. Acad. Sci. U. S. A.*, 2013, **110**, 18584–18589.

CHAPTER 11

Epilogue

ROBERT LANDICK*[a], TERENCE STRICK*[b] AND JUE WANG*[a]

[a]University of Wisconsin–Madison, USA; [b]École normale supérieure, France
*E-mail: landick@bact.wisc.edu, strick@biologie.ens.fr, wang@bact.wisc.edu

As we mark the diamond anniversary of the first purification of RNA polymerase in 1961, our understanding of the underlying mechanisms of transcription continues to deepen. New questions are arising, driven both by a refined appreciation of macromolecular function and by the development of new methods and approaches. We conclude this volume by asking what new trends and questions will we witness in the coming ten years. At the inevitable risk of not anticipating the happy surprises that are the true hallmark of fundamental science, we foresee several areas where research into transcriptional mechanisms is likely to head.

From a technical perspective, advances in detection systems and computing power will sustain continued success using cryo-electron microscopy (cryo-EM) to reconstruct more complex transcriptional machineries containing increasingly large numbers of subunits. The application of cryo-EM to active transcription reactions reconstituted *in vitro* and the development of time-resolved cryo-EM should make it possible to extract dynamics from the temporal progression of reaction intermediates. Single-molecule time-resolved Förster resonance energy transfer (FRET), reaching into the millisecond domain, may then be in a position to test kinetic models of these reactions using an independent set of spatial and temporal restraints. Advances in computing power will also drive more detailed and

Chemical Biology No. 19
RNA Polymerases as Molecular Motors: On the Road, 2nd Edition
Edited by Robert Landick, Terence Strick and Jue Wang

Published by the Royal Society of Chemistry, www.rsc.org

extensive all-atom simulations of transcriptional machineries. These predictions should both guide the development of thermodynamically realistic kinetic models and fill in gaps that are not readily captured by cryo-EM and single-molecule methods.

As progress is made in understanding stable, well-known partners of RNA polymerase, the field also can turn to the analysis of more weakly interacting molecular partners. RNA polymerase is clearly a target for dozens of competing and cooperating small-molecule and macromolecular regulators whose individual interactions with small distinct patches on the surface of RNA polymerase are transient and difficult to capture as stable complexes. Even when combined into relatively stable complexes comprised of multiple regulators [*e.g.* complexes formed by *Escherichia coli* N-utilization substance (Nus) factors], the web-like interactions among multiple partners and RNA polymerase indicate that only low-energy barriers govern changes in composition, structure, and thus function of these complexes. This view is consistent with the multitude of compositional and conformational states observed under the electron microscope. In the coming years, single-molecule approaches may become more central to characterizing, in solution conditions, the exchange rates for individual components in multicomponent complexes of well-defined stoichiometry. Use of cell extracts and even intact cells for single-molecule observations can extend these measurements to the complex milieu of natural systems. Cryo-electron tomography to characterize macromolecular structures *in vivo* will synergize with single-molecule approaches. A fruitful cycle of experiment and modeling combining atomic-resolution simulations, single-molecule functional assays, and single-molecule structural analysis, will probably emerge in the coming years.

We can also anticipate that evolutionarily diverse regulatory mechanisms, particularly in diverse bacteria, will become a huge area into which transcription research will expand in the coming decade. It is already abundantly clear that long-standing paradigms derived from the study of model bacteria like *Escherichia coli* and *Bacillus subtilis* represent only a limited set of solutions to the challenges of linking gene transcription to cellular and environmental conditions. New general initiation factors have been defined in Mycobacteria and even the well-known consensus −10 promoter element turns out to differ in some bacterial lineages. Researchers have barely begun when it comes to probing the diversity of mechanisms and regulators that await discovery as attention is turned to unexplored branches in the evolutionary tree of transcriptional regulatory mechanisms.

A greater appreciation and understanding of the connections between transcription of bacterial chromosomes and chromosomal structure and dynamics is likely to be another trend over the next decade. Although we also anticipate progress in the understanding of transcription of eukaryotic and archaeal chromatin, the more complex and heterogeneous composition of bacterial chromatin has long relegated understanding of these linkages in bacteria to a far more primitive state. New methodologies now make it possible to tackle these complexities and this trend is already evident in studies of R-loop formation, transcription–translation coupling, and DNA topology. We foresee increasing focus on understanding transcriptional regulation at the genome and cellular scales.

Although historically research has focused on the production of DNA-templated RNA and its presentation to the ribosome or to RNA-binding proteins *via* the RNA exit channel, there is also a growing interest in RNA-interacting proteins that monitor the secondary channel for the 3′ end of RNA that may appear as a result of RNA polymerase backtracking. Recent examples of this function include helicases that could assist the proofreading activities of viral RNA-dependent RNA polymerases by stabilizing backtracked states required for error correction. Surveillance of the secondary channel for nascent RNA is likely to be an important question in the coming years. In a similar vein, but regarding the RNA-exit channel, there is also evidence for modulation of the channel's function by partner proteins. λ Q-antitermination appears to involve prolongation of the RNA exit channel by a ring-forming protein that encircles the nascent RNA, precluding folding of RNA hairpins and thus preventing termination.

In addition, a growing body of evidence implicates RNA polymerase directly in a growing list of DNA damage and repair processes. This is true not only in bacteria but also in eukaryotes. Transcriptional processes have been associated with the generation of both single- and double-strand breaks *in vivo*. Although it is likely that many such lesions arise from replication–transcription conflicts, it is also possible that novel biochemical activities brought about by known or unknown partners of RNA polymerase could be responsible for DNA damage. Relatedly, RNA polymerase is also implicated in the signaling of DNA damage, and in particular, in the repair of DNA double-strand breaks. The propensity for RNA polymerase to interact with single-strand regions and with DNA ends made available by double-strand breaks may, in part, subtend this role.

We expect to see advances in all these areas but we suspect that the most significant advances may well emerge where we least expect them; many exciting and important discoveries remain to be made in the field of transcription. It will be illuminating to look back ten years hence and survey the steady progression of new knowledge about the most fundamental step in the expression of genetic information as well as the breakthroughs and discoveries that make it possible.

Robert Landick
Terence Strick
Jue Wang

Subject Index